Compendium of Clove

Welcome to the definitive guide *Compendium of Clove: Navigating Agriculture, Chemistry, Processing, and Health Benefits*, where centuries of tradition meet cutting-edge research on clove. Clove (*Syzygium aromaticum*) has a rich history dating back millennia, revered for its aromatic allure, medicinal properties, and economic significance across cultures. From the verdant plantations to the laboratory bench, each chapter in this book unfolds the intricate story of clove, bridging historical insights with contemporary studies, exploring its historical and botanical descriptions, community benefits, chemical composition, and diverse industrial applications.

This A–Z compendium not only consolidates existing knowledge but also pioneers new frontiers in clove research. It offers a panoramic view that caters to botanists, pharmacognosists, phytochemists, pharmacologists, food scientists, agriculturalists, industrialists, and policymakers alike.

KEY FEATURES:

- The book offers the origins and history of clove distribution, plant habits, and botanical descriptions.
- It provides insights into cultivation practices of clove, including good agricultural practices (GAP) and post-harvest management of clove.
- The book underlines how the biochemistry of plants, complete phytochemical screening, characterization, separation, and other factors affect the volatile oils of plants.
- It underlines clove's pharmacological and clinical aspects and highlights its usage in the food, pharmaceutical, and cosmetics industries.
- The book showcases market value, trade, and regulatory guidelines of clove in different countries.

Whether you seek a botanical expedition or a pharmacological breakthrough, whether your interest lies in chemistry or global economics, this book embarks on a journey that celebrates clove as not just a spice but a cornerstone of interdisciplinary research and industrial enterprise. Join us as we unearth the essence of clove—a testament to nature's bounty and human ingenuity, encapsulated within the pages of this definitive document.

Compendium of Clove

Navigating Agriculture, Chemistry, Processing, and Health Benefits

Edited by
Preet Amol Singh
Sukhvinder Singh Purewal

CRC Press
Taylor & Francis Group
Boca Raton London New York

CRC Press is an imprint of the
Taylor & Francis Group, an **informa** business

Designed cover image: Shutterstock

First edition published 2025
by CRC Press
2385 NW Executive Center Drive, Suite 320, Boca Raton FL 33431

and by CRC Press
4 Park Square, Milton Park, Abingdon, Oxon, OX14 4RN

CRC Press is an imprint of Taylor & Francis Group, LLC

Library of Congress Cataloging-in-Publication Data
Names: Singh, Preet Amol, editor. | Purewal, Sukhvinder Singh, editor.
Title: Compendium of clove : navigating agriculture, chemistry, processing, and health benefits / edited by Preet Amol Singh and Sukhvinder Singh Purewal.
Description: First edition | Boca Raton, FL : CRC Press, 2025. | Includes bibliographical references and index.
Identifiers: LCCN 2024041949 (print) | LCCN 2024041950 (ebook) | ISBN 9781032862026 (hardback) | ISBN 9781032862057 (paperback) | ISBN 9781003521785 (ebook)
Subjects: LCSH: Clove (Spice). | Clove tree. | Clove trade.
Classification: LCC SB307.C6 C66 2025 (print) | LCC SB307.C6 (ebook) | DDC 633.8/3—dc23/eng/20241015
LC record available at https://lccn.loc.gov/2024041949
LC ebook record available at https://lccn.loc.gov/2024041950

ISBN: 978-1-032-86202-6 (hbk)
ISBN: 978-1-032-86205-7 (pbk)
ISBN: 978-1-003-52178-5 (ebk)

DOI: 10.1201/9781003521785

Typeset in Times
by Apex CoVantage, LLC

Contents

Foreword

Throughout history, traditional and complementary medicine has played a vital role in promoting community and household health. Today, about 40% of pharmaceuticals including essential medications like aspirin and artemisinin were derived from natural sources, particularly plants. The World Health Organization (WHO) has received reports from 170 Member States on their use of traditional medicine, seeking evidence and data to shape policies, standards, and regulations that ensure its safe, affordable, and equitable application. This heritage has now sparked a global resurgence in the concept of **phytomedicine**, emphasizing rigorous scientific validation of herbal products and formulations for managing various health conditions.

Clove has a rich historical legacy as a medicinal herb. Over centuries, it has been valued for its therapeutic properties and has been used in various cultures around the world. In contemporary times, scientific research has validated many of these traditional uses of clove, attributing its health benefits to its rich content of antioxidants and active compounds such as eugenol. As a result, clove continues to be utilized in various forms including essential oils, extracts, and dietary supplements, maintaining its status as a versatile and effective natural remedy.

With new emerging global challenges such as climate change and its impact on the quality and quantity of the phytoconstituents present in medicinal plants, this book titled *Compendium of Clove: Navigating Agriculture, Chemistry, Processing, and Health Benefits*, published by CRC Press, Taylor & Francis, fulfils a crucial need by compiling and exploring diverse facets of clove, focusing on its geographical distribution, botanical characteristics, cultivation practices, processing techniques, phytochemical composition, trade dynamics, pharmaceutical applications, pharmacological effects, and regulatory considerations.

I believe this book will stimulate constructive dialogues on the optimum usage of clove, which can contribute to the development of resilient global policies on this topic.

Prof. (Dr.) Rajeev Sood
Vice Chancellor
Baba Farid University of Health Sciences
Faridkot, Punjab India

Editor Biographies

Preet Amol Singh, PhD, is a pharmacognosist by profession and is presently working as Associate Professor in Department of Phytomedicine, UCER, Baba Farid University of Health Sciences (BFUHS), Govt. of Punjab, Faridkot, Punjab, India. Dr. Singh was a researcher in the Forum on Indian Traditional Medicine (FITM) a platform established by Ministry of AYUSH at the Research and Information System for Developing Countries (RIS), a policy think-thank institute under the Ministry of External Affairs, New Delhi. His areas of interest include agro-climatology of medicinal plants; phyto-pharmacology; HPTLC finger-printing profiling; development of monographs and compendia of quality standards of medicinal plants; herbal product development; good agricultural and collection practices; the Schedule TA of Drug and Cosmetics Act, 1940; biodiversity products trade; pharmacovigilance of herbal medicine; and licensing and registration of medicinal plant products. Dr. Singh was a 'special invitee' to the Global AYUSH Investment and Innovation Summit 2022, chaired by Hon'ble Prime Minister of India Shri Narendra Modi and Mr. Tedros Adhanom Ghebreyesus, Director-General of the World Health Organization (WHO). He participated in discussions on medicinal plants organized by the Confederation of Indian Industries (CII), attended by the Hon'ble Vice President of India. Additionally, Dr. Singh contributed as a member in the drafting of Punjab's new Agricultural Policy by the Punjab State Farmers and Farm Workers Commission, Government of Punjab. Dr. Singh was awarded the FITM Doctoral Research Fellowship from the Ministry of AYUSH and Research and Information System for Developing Countries (RIS) for 2019–2020. He has authored numerous scientific articles with a cumulative impact factor exceeding 100, including publications in ADDR. Dr. Singh holds six patents (three international—two in Germany, one WIPO—and three national). He is currently preparing two books accepted by CRC Press, Taylor & Francis and has presented over 40 papers at prestigious international and national conferences. Dr. Singh serves as an academic editor for journals such as *Scientifica* and *Evidence-Based Complementary Alternative Medicine*. He has secured research projects from FITM, Ministry of AYUSH, RIS, DST, PSCST, and PBB and has received multiple research awards including from the American Chemical Society and Ministry of AYUSH. Dr. Singh is a life member of the Indian Science Congress, IPGA, and APTI, and he contributes regularly to newspapers like *Hindustan Times*, the *Tribune*, *Deccan Herald*, and the *Quint*.

Sukhvinder Singh Purewal, PhD, is presently working as Assistant Professor, University Centre for Research and Development (UCRD), Chandigarh University, Mohali, Punjab, India. In 2018, he was awarded with a major research project spanning from 2018 to 2022 from the SEED Division, Department of Science and Technology (DST), New Delhi. His research interests encompass solid-state fermentation, extraction of bioactive compounds from natural sources, antioxidants, and the processing of fruits. Having authored more than 45 research papers in internationally recognized journals, Dr. Sukhvinder Singh Purewal has been honored with best poster awards at both national and international conferences. He is an active life member of the Association of Microbiologists of India (AMI); Mycological Society of India (MSI); Association of Food Scientists and Technologists (AFSTI) in Mysore, India; and the Indian Science Congress Association (ISCA). In 2023, Dr. Purewal was honored with the prestigious Young Scientist Award by the renowned Association of Microbiologists of India (AMI) in recognition of his outstanding contributions to the field of food microbiology. Furthermore, Sukhvinder Singh Purewal has demonstrated his expertise by publishing an authored book titled *Millets: Properties, Processing, and Health Benefits*, along with edited books such as *Maize Nutritional Composition, Processing, and Industrial Uses* and *Chickpea and Cowpea: Nutritional Profile, Processing, Health Prospects, and Commercial Uses* with CRC Press, Taylor & Francis Group.

Contributors

Samuel Oluwatobi Agboola
Department of Pure and Applied Chemistry
Ladoke Akintola University of Technology
Ogbomoso, Nigeria

Iceu Agustinisari
Research Center for Agroindustry
Research Organization for Agriculture and Food
National Research and Innovation Agency
Cibinong, Indonesia

Fatemeh Jamshidi Alashti
Department of Atomic and Molecular Physics
Faculty of Science
University of Mazandaran
Babolsar, Iran

Mohammed Alrugaibah
Department of Food Science and Human Nutrition
College of Agriculture and Food
Qassim University
Buraydah, Saudi Arabia.

Toluwanimi Rachael Arowolo
Department of Pure and Applied Chemistry
Ladoke Akintola University of Technology
Ogbomoso, Nigeria

Samuel Adelani Babarinde
Department of Crop and Environmental Protection
Ladoke Akintola University of Technology
Ogbomoso, Nigeria

Neha Bajwa
Phytomedicine UCER
Baba Farid University of Health Sciences
Faridkot, Punjab, India

Hassan Barakat
Department of Food Science and Human Nutrition
College of Agriculture and Food
Qassim University
Buraydah, Saudi Arabia
Department of Food Technology
Faculty of Agriculture
Benha University
Moshtohor, Egypt

Olugbenga Solomon Bello
Department of Pure and Applied Chemistry
Ladoke Akintola University of Technology
Ogbomoso, Nigeria

Nurliani Bermawie
Research Center for Estate Crops
Research Organization for Agriculture and Food
National Research and Innovation Agency
Cibinong, Indonesia

Gustavo Adolfo Castillo-Herrera
Unidad de Tecnología Alimentaria
Centro de Investigación y Asistencia en Tecnología y Diseño del Estado de Jalisco, A.C.
Zapopan, Jalisco, México

Ireng Darwati
Research Center for Estate Crops
Research Organization for Agriculture and Food
National Research and Innovation Agency
Cibinong, Indonesia

Hugo Espinosa-Andrews
Unidad de Tecnología Alimentaria
Centro de Investigación y Asistencia en Tecnología y Diseño del Estado de Jalisco, A.C.
Zapopan, Jalisco, México

Zaira Yunuen Garcia-Carvajal
Unidad de Biotecnología Médica y Farmacéutica
Centro de Investigación y Asistencia en Tecnología y Diseño del Estado de Jalisco, A.C.
Guadalajara, Jalisco, México

R. Vitri Garvita
Research Center for Applied Botany
Research Organization for Life Sciences and Environment
National Research and Innovation Agency
Cibinong, Indonesia

José Nabor Haro-González
Unidad de Tecnología Alimentaria
Centro de Investigación y Asistencia en Tecnología y Diseño del Estado de Jalisco, A.C.
Zapopan, Jalisco, México

Subhajit Hazra
University Institute of Pharma Sciences
Chandigarh University
Chandigarh, Punjab, India

Hernani
Research Center for Agroindustry
Research Organization for Agriculture and Food
National Research and Innovation Agency
Cibinong, Indonesia

Tatang Hidayat
Research Center for Agroindustry
National Research and Innovation Agency
Indonesia

Anthony Temitope Idowu
Department of Biological Sciences
Faculty of Science and Engineering
University of Limerick
Limerick, Ireland

Nisha Khatri
University Institute of Pharma Sciences
Chandigarh University
Chandigarh, Punjab, India

Sneha Kumari
University Institute of Pharma Sciences
Chandigarh University
Chandigarh, Punjab, India

Heri Septya Kusuma
Department of Chemical Engineering
Faculty of Industrial Technology
Universitas Pembangunan Nasional "Veteran"
Yogyakarta, Indonesia

Ganing Irbah Al Lantip
Department of Chemical Engineering
Faculty of Industrial Technology
Universitas Pembangunan Nasional "Veteran"
Yogyakarta, Indonesia
Plasma Technology Research Core
Faculty of Science
University of Mazandaran
Babolsar, Iran

Xenna Mutiara
Department of Chemical Engineering
Faculty of Industrial Technology
Universitas Pembangunan Nasional "Veteran"
Yogyakarta, Indonesia

Olagoke Zachaeus Olatunde
Fujian Institute of Research on Structure of Matter
China

Devi Rusmin
Research Center for Estate Crops
Research Organization for Agriculture and Food
National Research and Innovation Agency
Cibinong, Indonesia

Bagem Br Sembiring
Research Center for Agroindustry
National Research and Innovation Agency
Indonesia

Adi Setiadi
Research Center for Estate Crops
Research Organization for Agriculture and Food
National Research and Innovation Agency
Cibinong, Indonesia

Ritika Sindhwani
University Institute of Pharma Sciences
Chandigarh University
Chandigarh, Punjab, India

Preet Amol Singh
Phytomedicine, UCER
Baba Farid University of Health Sciences
Faridkot, Punjab, India

Farshad Sohbatzadeh
Department of Atomic and Molecular Physics
Faculty of Science, University of Mazandaran
Babolsar, Iran
Plasma Technology Research Core, Faculty of Science
University of Mazandaran
Babolsar, Iran

Ankur Suri
Bhai Gurdas College of Pharmacy
Sangrur, Punjab, India

Rudi Suryadi
Research Center for Estate Crops
Research Organization for Agriculture and Food
National Research and Innovation Agency
Cibinong, Indonesia

Sitti Fatimah Syahid
Research Center for Estate Crops
Research Organization for Agriculture and Food
National Research and Innovation Agency
Cibinong, Indonesia

Octivia Trisilawati
Research Center for Estate Crops
Research Organization for Agriculture and Food
National Research and Innovation Agency
Cibinong, Indonesia

Shiva Tushir
Panipat Institute of Engineering and Technology
Samalkha, Panipat, Haryana, India

Moisés Martínez Velázquez
Unidad de Biotecnología Médica y Farmacéutica
Centro de Investigación y Asistencia en Tecnología y Diseño del Estado de Jalisco, A.C.
Guadalajara, Jalisco, México

Sri Wahyuni
Research Center for Estate Crops
Research Organization for Agriculture and Food
National Research and Innovation Agency
Cibinong, Indonesia

Christina Winarti
Research Center for Agroindustry
Research Organization for Agriculture and Food
National Research and Innovation Agency
Cibinong, Indonesia

Muchamad Yusron
Research Center for Estate Crops
Research Organization for Agriculture and Food
National Research and Innovation Agency
Cibinong, Indonesia

Abbreviations

Abbreviations	Full form
ABTS	2,2-azino-bis (3-ethylbenzothiazoline-6-sulphonic acid
DPPH	2,2-diphenyl-1-picrylhydrazyl
MTT	3-(4,5-dimethylthiazol-2-yl)-2,5-diphenyltetrazolium bromide
ACh	Acetylcholine
AChE	Acetylcholinesterase
AGEP	Advanced glycation end products
APEDA	Agricultural and processed food products export development authority
AGMARK	Agricultural marketing certification
ALT	Alanine transaminase
ALP	Alkaline phosphatase
ASTA	American Spice Trade Association
ACE2	Angiotensin converting enzyme 2
BMI	Body mass index
BSLT	Brine shrimp lethality test
BIS	Bureau of Indian Standards
CFIA	Canada Food Inspection Agency
CO2	Carbon dioxide
CAT	Catalase
CVM	Center for Veterinary Medicine
CCC	China Compulsory Certification
CNCA	China National Certification and Accreditation Administration
CBP	Clove bud powder
CEO	Clove essential oil
CAC	Codex Alimentarius Commission
CCSCH	Codex Committee on Spices and Culinary Herbs
CXS	Codex standard
CFR	Code of federal regulations
CE	Common era
COMESA	Common Market for Eastern and Southern Africa
COLP	Conditioned open-label placebo
COVID-19	Coronavirus Disease-19
COX	Cyclooxygenase
DDT	Dichlorodiphenyltrichloroethane
DM	Diabetes mellitus
DW	Dry weight basis
EIC	Export Inspection Council

Abbreviations	Full form
TE-13	Esophageal cancer cell lines
EO	Essential oil
EFSA	European Food Safety Authority
EMA	European Medicines Agency
ESA	European Spice Association
EY	Extraction yield
BVL	Federal Office of Consumer Protection and Food Safety
FAO	Food and Agriculture Organization
FDA	Food and Drug Administration
FSSAI	Food Safety and Standards Authority of India
FSA	Food Standards Agency
ANSES	French Agency for Food, Environmental and Occupational Health and Safety
FDCH	Fully deacetylated chitosan
GABA	Gamma-aminobutyric acid
GC	Gas chromatography
GC-MS	Gas chromatography-mass spectrometry
GRAS	Generally recognized as safe
GHz	Gigahertz
GFR	Glomerular filtration rate
G6Pase	Glucose-6-phosphatase
GAP	Good agriculture practices
GHP	Good hygienic practices
GMP	Good manufacturing practices
IC50	Half-maximal inhibitory concentration
HS	Harmonized system
HeLa	Henrietta Lacks-cervical cancer cell lines
HSV	Herpes simplex virus
HPLC	High-performance liquid chromatography
His41	Histidine41
DU-145	Human prostate cancer line
PPAR-c	Human peroxisome proliferator-activated receptor
IL	Interleukin
IL-1β	Interleukin-1 beta
IL-10	Interleukin-10
IL-6	Interleukin-6
ISO	International Organization for Standardization
IUPAC	International Union of Pure and Applied Chemistry
IBS	Irritable bowel syndrome
JETRO	Japan External Trade Organization
JECFA	Joint FAO/WHO Expert Committee on Food Additives

Abbreviations	Full form
kcal	Kilocalorie
kHz	Kilohertz
KA	Kojic acid
LC50	Lethal concentration 50
LC90	Lethal concentration 90
LC-MS	Liquid chromatography-mass spectrometry
LC-MS/MS	Liquid chromatography with tandem mass spectrometry
Mpro	Main protease
MDA	Malonaldehyde
MRL	Maximum residue level
MRLs	Maximum residue limits
MDA-MB-231	MD Anderson-Metastatic breast cancer
MHz	Megahertz
MPa	Megapascal
MMP-2	Metalloproteinase-2
MMP-9	Metalloproteinase-9
MAD	Microwave-assisted drying
MAE	Microwave-assisted extraction
MAHD	Microwave-assisted hydrodistillation
MHLW	Ministry of Health, Labour and Welfare
MRSA	Multidrug-resistant *Staphylococcus aureus*
NAM	National AYUSH Mission
NHC	National Health Commission
BSN	National Standardization Agency of Indonesia
NTP	National Toxicology Program
N	Nitrogen
NF-$k\beta$	Nuclear factor kappa beta
ONN	Office National des Normes
OTC	Over-the-counter
PEPCK	Phosphoenolpyruvate carboxy kinase
P	Phosphor
P_2O_5	Phosphorus pentoxide
PMNs	Polymorphonuclear neutrophils
KCl	Potassium chloride
SP36	Potassium sulphate
PCA	Principal component analysis
Akt	Protein kinase B
QE	Quercetin equivalent
RASFF	Rapid alert system for food and feed
DPPH	Reactive oxygen-nitrogen species
RCMC	Registration-Cum-Membership Certificate

Abbreviations	Full form
SAMR	State Administration for Market Regulation
AQSIQ	State Administration for Quality Supervision, Inspection, and Quarantine
SLSI	Sri Lanka Standards Institution
SCE	Supercritical carbon dioxide extraction
SF	Supercritical fluid
SFE	Supercritical fluid extraction
SC-CO_2	Supercritical fluid extraction using carbon dioxide
TBT	Technical barriers to trade
TBHQ	Tertiary butylhydroquinone
TAC	Total antioxidant capacity
TFC	Total flavonoid compound
TPC	Total phenolic content
TRP	Transient receptor potential
TRPV	Transient receptor potential cation channel
TNF-α	Tumor necrosis factor alpha
USDA	U.S. Department of Agriculture
UV	Ultraviolet
UVB	Ultraviolet B
UAE	Ultrasonic-assisted extraction
UAHD	Ultrasound-assisted hydrodistillation
USC-CO_2	Ultrasound-assisted supercritical carbon dioxide
UAE	United Arab Emirates
UK	United Kingdom
USD	United State Dollar
U.S.	United States
USA	United States of America
WHO	World Health Organization
WTO	World Trade Organization

1 A Precise Spice

Anthony Temitope Idowu

1.1 INTRODUCTION

Spices are known to be used before civilization as a food additive for a several reasons ranging from appetite stimulation to attracting consumer appearance and to garnish the taste of food in diverse culinary preparation across the globe (Raghavan, 2006). Spices and herbs are sometimes misconstrued to mean the same thing, but in botany they have a defined terminology. A spice may be a flower stigma (saffron), bark (cinnamon), aromatic seed (cumin), root (ginger) and bud (clove) of a plant (Kaefer & Milner, 2012). While herbs reserve their flavour constituents in the leaves, it was reported that spices reserve theirs in the bark, seed and roots (Ogbunugafor et al., 2017). Some of the common groups of spices are fenugreek, coriander, cumin, thyme, garlic, onion, ginger, black pepper, turmeric, rosemary, oregano and clove etc. (Shahidi & Ambigaipalan, 2015; Srinivasan, 2005; Rahman & Husen, 2024). As scientific evolution occurs, a broad range of scientific investigation increased on the composition and significance of clove spices which, among others, gained prominence to fully explore their applications (Singh et al., 2021a).

Clove (*Syzygium aromaticum*) is a plant that was believed to have originated from the *Mirtaceae* family native located in the Maluku islands of the east Indonesia (Kamatou et al., 2012). Commercial harvesting of cloves is common in countries such as Pakistan, Madagascar, Tanzania, Zanzibar, Malaysia, India, China, Sri Lanka and Indonesia (Uddin et al., 2017; Singletary, 2014a). The flower bud of clove has often been dried and used as a spice since antiquity in delicacies of different cultures around the world (Sultana et al., 2014). In these delicacies, clove was added to give a scintillating flavour (Modupalli et al., 2022), to enhance the appearance of foods to attract consumers (Darriet, 2007) and also as a preservative to extend shelf life of foods (Ibrahium et al., 2013). These applications may be due to its pungent and fragrant smell coupled with a blend of sweet and astringent taste (Milind & Deepa, 2011). Apart from food application, clove spice was also reportedly used for medicinal purposes in ancient times (Parthasarathy et al., 2008; Singletary, 2014a). Currently, clove has been introduced into household products such as soaps, oral products, tobacco (kreteks) and perfumes (Singletary, 2014a).

In general, prevalent circumstances such as food, agricultural, medical, cosmetics and pharmaceutical industries are targeting bio-compounds (phytochemicals) from natural sources for use in new or existing food ingredients, colorants, preservatives and flavorant products. In addition, pharmaceutical industries are interested in bioactive compounds from plants for use in drug formulations, testing and applications. For instance, countries such as Norway and Australia have disallowed the use of tartrazine (E102), a synthetic yellow dye (Deville, 2018). Furthermore, there are strict

DOI: 10.1201/9781003521785-1

regulations by food safety agencies such as per batch certification to allow the use of synthetic yellow dye in the United States (Simon et al., 2017), thus creating opportunities for use of natural colourant such as cloves.

Furthermore, clove bud consists of essential oil (EO) whose composition and yield may vary based on the growth conditions, nutritional uptake of the plant geographic origins, different chemotypes and genetic factors (Hanif et al., 2019; Singh et al., 2022). The fragrant aroma in clove buds has been attributed to the presence of these EOs (Mittal et al., 2014). Elsewhere, it was documented that the EO contributed up to 17% of the fragrance and flavour market size worldwide which was valued to be around US$ 8 billion (Riaz et al., 2021). Global production of the EO was around 40,000 to 60,000 tons per annum with an expected demand of about 2,000 tonnes per year (Riaz et al., 2021). These EO have a range of functionality in food and pharmaceutical industries. Several studies have also reported the biological functions of dried, milled, EO and other extracted compounds from clove buds which appear to be promising for therapeutic applications. Therefore, this review aims to present an overview of clove components, some extraction techniques that are useful to obtain bioactive compounds from clove as well as the various biological functions reported in literature to further broaden the potential functionality of clove in the global market space.

1.2 BIOACTIVE COMPOUNDS/CHEMICAL CONSTITUENTS OF CLOVES

Natural bioactive compounds are clusters of phytochemicals that are useful for human healthy living (Li et al., 2019; Singh et al., 2021; Riar & Panesar, 2024). Xiao et al. (2018) reported that 40% of the drugs often used in medicine are obtained from natural compounds. These compounds are categorized into secondary (flavonoids, polyphenols, alkaloids, etc.) and primary (protein, fats, ash, vitamins, minerals and carbohydrate) metabolites in plants (Xiang et al., 2022). The secondary metabolites are known to contribute to the plant adaptability to the environment while the primary metabolites (chemical constituent) promote the growth and development of the plant (Qin & Xi, 2021; Choudhury et al., 2023). These bioactive compounds are known to contribute to the different biological functions that emanate from the plants.

Specifically, clove bud is made up of bioactive compounds such as eugenol, eugenol acetate and gallic acid (Idowu et al., 2021b). These bioactive compounds have been widely investigated and linked to their various biological functions such as antimicrobial, antioxidant, antiglycation, antinociceptive, antidiabetic, anticarcinogenic, anticancer and insecticidal (Rubió et al., 2013; Ryu et al., 2016; Pandey et al., 2024; Tungare et al., 2024; Shahzadiand & Sonmez, 2023). Other notable compounds in clove bud are α-humulene and β-caryophyllene, cinnamic and flavonoids (Haro-González et al., 2021). In Figure 1.1, the constituents of clove buds are briefly illustrated.

Chemical constituents such as ash, crude fat, crude protein, crude fibre, moisture and carbohydrates as well as the mineral profile of clove bud powders have been reported in some studies, some of which are presented in Table 1.1. The variations in the chemical composition may be due to the different drying conditions used,

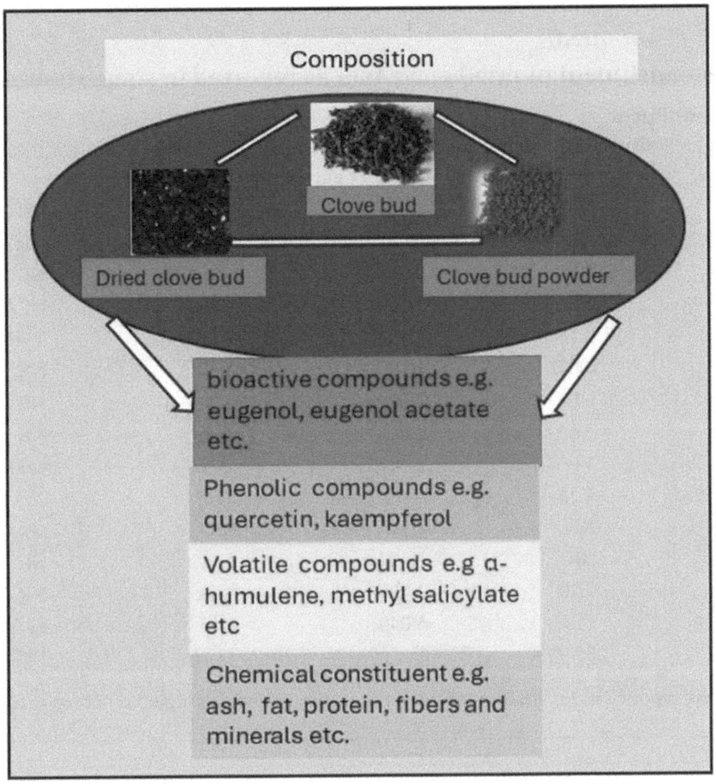

FIGURE 1.1 A brief overview of clove bud constituents.

TABLE 1.1
Chemical constituent of dried clove bud as reported in some studies.

Chemical constituent (proximate in % dry weight basis)	(Al-Jasass & Al-Jasser, 2012)	(Kaur et al., 2019)	(Lartey et al., 2023)	(Moulick et al., 2023)
Ash	5.96 ± 0.1	5.29 ± 0.08	NR	6.11 ± 0.07
Crude fat	16.63 ± 0.3	5.86 ± 0.03	16.33 ± 0.00	13.75 ± 0.46
Crude protein	6.9 ± 0.4	6.91 ± 0.37	7.24 ± 0.12	6.29 ± 0.20b
Crude fibre	11.47 ± 0.5	14.37 ± 0.04	10.57 ± 0.22	9.37 ± 0.03
Moisture	7.74 ± 0.2	29.47 ± 0.08	24.56 ± 0.07	10.48 ± 0.39
carbohydrate	51.3 ± 2.7	32.1 ± 0.05	36.02 ± 0.24	53.99 ± 0.17
Mineral composition (concentration in mg/kg)	(Al-Jasass & Al-Jasser, 2012)	(Kaur et al., 2019)	(Lartey et al., 2023)	(Moulick et al., 2023)
Cadmium	NR	0.03	NR	NR

(*Continued*)

TABLE 1.1 *(Continued)*
Chemical constituent of dried clove bud as reported in some studies.

Chemical constituent (proximate in % dry weight basis)	(Al-Jasass & Al-Jasser, 2012)	(Kaur et al., 2019)	(Lartey et al., 2023)	(Moulick et al., 2023)
Arsenic	NR	0.13	NR	NR
Nickel	NR	0.65	NR	NR
Lead	NR	0.90	NR	NR
Chromium	NR	1.16	NR	NR
Copper	0.08	5.42	2.27	2.304
Zinc	0.07	15.3	2.36	NR
Boron	NR	22.9	NR	NR
Iron	3.60	235	3.46	NR
Sodium	NR	244.45	29.11	6.90
Manganese	4.38	531	NR	265.10
Sulphur	NR	1,371	NR	NR
Phosphorus	NR	1,504.5	17.41	NR
Magnesium	9.70	2,504	0.03	NR
Calcium	27.00	5,040	0.15	42.30
Potassium	65.00	1,1560	129.6	145.29

* Values were reported as the mean standard deviation ($n = 3$) and NR signifies not reported.

differences in varieties, harvesting times, growth conditions and geographical locations, length of drying times, seasonality of harvest, plant variety, environmental conditions and method used during analysis etc. (Lartey et al., 2023).

1.3 COMMON DRYING AND EXTRACTION TECHNIQUES USED TO FRACTIONATE BIOACTIVE COMPOUNDS FROM CLOVES

1.3.1 Drying

Drying is often employed to reduce the amount of water, in some cases to about < 15%, in order to prevent microbial activity and also limit the biochemical changes (Hamrouni-Sellami et al., 2013; Orphanides et al., 2016). Common drying techniques are depicted in Figure 1.2 and some of them have been applied to obtain dried spices and herbs.

Some examples are hot air drying (Demiray & Tulek, 2014), microwave drying (Arslan et al., 2010), solar drying (Özcan et al., 2005) and freeze-drying (Gümüşay et al., 2015). Mild temperatures are used because of the heat sensitivity of bioactive compounds in cloves. Milling is basically employed to reduce the particle size of a plant material (Chimwani, 2021; Haros & Wronkowska, 2017). For instance, Yang et al. (2012) dried clove and cinnamon at a low temperature of 50°C for a period of 24 h.

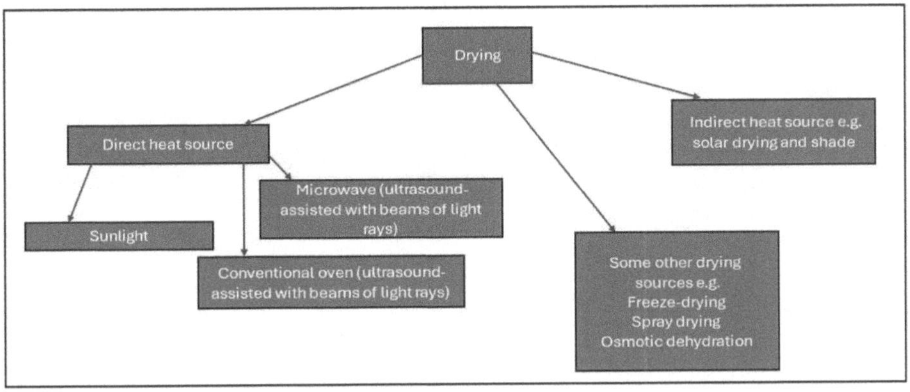

FIGURE 1.2 Some common drying techniques.

1.3.2 Extraction Methods for Fractionation of Bioactive Compound in Cloves

Extraction plays a vital role in obtaining bioactive ingredients from plant material both quantitatively and qualitatively (Huie, 2002). It is known that simple operational equipment such as heat reflux extraction, Soxhlet extraction, water batch extraction and maceration have been employed to obtain different bioactive compounds from plant biomass (Li et al., 2019). The limitations of this equipment is that it has high energy consumption and longer extraction time (Ramos et al., 2019).

Given a wide range of the bioactive constituents of cloves, a comprehensive method of extraction is thereby necessary to obtain an extract with useful biological functions. Therefore, judicious extraction procedures are necessary to enhance further separation, identification, purification and characterization of bioactive constituents. In some instances, a combination of different extraction methods rather than a single method may be considered to obtain an extract with enhanced active constituents from plant material (e.g. clove bud). In Figure 1.3, a brief illustration of the mechanism of extraction was shown.

Usually, the plant material biomass or milled powder is suspended in either an aqueous medium (e.g., distilled water), other polar solvent (methanol, acetone etc.) or non-polar solvent (chloroform, ether, hexane etc.) followed by either a single and/or a combination of other approaches such as hot plate continuous stirring, vacuum heating, hydrodistillation or irradiation treatment mediation (microwaves extraction). During this process, kinetic energy and pressure increase in the plant cell at a pre-set temperature, with stirring or irradiation to allow gradual permeability of the solvent. This ultimately leads to the collapse of the cell wall and tissues and then the release of the intact chemical constituents as depicted in Figure 1.3. However, further separation and fractionation techniques may be required to obtain the desired active constituent.

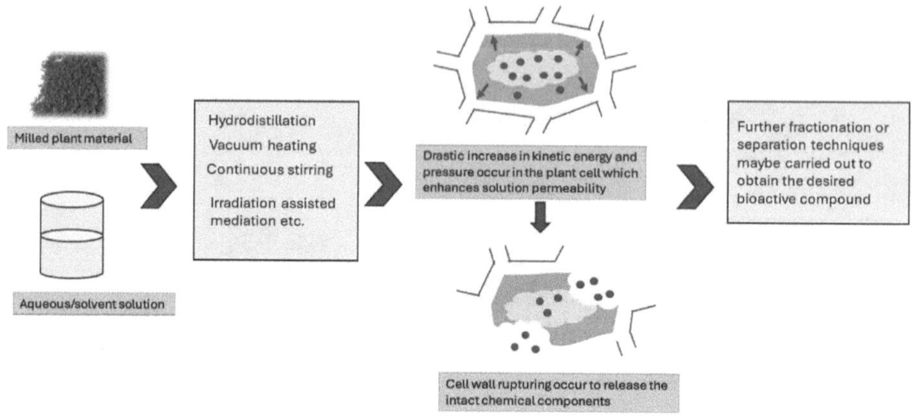

FIGURE 1.3 A basic schematic diagram to show the mechanism of extraction.

1.3.2.1 Aqueous/Solvent Extraction

Aqueous/solvent extraction is a cheap and common technique that is used to obtain bioactive compounds such as essential oil from plant material (Pandey et al., 2024). Most green extraction of bioactive compound from plant material uses an aqueous medium such as distilled water (Awad et al., 2021). Other polar solvents (methanol, acetone, ethanol, isopropanol acetonitrile etc.) or non-polar solvents (chloroform, ether, acetic acid, hexane etc.) are also considered for extraction of bioactive compounds from plant material. The use of these solvents appears to alter the desirable food flavour, and sometimes chemical residue is deposited in the extract (Chaichi et al., 2021). From the literature, some of these methods have been used to obtain eugenol and EO from clove bud. Ideally, milled clove buds are weighed on a filter paper and placed in a thimble followed by insertion in a reflux. Thereafter, a suitable solvent is selected and then extraction is carried out in a Soxhlet apparatus and the obtained extract in subjected to drying in a vacuum evaporator (Khalil et al., 2017; Quan et al., 2004). In a separate study, El-Maati et al. (2016) reported the use of aqueous solvent to obtain flavonoids and phenolic-rich extract from clove which possess good antimicrobial and antioxidant properties.

1.3.2.2 Hydrodistillation

This method is practically preferred in order to obtain essential oil (El Kharraf et al., 2020). A typical illustration for carrying out hydrodistillation extraction is shown in Figure 1.4.

This is usually carried out by dispersing a known amount of clove powder (between 50–100g) in a known amount of distilled water placed in a volumetric flask, and then the hydrodistillation is carried out for about 4–6 hrs. Organic solvents or petroleum are added before collection and saturation of the of the volatile distillate with sodium chloride. Afterwards, the hydro and ether constituents are separated using sodium sulphate. This is followed by heating the sample in a water bath at 60°C to obtain the

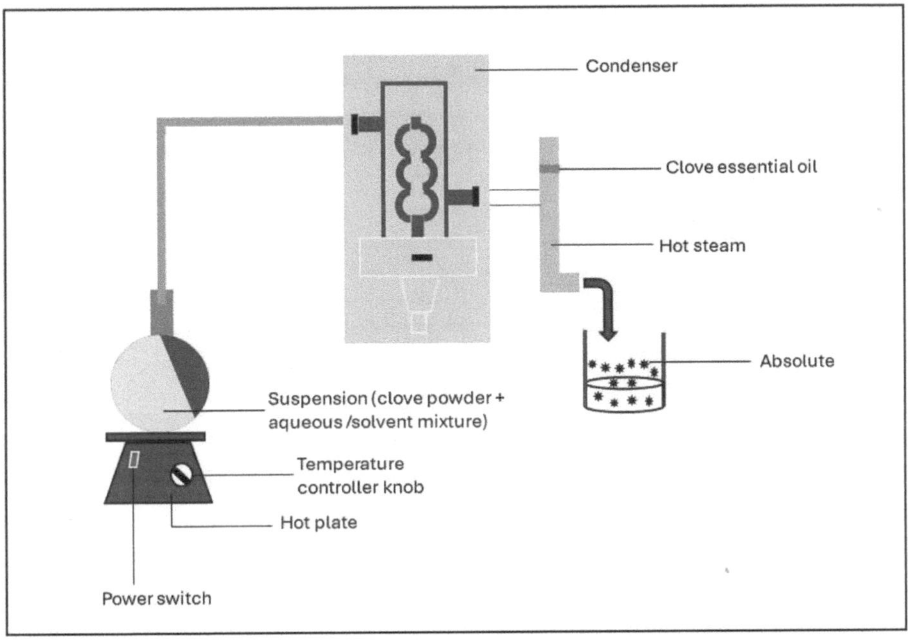

FIGURE 1.4 Brief schematic illustration of hydrodistillation procedure.

extract concentrate and the ether constituents. Guan et al. (2006) stated the possibility to obtain up to 11.5% of EO that contains around 50% of eugenol compound using the hydrodistillation procedure. By further reduction of milled clove bud particle size, the yield of oil obtained can be enhanced (Mostafa Khajeh et al., 2004).

1.3.2.3 Microwave-assisted Extraction

Microwave-assisted extraction has been identified as a prominent green approach that can be used to obtain eugenol and other EO whose sensory attribute is within the range of those obtained by conventional methods (Liu et al., 2020). A typical illustration of microwave-assisted extraction is shown in Figure 1.5.

Different configurations have emerged such as microwave-assisted hydro-distillation, microwave steam distillation, coaxial microwave-assisted hydro-distillation and microwave-assisted hydro-diffusion in a bit to extract bio-compounds from plant tissues. From literature, the use of coaxial microwave-assisted hydrodistillation (CMH-D) extraction was used to extract EO from different herbs, for example sage, fennel, lavender and clove, and this was compared with the crude hydrodistillation procedure (González-Rivera et al., 2016). Findings from the study showed that microwave extraction was faster than the hydrodistillation procedure under similar incubation time for extraction of bio-compounds from cloves. Elsewhere, it was stated that CMH-D extraction could help to save energy demand (30%) and heating time (400%) in comparison to other microwave extraction procedures (Khalil et al., 2017).

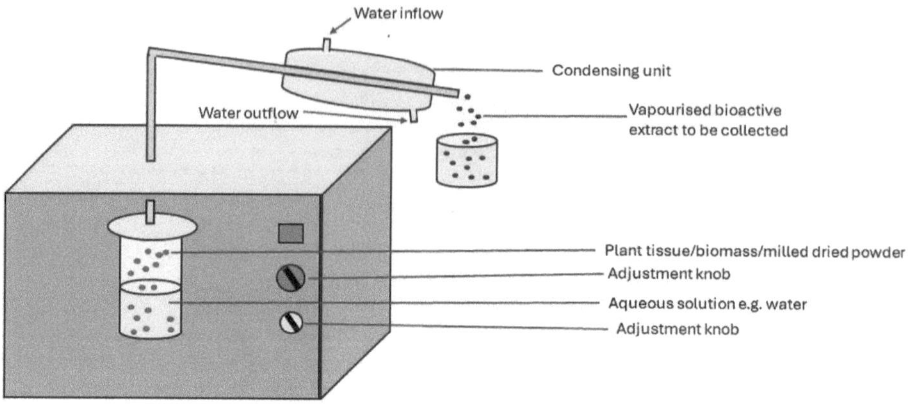

FIGURE 1.5 A brief schematic illustration of a microwave oven set up.

1.3.2.4 Ultrasound-assisted Extraction (UAE)

This is another technique that has been introduced to accelerate the extraction processes. This technique has been used to recover bioactive ingredients from herbs and spices (Benarfa et al., 2020). Some notable benefits of UAE in comparison to conventional methods is that it is less tedious, no residue is deposited in the extract, it is eco-friendly, it provides enhanced quality, it allows for prevention of extract degradation and it facilitates easy handling of extracts (Chemat & Khan, 2011). In a separate study, Alexandru et al. (2013) obtained polyphenol rich extract from clove bud. This was achieved by suspending the milled clove in a solution of ethanol and water assisted with UAE. Elsewhere, Tekin et al. (2015) obtained EO from clove using UAE techniques.

1.3.2.5 Supercritical Carbon Dioxide Extraction (SCE)

This methodical approach has been investigated for extraction of bioactive compounds, e.g. eugenol from clove buds. For instance, Reverchon and Marrone (1997) obtained eugenol from clove bud using SCE. It was concluded that the yield of eugenol obtained can be enhanced by increasing the flow rate of CO_2 during SCE. Elsewhere, Guan et al. (2007) obtained EO using SCE in comparison with steam distillation, Soxhlet extraction and hydrodistillation methods. The study showed that the EO obtained using SCE was of higher quality in comparison to the other method used. Similar findings were reported by Yu et al. (2009) whereby EO was obtained by SCE which showed higher quality and required lower extraction time than those obtained by organic solvent reflux method and hydrodistillation.

1.4 BIOLOGICAL FUNCTIONS OF CLOVES BLENDS, POWDER, AND ASSOCIATED EXTRACTED BIOACTIVE COMPOUNDS

1.4.1 Antimicrobial

It is known that foodborne diseases, food poisoning and food spoilage do occur when food is contaminated by either gram positive and gram negative bacteria (Mostafa et al., 2018). Hence, extracts from plants are often used to counter microbial activities to reduce foodborne disease and spoilage. Some studies have tested and reported the potency of clove extracts, eugenol or EO to act as antimicrobials against bacterial and fungi strain. Xu et al. (2016) attributed the antimicrobial function of clove to the presence of bioactive compounds which function to disrupt the cellular membrane of the microorganisms and sometimes loss of cellular materials as well as cell death. Some of the antimicrobial properties of clove bud powder, essential oil and extracts are enumerated in Table 1.2.

1.4.2 Antioxidant

Antioxidants help to prevent the oxidative destruction of the human cells or body tissues from free radicals (Idowu et al., 2021a; Idowu et al., 2021b). The use of synthetic antioxidant is currently downplayed for use because it is reported to be carcinogenic (Sindhi et al., 2013), thus shifting attention to natural antioxidants such as plant phenols that are known to be high in antioxidant compound and polyphenolic (Cortés-Rojas et al., 2014). Some of the of the studies confirming the potential of cloves as antioxidants are reported in Table 1.2. Some of the tests used include chelating metal, scavenging and reducing power assay to ascertain the potency of cloves.

1.4.3 Antiglycation

Glycation leads to the generation of irreversible compounds that are called advanced glycation end products (AGEP) (Singh et al., 2001). Excessive tissue accumulation of AGEP can proliferate some health disorders and associated complications such as diabetes, Alzheimer's disease, kidney diseases, atherosclerosis and chronic heart failure (Uribarri et al., 2015; Sadowska-Bartosz & Bartosz, 2016). Some investigations on the antiglycation properties of cloves are showed in Table 1.2. In some instances, the antiglycation function of cloves was linked to their phenolic constituents.

1.4.4 Antinociceptive

Antinociceptive characteristics are attributed to an increase in tolerance and its ability to lower sensitivity to a harmful sensation or pain (Ginzburg et al., 2015). From literature, various studies have reported the possibility of cloves to perform an analgesic function by reducing toothache, as an antispasmodic and in relieving joint pain. This therapeutic attribute was associated with the presence of eugenol in clove (Cortés-Rojas et al., 2014). The mechanism entails the activation of calcium and sodium channels in ganglionar cells (Li et al., 2008). Elsewhere, it was stated that the analgesic attribute of cloves was due to its function as capsaicin agonist (Ohkubo & Shibata, 1997). Investigations on the antinociceptive characteristics of cloves are showed in Table 1.2.

1.4.5 Antidiabetic

Abnormalities in the body glucose metabolism leads to diabetes mellitus (DM) which is a degenerative disease (Khalil et al., 2017). This occurrence is related to a reduction in production of pancreatic insulin and/or body passivity which result in hyperglycaemia, a pre-condition to for diabetes (Veerapur et al., 2012). Instability of glucose metabolism poses a threat to the endocrine system-secreted hormones which eventually cause a disturbance in endocrine control (Veerapur et al., 2012). Diet has been identified as one of the strategies that can be used to combat diabetes since it is instrumental in controlling the hyper-glycosylation of biologically active molecules linked with various metabolic pathways, blood sugar levels as well as pathogen inhibition (Khalil et al., 2017). The rich polyphenol contains of cloves contributed to its exploitation as a potential antidiabetic agent. Some of the studies reported that clove performs similar functions to insulin by controlling the gene expression linked to diabetes such as phosphoenolpyruvate carboxykinase (PEPCK) and glucose-6-phosphatase (G6Pase) (Prasad et al., 2005). Some of the various studies on the antidiabetic potential of cloves are enumerated in Table 1.2.

1.4.6 Anticancer

Cancer occur when there is a rapid cell division that leads to cell aggregation and then to tumour formation (Khalil et al., 2017). Numerous causes of cancer at the onset are ingestion of heavy metals, pesticides, smoking, lack of safe and healthy diet, genetic mutations etc. (Yousefzadi et al., 2014). Amelioration of cancer is achieved by inhibiting the cells' propagation and obliteration of the malicious cells (Jaganathan & Supriyanto, 2012). Scientific investigations have reported the chemo-protective viability of eugenol to inhibit the proliferation of different cancerous cell activities. In Table 1.2, some of the studies on the anticancer effects using cell cultures and rat studies focused on thyroid, colon and breast cancer.

Overall, clove constituents (eugenol, EO, volatile compounds etc.) may need to be explored in community-based/clinical or human trials in order to ascertain the dosage fit for human consumption as well as to explore the possibility of usage in other numerous unknown biological functions. Furthermore, the absorption mechanism in the gastrointestinal tract and bioavailability studies is recommended.

1.4.7 Insecticidal Function

The use of chemicals and additives for controlling parasitic anthropoids are facing the problems of environmental pollution and drug resistance. Meanwhile, *Eugenia caryophyllata* compound has been identified for its biological functions against parasites. Studies have also demonstrated its insecticidal function against *Culex pipiens larvae*, *Pediculus capitis*, *Anopheles dirus* mosquitoes, *Tribolium castaneum* and also to decrease the reproduction cycle of *Sitophilus zeamais* when used with isoeugenol (Ho et al., 1994). Some of the other findings on the insecticidal functions are briefly enumerated in Table 1.2

TABLE 1.2
Some of the biological functions of clove that were investigated and reported in different studies.

Test sample	Biological function tested	Experiment	Brief outcome of study	References
EO from Oregano and clove	Antimicrobial	EOs from test samples were tested for antifungal efficacy against *Zygosaccharomyces bailii*	Showed efficacy as inhibitors against the tested fungi	(Ribes et al., 2019)
EO from clove	Antimicrobial	EO was tested on gram-negative bacteria *Pseudomonas aeruginosa* and *Escherichia coli* and on *Bacillus cereus* and *Staphylococcus aureus*	EO exhibited antibacterial properties against the tested bacteria strain	(Ogunwande et al., 2005)
Eugenol and its isomer isoeugenol	Antimicrobial	Antibacterial test against *Staphylococcus aureus, Listeria monocytogenes* and *Bacillus subtilis*	Good antibacterial potentials on all tested bacteria strains	(Zhang et al., 2017b)
Clove oil	Antimicrobial	Antibacterial test on gram-negative *Erminia carotovora pv, Agrobacterium tumefaciens, Xanthomonas campestris pv* and gram-positive *Rhodococcus fascians* and *Streptomyces sp.*	Functioned as antibacterial against tested bacteria strains	(Huang & Lakshman, 2010)
EO from clove	Antimicrobial	Antifungal efficacy against *Candida sp*	EO exhibited powerful antifungal activity against the fungi	(Chaieb et al., 2007)
Clove oil combined with a chitosan coating	Antimicrobial	Possibility to extend shelf life against bacterial strain e.g., *Pseudomonas sp.* and *Enterobacteriaceae*	Reduction in bacterial activity	(Li et al., 2017)
Clove oil	Antimicrobial	Tested against *Aspergillus favus* in peanuts to investigate antifungal effect	Reduction in fungi population was observed	(Matan & Jantamas, 2011)
EO from aniseed, Lemongrass and clove	Antimicrobial	Antibacterial effect on wilt (*Ralstonia solanacearum*)	Showed potency against *R. solanacearum*	(Deberdt et al., 2018)
Thymol, menthol, carvacrol and eugenol	Antimicrobial	Antifungal test against *Aspergillus sp* and *Cladosporium sp*	Compounds tested showed antifungal activity	(Abbaszadeh et al., 2014)

(Continued)

TABLE 1.2 ***(Continued)***
Some of the biological functions of clove that were investigated and reported in different studies.

Test sample	Biological function tested	Experiment	Brief outcome of study	References
EO from clove	Antimicrobial	Antifungal properties against pathogens such as *Colletotrichum gloeosporioides, Alternaria citrii, Lasiodiplodia theobromae, Penicillium digitatum* and *Botrytis cinerea*	Good fungicide activity was observed	(Combrinck et al., 2011)
Oregano, geranium, nutmeg, black pepper, thyme and clove	Antimicrobial	Antibacterial activity against some gram positive and negative bacteria strain	Oregano, thyme and clove showed the widest antibacterial range	(Dorman & Deans, 2000)
Cinnamon, mustard, mint, garlic and clove	Antimicrobial	Efficiency against pathogens (Staphylococcus aureus, Escherichia coli and Bacillus cereus) were tested	At 3% concentration antibacterial activity for clove was observed.	(Sofia et al., 2007)
Clove EO, Oregano (*Origanum vulgare*), bay (*Pimenta racemosa*) and thyme (*Thymus vulgaris*)	Antimicrobial	Tested against *Escherichia coli*	Inhibition against the micro-organism was observed	(Burt & Reinders, 2003)
Eugenol	Antimicrobial	Tested against *Salmonella typhi*	Eugenol deformed and disrupted the cellular membrane of *Salmonella typhi*	(Devi et al., 2010)
Clove oil	Antimicrobial	Test against dermatophytes such as *Epidermophyton foccosum, Trichophyton rubrum, Trichophyton mentagrophytes, Microsporum canis* and *Microsporum gypseum*	At concentration of 0.2 mg/mL a maximum activity with up to 60% effectiveness was observed	(Park et al., 2007)
Aqueous, methanolic and acetone clove extract	Antioxidant	DPPH activity	DPPH activity: acetone > methanol > water	(Adaramola & Onigbinde, 2016)
Clove bud oil	Antioxidant	DPPH activity	Positive DPPH activity	(Alfikri et al., 2020)

TABLE 1.2 ***(Continued)***
Some of the biological functions of clove that were investigated and reported in different studies.

Test sample	Biological function tested	Experiment	Brief outcome of study	References
Essential oil from clove	Antioxidant	DPPH activity	Positive DPPH activity	(El-Mesallamy et al., 2012)
Essential oil from clove	Antioxidant	Inhibition against oxidation of aliphatic aldehyde hexanal to carboxylic acid	Moderate inhibition against oxidation of hexanal	(Misharina & Samusenko, 2008)
Clove bud oil	Anti-tyrosinase and antioxidant activity	Anti-tyrosinase: antimelanogenic effect of clove bud oil using tyrosinase as substrate. Antioxidant: DPPH and hydrogen peroxide scavenging activity	Clove bud oil showed antityrosinase activity and antioxidant activity	(Ahmed, 2016)
Ethanolic extract	Antioxidant	DPPH activity	Extract exhibited antioxidant activity	(Baghshahi et al., 2014)
Methanolic extract from clove bud	Antioxidant	Estimation of total antioxidant capacity by the $ABTS^+$ method	Antioxidant activity (168.660 ± 0.024 in mmol of Trolox/100 g dry weight)	(Shan et al., 2005)
Water and ethanolic extracts of clove buds	Antioxidant	Free radical scavenging, reductive potential, metal chelating, superoxide anion radical scavenging activities	Antioxidant activity of more than 80% was observed	(Gülçin et al., 2004)
Aqueous extract	Antioxidant	Inhibitory effect toward oxidation of aldehyde to acid	Inhibited effect of more than 80% was observed	(Lee & Shibamoto, 2001)
Clove oil	Antioxidant	Nitric oxide free-radical scavenging and DPPH assay	Activity with IC_{50} of about 10.87 to 31.63 µg/mL	(Kaur et al., 2019)
Methanolic extract	Antioxidant	Reducing power, DPPH, β-carotene-linoleate, hydroxyl radical and ferric thiocyanate	Similar activity was obtained with butylated hydroxy toluene (BHT) synthetic antioxidant	(Bamdad et al., 2006)
Aqueous and methanol extract	Antioxidant	Inhibition of linoleic acid and DPPH radical scavenging capacity	Inhibition of linoleic acid: 54.96–86.89% and DPPH activity: 71.16–94.58%	(Sultana et al., 2014)
Ethanolic extract	Antiglycation	Tested for its ability to inhibit glycation of albumin	Extract exhibited antiglycation effect	(Dearlove et al., 2008)

(Continued)

TABLE 1.2 ***(Continued)***
Some of the biological functions of clove that were investigated and reported in different studies.

Test sample	Biological function tested	Experiment	Brief outcome of study	References
Aqueous/ethanol extract	Antiglycation	Inhibitory effect exerted against glycation in the presence or absence of a substance or an extract using a reaction model system	Positive inhibitory activity was observed	(Starowicz & Zieliński, 2019)
Eugenol	Antinociceptive	Anti-inflammatory and analgesic effect	Moderate possibility to relief acute pain and greater efficiency to relieve inflammatory-type pain	(Lugo-Lugo et al., 2019)
Eugenol	Antinociceptive	Inflammatory effect	Positive inflammatory activity	(Daniel et al., 2009)
Methanolic and aqueous extract of clove	Antinociceptive	Analgesic effect	Aqueous extract of clove has analgesic effect	(Asl et al., 2013)
Clove bud extract	Antioxidant	DPPH, 2,2′-azino-bis(3-ethylbenzthiazoline-6-sulfonic acid) (ABTS), ferrous reducing antioxidant power (FRAP)	Moderate antioxidant activity	(Dev et al., 2021)
Clove bud extract	Antioxidant	DPPH and reducing power	Good antioxidant capacity	(Adefegha et al., 2016)
Clove bud powder	Nitrite scavenging abilities, antimicrobial and antioxidant	Nitrite scavenging abilities, antimicrobial and antioxidant were investigated at (0.1 and 0.2%) incorporation in emulsified pork sausage	Positive nitrite scavenging activity but limited antioxidant and antimicrobial activity	(Jin et al., 2016)
Clove extract	Antioxidant	DPPH, ABTS and hydrogen peroxide scavenging	Good antioxidant activity that is in comparison with the synthetic antioxidant	(Gülçin et al., 2012)
Clove bud extract	Antioxidant and antimicrobial	Antioxidant: DPPH Antimicrobial: test against gram-positive bacteria (*Staphyloccocus aureus* and *Bacillus subtilis*). Gram negative bacteria (*Vibrio cholerae* and *Klebsiella pneumonia*)	Good antioxidant and antimicrobial activity	(Hemalatha et al., 2016)

TABLE 1.2 ***(Continued)***
Some of the biological functions of clove that were investigated and reported in different studies.

Test sample	Biological function tested	Experiment	Brief outcome of study	References
Essential oil from clove bud	Antioxidant and antimicrobial	Antioxidant: DPPH Antimicrobial: tested against activity of bacterial pathogen such as *S. aureus, B. cereus, E. coli* and *K. pneumoniae*	Positive DPPH activity and inhibitory effect against tested bacterial pathogen	(Nada et al., 2022)
Clove extract	Antioxidant	DPPH and total reduction power	Extract demonstrated good antioxidant activity	(Nikousaleh & Prakash, 2016)
Clove extract	Antioxidant	DPPH and total antioxidant power	Good antioxidant capacity	(Moulick et al., 2023)
Clove bud powder (CBP)	Antidiabetic	Test on biochemical effect of CBP on a type 2 diabetes rat	Reduction in blood glucose level in diabetic rat in comparison to the rat that was not fed the CBP	(Adefegha et al., 2014)
Phenolic extract from clove bud	Antidiabetic	The extract was tested for α-glucosidase and α-amylase inhibitory activity	Extracts showed potency against α-amylase and α-glucosidase activity	(Adefegha & Oboh, 2012)
Isoeugenol from clove oil	Antidiabetic and anticholinergic	Antidiabetic: inhibitory activity against α-glycosidase, and α-amylase Anticholinergic: test against acetylcholinesterase activity	Isoeugenol showed excellent activity against α-glycosidase and α-amylase as well as against acetylcholinesterase	(Topal, 2019)
Clove bud essential oil	Anticancer and antibacterial	Anticancer: using thyroid cancer cell culture (HTh-7) and non-tumorigenic (cell HeK-293). Antibacterial: test against *S. aureus* activity	Anticancer: reduction in cancer cell proliferation Antibacterial: positive inhibitory effect against *S. aureus* activity	(Nirmala et al., 2019)
Clove bud extract	Anticancer	In vivo and in vitro mammary carcinoma model	Study showed antiproliferative effect of the extract	(Kubatka et al., 2017)
Clove bud extract	Anticancer	Inhibition of cell migration and induction of apoptosis	Positive anticancer activity	(Ali et al., 2019)

(Continued)

TABLE 1.2 ***(Continued)***
Some of the biological functions of clove that were investigated and reported in different studies.

Test sample	Biological function tested	Experiment	Brief outcome of study	References
Clove essential oil nanoemulsion	Anticancer	Apoptotic effect and anti tumour impact against colon cancer cells (HT-29)	Moderate anticancer activity	(Abadi et al., 2022)
Volatile oil from clove	Anticancer	Anticancer effect on breast carcinoma cell line	Oil inhibited the growth of MCF7 cell line at increasing concentration	(El-Darier et al., 2018)
EO from clove and garlic	Insecticidal function	Fumigant test on Japanese terminate (*Reticulitermes speratus*)	Potent insecticidal activity was observed	(Park & Shin, 2005a)
Eugenol	Insecticidal function	Acaricidal effects on *Dermatophagoides pteronyssinus* and *Dermatophagoides farinae*	Exhibited acaricidal effects	(Kim et al., 2003)
Clove oil	Insecticidal function	Tested against *Leptotrombidium* chiggers (larvae)	Result showed the repellency of clove oil against *Leptotrombidium* chiggers (larvae)	(Eamsobhana et al., 2009)
Clove oil	Insecticidal function	Nematicidal effect on pine wood nematode *(Bursaphelenchus xylophilus)*	Potent nematicidal effect	(Park & Shin, 2005b)
Clove oil	Insecticidal function	Repellent effect against mosquitoes such as *Anopheles dirus Peyton* and *Harrion* as well as *Aedes aegypti (L.)*	Result showed positive repellent effect against the mosquitoes	(Sritabutra et al., 2011)
Eugenol acetate, eugenol and β-caryophyllene	Insecticidal function	Repellent effect against free ants (*Solenopsis Invicta*)	Exhibited repellent activity	(Kafle & Shih, 2013)
Clove oil	Insecticidal function	Repellent effect against paper wasps e.g. social wasps, *Vespula pensylvanica* (Saussure) and *Polistes dominulus*	Moderate repellent effect	(Zhang et al., 2014)
EO from cinnamon and cloves	Insecticidal function	Larvicidal efficacy against *Anopheles* larvae	Mortality rate of the larvae increased	(Thomas et al., 2017)

1.5 TOXICITY AND DOSAGE

An investigations into the cellular activities of clove compounds by assessing their toxicity helps to ascertain their safety (Nagaraju et al., 2021). According to Vijayasteltar et al. (2016), the food and drug administration (FDA) of the United States has corroborated that eugenol, clove oil, oleoresins and clove bud are safe as a food supplement. The European Food Safety Authority has highlighted that eugenol is not carcinogenic and did not show any neurotoxic and teratogenic effects (Singletary, 2014b). However, the World Health Organization has recommended that the acceptable daily intake of clove for human consumption should be around 2.5 mg/kg body weight (Ogunwande et al., 2005). Clove oil was reported to be safe as an additive in small proportions (< 1,500 ppm) but may result in severe complications such as liver failure, nervous system depression and seizures, respiratory distress and systemic poisoning (Janes et al., 2005; Dyrbye et al., 2012; Brown et al., 1992; Lane et al., 1991).

Elsewhere, Prashar et al. (2006) investigated the in vitro cytotoxicity effect of eugenol and clove EO against endothelial cells and human fibroblasts and the outcome of the study indicated that they are safe. In other instances, studies have also reported some side effects of clove compounds, for instance, allergic reaction to eugenol when utilized for dentistry purposes (Sarrami et al., 2002; Anuj & Sanjay, 2010). Also, clove EO and eugenol when tested in an in vitro study were observed to have a spermicidal effect on six male partners of infertile couples (mice). However, further studies affirmed the cytotoxicity of clove.

1.6 FUTURE OUTLOOK ON CLOVE AS A FUNCTIONAL FOOD

Functional foods are known to be consumed for nutritional purposes as well as for their targeted function in the body system (Viuda-Martos et al., 2010). In the current review, the presence of bioactive compounds and various biological functions of clove (Table 1.2) reported in various studies and presented herein is a testament to the possibility of clove to serve as a functional food.

For future trends and considerations, the demand for clove as a spice or condiment to enhance flavour, colour, shelf life and aroma is expected to increase in the food and spice industry. So far, some studies have elucidated the application of clove oil, extracts and blends for use in cheese (Ratiba et al., 2006), pasteurized milk (Cava et al., 2007), sausages (Zhang et al., 2017a), meat products (Wang et al., 2018; Pateiro et al., 2021; Khaleque et al., 2016) and food packaging (Sharma et al., 2020) as well as improvements in oxidative stability, natural preservatives, colour and shelf life. Hence, there is a possibility of continuous demand for clove in the global market space in food processing.

Furthermore, bioactive components extracted from clove are also applicable for use in the pharmaceutical industry for use in therapeutic drug development owing to its numerous biological functions. However, optimization of the extraction techniques may need to be investigated to obtain the consistent high-quality active ingredient in cloves. In addition, waste or residue (by-product) generated after extraction may need to be valorised for use in the food or agro sector to fully maximize its potential application. For instance, clove bud residue after extraction may be investigated for potential use as a fibre source.

1.7 CONCLUSION

In summary, the information presented herein in terms of the composition and biological functions of clove are pointers to the potential significant value of cloves for use in the food and pharmaceutical industry. However, appropriate selection of single or combined extraction techniques is necessary to fully utilize cloves for use in the food and pharmaceutical industry. Overall, it can be inferred that the clove is a potential functional food that goes beyond its application as a spice.

REFERENCES

Abadi, A. V. M., Karimi, E., Oskoueian, E., Mohammad, G. R. K. S. & Shafaei, N. (2022). Chemical investigation and screening of anti-cancer potential of *Syzygium aromaticum* L. bud (clove) essential oil nanoemulsion. *3 Biotechnology*, 12, 49.

Abbaszadeh, S., Sharifzadeh, A., Shokri, H., Khosravi, A. & Abbaszadeh, A. (2014). Antifungal efficacy of thymol, carvacrol, eugenol and menthol as alternative agents to control the growth of food-relevant fungi. *Journal de Mycologie Medicale*, 24, e51–e56.

Adaramola, B. & Onigbinde, A. (2016). Effect of extraction solvent on the phenolic content, flavonoid content and antioxidant capacity of clove bud. *IOSR Journal of Pharmaceutical and Biological Science*, 11, 33–38.

Adefegha, S. A. & Oboh, G. (2012). *In vitro* inhibition activity of polyphenol-rich extracts from Syzygium aromaticum (L.) Merr. & Perry (Clove) buds against carbohydrate hydrolyzing enzymes linked to type 2 diabetes and Fe2+-induced lipid peroxidation in rat pancreas. *Asian Pacific Journal of Tropical Biomedicine*, 2, 774–781.

Adefegha, S. A., Oboh, G., Adefegha, O. M., Boligon, A. A. & Athayde, M. L. (2014). Antihyperglycemic, hypolipidemic, hepatoprotective and antioxidative effects of dietary clove (*Szyzgium aromaticum*) bud powder in a high-fat diet/streptozotocin-induced diabetes rat model. *Journal of the Science of Food and Agriculture*, 94, 2726–2737.

Adefegha, S. A., Oboh, G., Oyeleye, S. I. & Osunmo, K. (2016). Alteration of starch hydrolyzing enzyme inhibitory properties, antioxidant activities, and phenolic profile of clove buds (*Syzygium aromaticum L.*) by cooking duration. *Food Science & Nutrition*, 4, 250–260.

Ahmed, W. (2016). Monitoring antioxidant and antityrosinase activity of clove aromatic flower buds. *Journal of Medicinal Plants Studies*, 4, 163–169.

Alexandru, L., Cravotto, G., Giordana, L., Binello, A. & Chemat, F. (2013). Ultrasound-assisted extraction of clove buds using batch-and flow-reactors: A comparative study on a pilot scale. *Innovative Food Science & Emerging Technologies*, 20, 167–172.

Alfikri, F. N., Pujiarti, R., Wibisono, M. G. & Hardiyanto, E. B. (2020). Yield, quality, and antioxidant activity of clove (*Syzygium aromaticum L.*) bud oil at the different phenological stages in young and mature trees. *Scientifica*, 2020.

Ali, I., Naqshbandi, M. F. & Husain, M. (2019). Cell migration and apoptosis in human lung cancer cells by Clove (Syzygium aromaticum) dried flower buds extract. *Journal of Taibah University for Science*, 13, 1163–1174.

Al-Jasass, F. M. & Al-Jasser, M. S. (2012). Chemical composition and fatty acid content of some spices and herbs under Saudi Arabia conditions. *The Scientific World Journal*, 2012.

Anuj, G. & Sanjay, S. (2010). Eugenol: A potential phytochemical with multifaceted therapeutic activities. *Pharmacologyonline*, 2, 108–120.

Arslan, D., Özcan, M. M. & Mengeş, H. O. (2010). Evaluation of drying methods with respect to drying parameters, some nutritional and colour characteristics of peppermint (*Mentha x piperita L.*). *Energy Conversion and Management*, 51, 2769–2775.

Asl, M. K., Nazariborun, A. & Hosseini, M. (2013). Analgesic effect of the aqueous and ethanolic extracts of clove. *Avicenna Journal of Phytomedicine*, 3, 186.

Awad, A. M., Kumar, P., Ismail-Fitry, M. R., Jusoh, S., Ab Aziz, M. F. & Sazili, A. Q. (2021). Green extraction of bioactive compounds from plant biomass and their application in meat as natural antioxidant. *Antioxidants*, 10, 1465.

Baghshahi, H., Riasi, A., Mahdavi, A. & Shirazi, A. (2014). Antioxidant effects of clove bud (*Syzygium aromaticum*) extract used with different extenders on ram spermatozoa during cryopreservation. *Cryobiology*, 69, 482–487.

Bamdad, F., Kadivar, M. & Keramat, J. (2006). Evaluation of phenolic content and antioxidant activity of Iranian caraway in comparison with clove and BHT using model systems and vegetable oil. *International Journal of Food Science & Technology*, 41, 20–27.

Benarfa, A., Gourine, N., Hachani, S., Harrat, M. & Yousfi, M. (2020). Optimization of ultrasound-assisted extraction of antioxidative phenolic compounds from Deverra scoparia Coss. & Durieu (flowers) using response surface methodology. *Journal of Food Processing and Preservation*, 44, e14514.

Brown, S., Biggerstaff, J. & Savidge, G. (1992). Disseminated intravascular coagulation and hepatocellular necrosis due to clove oil. *Blood Coagulation & Fibrinolysis*, 3, 665–668.

Burt, S. A. & Reinders, R. D. (2003). Antibacterial activity of selected plant essential oils against Escherichia coli O157: H7. *Letters in Applied Microbiology*, 36, 162–167.

Cava, R., Nowak, E., Taboada, A. & Marin-Iniesta, F. (2007). Antimicrobial activity of clove and cinnamon essential oils against *Listeria monocytogenes* in pasteurized milk. *Journal of Food Protection*, 70, 2757–2763.

Chaichi, M., Mohammadi, A., Badii, F. & Hashemi, M. (2021). Triple synergistic essential oils prevent pathogenic and spoilage bacteria growth in the refrigerated chicken breast meat. *Biocatalysis and Agricultural Biotechnology*, 32, 101926.

Chaieb, K., Zmantar, T., Ksouri, R., Hajlaoui, H., Mahdouani, K., Abdelly, C. & Bakhrouf, A. (2007). Antioxidant properties of the essential oil of *Eugenia caryophyllata* and its antifungal activity against a large number of clinical Candida species. *Mycoses*, 50, 403–406.

Chemat, F. & Khan, M. K. (2011). Applications of ultrasound in food technology: Processing, preservation and extraction. *Ultrasonics Sonochemistry*, 18, 813–835.

Chimwani, N. (2021). A review of the milestones reached by the attainable region optimisation technique in particle size reduction. *Minerals*, 11, 1280.

Choudhury, A., Singh, P. A., Bajwa, N., Dash, S. & Bisht, P. (2023). Pharmacovigilance of herbal medicines: Concerns and future prospects. *Journal of Ethnopharmacology*, 309, 116383.

Combrinck, S., Regnier, T. & Kamatou, G. P. (2011). *In* vitro activity of eighteen essential oils and some major components against common postharvest fungal pathogens of fruit. *Industrial Crops and Products*, 33, 344–349.

Cortés-Rojas, D. F., De Souza, C. R. F. & Oliveira, W. P. (2014). Clove (*Syzygium aromaticum*): A precious spice. *Asian Pacific Journal of Tropical Biomedicine*, 4, 90–96.

Daniel, A. N., Sartoretto, S. M., Schmidt, G., Caparroz-Assef, S. M., Bersani-Amado, C. A. & Cuman, R. K. N. (2009). Anti-inflammatory and antinociceptive activities A of eugenol essential oil in experimental animal models. *Revista Brasileira de Farmacognosia*, 19, 212–217.

Darriet, A. (2007). Herbs, spices and essential oils. In *Handbook of Food Products Manufacturing*. Hui, Y. H., Ed. John Wiley & Sons, 205–220.

Dearlove, R. P., Greenspan, P., Hartle, D. K., Swanson, R. B. & Hargrove, J. L. (2008). Inhibition of protein glycation by extracts of culinary herbs and spices. *Journal of Medicinal Food*, 11, 275–281.

Deberdt, P., Davezies, I., Coranson-Beaudu, R. & Jestin, A. (2018). Efficacy of leaf oil from *Pimenta racemosa* var. racemosa in controlling bacterial wilt of tomato. *Plant Disease*, 102, 124–131.

Demiray, E. & Tulek, Y. (2014). Drying characteristics of garlic (*Allium sativum L*) slices in a convective hot air dryer. *Heat and Mass Transfer*, 50, 779–786.

Dev, M., Bhattacharyya, D. & Ghosh, M. (2021). Composition and antioxidant activity of vapours from clove bud during roasting. In *Advances in Bioprocess Engineering and Technology: Select Proceedings ICABET 2020*. Springer, 151–158.

Devi, K. P., Nisha, S. A., Sakthivel, R. & Pandian, S. K. (2010). Eugenol (an essential oil of clove) acts as an antibacterial agent against *Salmonella typhi* by disrupting the cellular membrane. *Journal of Ethnopharmacology*, 130, 107–115.

Deville, L. (2018). *The Holistic Gut Prescription: Create Your Own Personal Path to Optimal Digestive Wellness*. Morgan James Publishing.

Dorman, H. D. & Deans, S. G. (2000). Antimicrobial agents from plants: Antibacterial activity of plant volatile oils. *Journal of Applied Microbiology*, 88, 308–316.

Dyrbye, B., Dubois, L., Vink, R. & Horn, J. (2012). A patient with clove oil intoxication. *Anaesthesia and Intensive Care*, 40, 365–367.

Eamsobhana, P., Yoolek, A., Kongkaew, W., Lerdthusnee, K., Khlaimanee, N., Parsartvit, A., Malainual, N. & Yong, H.-S. (2009). Laboratory evaluation of aromatic essential oils from thirteen plant species as candidate repellents against *Leptotrombidium* chiggers (Acari: Trombiculidae), the vector of scrub typhus. *Experimental and Applied Acarology*, 47, 257–262.

El Kharraf, S., Farah, A., Miguel, M. G., El-Guendouz, S. & El Hadrami, E. M. (2020). Two extraction methods of essential oils: Conventional and non-conventional hydrodistillation. *Journal of Essential Oil Bearing Plants*, 23, 870–889.

El-Darier, S. M., El-Ahwany, A. M., Elkenany, E. T. & Abdeldaim, A. A. (2018). An *in vitro* study on antimicrobial and anticancer potentiality of thyme and clove oils. *Rendiconti Lincei. Scienze Fisiche e Naturali*, 29, 131–139.

El-Maati, M. F. A., Mahgoub, S. A., Labib, S. M., Al-Gaby, A. M. & Ramadan, M. F. (2016). Phenolic extracts of clove (*Syzygium aromaticum*) with novel antioxidant and antibacterial activities. *European Journal of Integrative Medicine*, 8, 494–504.

El-Mesallamy, A. M., El-Gerby, M., Azim, M. H. A. E. & Awad, A. (2012). Antioxidant, antimicrobial activities and volatile constituents of clove flower buds oil. *Journal of Essential Oil Bearing Plants*, 15, 900–907.

Ginzburg, K., Tsur, N., Karmin, C., Speizman, T., Tourgeman, R. & Defrin, R. (2015). Body awareness and pain habituation: The role of orientation towards somatic signals. *Journal of Behavioral Medicine*, 38, 876–885.

González-Rivera, J., Duce, C., Falconieri, D., Ferrari, C., Ghezzi, L., Piras, A. & Tine, M. R. (2016). Coaxial microwave assisted hydrodistillation of essential oils from five different herbs (lavender, rosemary, sage, fennel seeds and clove buds): Chemical composition and thermal analysis. *Innovative Food Science & Emerging Technologies*, 33, 308–318.

Guan, W., Li, S., Yan, R., Tang, S. & Quan, C. (2007). Comparison of essential oils of clove buds extracted with supercritical carbon dioxide and other three traditional extraction methods. *Food Chemistry*, 101, 1558–1564.

Guan, W., Li Shufen, L. S., Yan Ruixiang, Y. R., Tang Shaokun, T. S. & Quan Can, Q. C. (2006). Comparison of essential oils of clove buds extracted with supercritical carbon dioxide and other three traditional extraction methods. *Food Chemistry*, 101, 1558–1564.

Gulcin, I., Elmastaş, M. & Aboul-Enein, H. Y. (2012). Antioxidant activity of clove oil–A powerful antioxidant source. *Arabian Journal of Chemistry*, 5, 489–499.

Gulcin, I., Şat, I. G., Beydemir, S., Elmastas, M. & Küfrevioğlu, Ö. İ. (2004). Comparison of antioxidant activity of clove (Eugenia caryophylata Thunb) buds and lavender (Lavandula stoechas L.). *Food Chemistry*, 87, 393–400.

Gümüşay, Ö. A., Borazan, A. A., Ercal, N. & Demirkol, O. (2015). Drying effects on the antioxidant properties of tomatoes and ginger. *Food Chemistry*, 173, 156–162.

Hamrouni-Sellami, I., Rahali, F. Z., Rebey, I. B., Bourgou, S., Limam, F. & Marzouk, B. (2013). Total phenolics, flavonoids, and antioxidant activity of sage (Salvia officinalis L.) plants as affected by different drying methods. *Food and Bioprocess Technology*, 6, 806–817.

Hanif, M. A., Nisar, S., Khan, G. S., Mushtaq, Z. & Zubair, M. (2019). Essential oils. In *Essential Oil Research: Trends in Biosynthesis, Analytics, Industrial Applications and Biotechnological Production*, Springer, 3–17.

Haro-González, J. N., Castillo-Herrera, G. A., Martínez-Velázquez, M. & Espinosa-Andrews, H. (2021). Clove essential oil (*Syzygium aromaticum L.* Myrtaceae): Extraction, chemical composition, food applications, and essential bioactivity for human health. *Molecules*, 26, 6387.

Haros, C. M. & Wronkowska, M. (2017). Pseudocereal dry and wet milling: Processes, products and applications. *Pseudocereals: Chemistry and Technology*, 140–162.

Hemalatha, R., Nivetha, P., Mohanapriya, C., Sharmila, G., Muthukumaran, C. & Gopinath, M. (2016). Phytochemical composition, GC-MS analysis, *in vitro* antioxidant and antibacterial potential of clove flower bud (*Eugenia caryophyllus*) methanolic extract. *Journal of Food Science and Technology*, 53, 1189–1198.

Ho, S., Cheng, L., Sim, K. & Tan, H. (1994). Potential of cloves (Syzygium aromaticum (L.) Merr. and Perry as a grain protectant against *Tribolium castaneum* (Herbst) and Sitophilus zeamais Motsch. *Postharvest Biology and Technology*, 4, 179–183.

Huang, Q. & Lakshman, D. (2010). Effect of clove oil on plant pathogenic bacteria and bacterial wilt of tomato and geranium. *Journal of Plant Pathology*, 701–707.

Huie, C. W. (2002). A review of modern sample-preparation techniques for the extraction and analysis of medicinal plants. *Analytical and Bioanalytical Chemistry*, 373, 23–30.

Ibrahium, M., El-Ghany, M. & Ammar, M. (2013). Effect of clove essential oil as antioxidant and antimicrobial agent on cake shelf life. *World Journal of Dairy and Food Sciences*, 8(2), 140–146.

Idowu, A. T., Igiehon, O. O., Idowu, S., Olatunde, O. O. & Benjakul, S. (2021a). Bioactivity potentials and general applications of fish protein hydrolysates. *International Journal of Peptide Research and Therapeutics*, 27, 109–118.

Idowu, S., Adekoya, A. E., Igiehon, O. O. & Idowu, A. T. (2021b). Clove (*Syzygium aromaticum*) spices: A review on their bioactivities, current use, and potential application in dairy products. *Journal of Food Measurement and Characterization*, 15, 3419–3435.

Jaganathan, S. K. & Supriyanto, E. (2012). Antiproliferative and molecular mechanism of eugenol-induced apoptosis in cancer cells. *Molecules*, 17, 6290–6304.

Janes, S. E., Price, C. S. & Thomas, D. (2005). Essential oil poisoning: N-acetylcysteine for eugenol-induced hepatic failure and analysis of a national database. *European Journal of Pediatrics*, 164, 520–522.

Jin, S. K., Choi, J. S., Jeong, J. Y. & Kim, G. D. (2016). The effect of clove bud powder at a spice level on antioxidant and quality properties of emulsified pork sausage during cold storage. *Journal of the Science of Food and Agriculture*, 96, 4089–4097.

Kaefer, C. M. & Milner, J. A. (2012). *Herbs and Spices in Cancer Prevention and Treatment.* CRC Press/Taylor & Francis.

Kafle, L. & Shih, C. J. (2013). Toxicity and repellency of compounds from clove (*Syzygium aromaticum*) to red imported fire ants Solenopsis invicta (Hymenoptera: Formicidae). *Journal of Economic Entomology*, 106, 131–135.

Kamatou, G. P., Vermaak, I. & Viljoen, A. M. (2012). Eugenol—from the remote Maluku Islands to the international market place: A review of a remarkable and versatile molecule. *Molecules*, 17, 6953–6981.

Kaur, K., Kaushal, S. & Rani, R. (2019). Chemical composition, antioxidant and antifungal potential of clove (*Syzygium aromaticum*) essential oil, its major compound and its derivatives. *Journal of Essential Oil Bearing Plants*, 22, 1195–1217.

Khaleque, M., Keya, C., Hasan, K., Hoque, M., Inatsu, Y. & Bari, M. (2016). Use of cloves and cinnamon essential oil to inactivate *Listeria monocytogenes* in ground beef at freezing and refrigeration temperatures. *LWT*, 74, 219–223.

Khalil, A. A., Ur Rahman, U., Khan, M. R., Sahar, A., Mehmood, T. & Khan, M. (2017). Essential oil eugenol: Sources, extraction techniques and nutraceutical perspectives. *RSC Advances*, 7, 32669–32681.

Kim, E.-H., Kim, H.-K. & Ahn, Y.-J. (2003). Acaricidal activity of clove bud oil compounds against Dermatophagoides farinae and Dermatophagoides pteronyssinus (Acari: Pyroglyphidae). *Journal of Agricultural and Food Chemistry*, 51, 885–889.

Kubatka, P., Uramova, S., Kello, M., Kajo, K., Kruzliak, P., Mojzis, J., Vybohova, D., Adamkov, M., Jasek, K. & Lasabova, Z. (2017). Antineoplastic effects of clove buds (*Syzygium aromaticum L.*) in the model of breast carcinoma. *Journal of Cellular and Molecular Medicine*, 21, 2837–2851.

Lane, B. W., Ellenhorn, M. J., Hulbert, T. V. & Mccarron, M. (1991). Clove oil ingestion in an infant. *Human & Experimental Toxicology*, 10, 291–294.

Lartey, G. S. T. J., Tachie-Menson, J. W. & Adu, S. (2023). Nutritional composition of cinnamon and clove powder and the evaluation of the antimicrobial properties of their extracts: A comparison between Ghana and other countries. *African Journal of Plant Science*, 17, 11–17.

Lee, K.-G. & Shibamoto, T. (2001). Antioxidant property of aroma extract isolated from clove buds [*Syzygium aromaticum* (L.) Merr. et Perry]. *Food Chemistry*, 74, 443–448.

Li, H. Y., Lee, B. K., Kim, J. S., Jung, S. J. & Oh, S. B. (2008). Eugenol inhibits ATP-induced P2X currents in trigeminal ganglion neurons. *The Journal of the Korean Physiological Society and the Korean Society of Pharmacology*, 12, 315.

Li, X. P., Zhou, M. Y., Liu, J. F., Xu, Y. X., Mi, H. B., Yi, S. M., Li, J. R. & Lin, H. (2017). Shelf-life extension of chilled olive flounder (*Paralichthys olivaceus*) using chitosan coatings containing clove oil. *Journal of Food Processing and Preservation*, 41, e13204.

Li, Z., Fan, Y. & Xi, J. (2019). Recent advances in high voltage electric discharge extraction of bioactive ingredients from plant materials. *Food Chemistry*, 277, 246–260.

Liu, X., Zhang, C., Liu, S., Gao, J., Cui, S. W. & Xia, W. (2020). Coating white shrimp (*Litopenaeus vannamei*) with edible fully deacetylated chitosan incorporated with clove essential oil and kojic acid improves preservation during cold storage. *International Journal of Biological Macromolecules*, 162, 1276–1282.

Lugo-Lugo, D. E., Pozos-Guillén, A. D. J., Zapata-Morales, J. R., Rodríguez-Chong, A., Rangel-López, A. D. J., Saavedra-Leos, M. Z. & Vértiz-Hernández, A. A. (2019). Antinociceptive local activity of 4-allyl-1-hydroxy-2-methoxybenzene (eugenol) by the formalin test: An anti-inflammatory effect. *Brazilian Journal of Pharmaceutical Sciences*, 55, e18022.

Matan, N. & Jantamas, S. (2011). Efficacy of clove oil in the inactivation of *Aspergillus flavus* on peanuts. *International Symposium on Medicinal and Aromatic Plants*, 1023, 301–306.

Milind, P. & Deepa, K. (2011). Clove: A champion spice. *International Journal of Research in Ayurveda Pharmacy*, 2, 47–54.

Misharina, T. & Samusenko, A. (2008). Antioxidant properties of essential oils from lemon, grapefruit, coriander, clove, and their mixtures. *Applied Biochemistry and Microbiology*, 44, 438–442.

Mittal, M., Gupta, N., Parashar, P., Mehra, V. & Khatri, M. (2014). Phytochemical evaluation and pharmacological activity of *Syzygium aromaticum*: A comprehensive review. *International Journal of Pharmacy and Pharmaceutical Sciences*, 6, 67–72.

Modupalli, N., Devraj, L. & Natarajan, V. (2022). Plant extracts as flavoring agents. In *Plant Extracts: Applications in the Food Industry*. Elsevier.

Mostafa, A. A., Al-Askar, A. A., Almaary, K. S., Dawoud, T. M., Sholkamy, E. N. & Bakri, M. M. (2018). Antimicrobial activity of some plant extracts against bacterial strains causing food poisoning diseases. *Saudi Journal of Biological Sciences*, 25, 361–366.

Mostafa Khajeh, M. K., Yadollah Yamini, Y. Y., Fatemeh Sefidkon, F. S. & Naader Bahramifar, N. B. (2004). Comparison of essential oil composition of *Carum copticum* obtained by supercritical carbon dioxide extraction and hydrodistillation methods. *Food Chemistry*, 86, 587–591.

Moulick, S. P., Jahan, F., Al Mamun, M. Z. U., Hossain, M. I. S., Waliullah, M. & Sathee, R. A. (2023). Analysis of indigenous spices widely consumed in Bangladesh: An assessment to explore its proximate contents, minerals, phytochemical compositions, and antioxidant activities. *Journal of Agriculture and Food Research*, 14, 100720.

Nada, H. G., Mohsen, R., Zaki, M. E. & Aly, A. A. (2022). Evaluation of chemical composition, antioxidant, antibiofilm and antibacterial potency of essential oil extracted from gamma irradiated clove (*Eugenia caryophyllata*) buds. *Journal of Food Measurement and Characterization*, 16, 673–686.

Nagaraju, P. G., Sengupta, P., Chicgovinda, P. P. & Rao, P. J. (2021). Nanoencapsulation of clove oil and study of physicochemical properties, cytotoxic, hemolytic, and antioxidant activities. *Journal of Food Process Engineering*, 44, e13645.

Nikousaleh, A. & Prakash, J. (2016). Antioxidant components and properties of dry heat treated clove in different extraction solvents. *Journal of Food Science and Technology*, 53, 1993–2000.

Nirmala, M. J., Durai, L., Gopakumar, V. & Nagarajan, R. (2019). Anticancer and antibacterial effects of a clove bud essential oil-based nanoscale emulsion system. *International Journal of Nanomedicine*, 6439–6450.

Ogbunugafor, H. A., Ugochukwu, C. G. & Kyrian-Ogbonna, A. E. (2017). The role of spices in nutrition and health: A review of three popular spices used in Southern Nigeria. *Food Quality and Safety*, 1, 171–185.

Ogunwande, I., Olawore, N., Ekundayo, O., Walker, T., Schmidt, J. & Setzer, W. (2005). Studies on the essential oils composition, antibacterial and cytotoxicity of *Eugenia uniflora L. International Journal of Aromatherapy*, 15, 147–152.

Ohkubo, T. & Shibata, M. (1997). The selective capsaicin antagonist capsazepine abolishes the antinociceptive action of eugenol and guaiacol. *Journal of Dental Research*, 76, 848–851.

Orphanides, A., Goulas, V. & Gekas, V. (2016). Drying technologies: Vehicle to high-quality herbs. *Food Engineering Reviews*, 8, 164–180.

Özcan, M., Arslan, D. & Ünver, A. (2005). Effect of drying methods on the mineral content of basil (*Ocimum basilicum L.*). *Journal of Food Engineering*, 69, 375–379.

Pandey, V. K., Srivastava, S., Dash, K. K., Singh, R., Dar, A. H., Singh, T., Farooqui, A., Shaikh, A. M. & Kovacs, B. (2024). Bioactive properties of clove (*Syzygium aromaticum*) essential oil nanoemulsion: A comprehensive review. *Heliyon*, 10, e22437.

Park, I.-K. & Shin, S.-C. (2005a). Fumigant activity of plant essential oils and components from garlic (Allium sativum) and clove bud (*Eugenia caryophyllata*) oils against the Japanese termite (Reticulitermes speratus Kolbe). *Journal of Agricultural and Food Chemistry*, 53, 4388–4392.

Park, I.-K. & Shin, S.-C. (2005b). Nematicidal activity of plant essential oils and components from Garlic (*Allium sativum*) and Cinnamon (*Cinnammon verum*) oils against Pine Wood Nematode (*Bursaphelenchus xylophilus*). *Nematology*, 103–103.

Park, M.-J., Gwak, K.-S., Yang, I., Choi, W.-S., Jo, H.-J., Chang, J.-W., Jeung, E.-B. & Choi, I.-G. (2007). Antifungal activities of the essential oils in *Syzygium aromaticum* (L.) Merr. Et Perry and Leptospermum petersonii Bailey and their constituents against various dermatophytes. *Journal of Microbiology*, 45, 460–465.

Parthasarathy, V. A., Chempakam, B. & Zachariah, T. J. (2008). *Chemistry of Spices*. Cabi.

Pateiro, M., Munekata, P. E., Sant'ana, A. S., Domínguez, R., Rodríguez-Lázaro, D. & Lorenzo, J. M. (2021). Application of essential oils as antimicrobial agents against spoilage and pathogenic microorganisms in meat products. *International Journal of Food Microbiology*, 337, 108966.

Prasad, R. C., Herzog, B., Boone, B., Sims, L. & Waltner-Law, M. (2005). An extract of Syzygium aromaticum represses genes encoding hepatic gluconeogenic enzymes. *Journal of Ethnopharmacology*, 96, 295–301.

Prashar, A., Locke, I. C. & Evans, C. S. (2006). Cytotoxicity of clove (*Syzygium aromaticum*) oil and its major components to human skin cells. *Cell Proliferation*, 39, 241–248.

Qin, D. & Xi, J. (2021). Flash extraction: An ultra-rapid technique for acquiring bioactive compounds from plant materials. *Trends in Food Science & Technology*, 112, 581–591.

Quan, L., Li, S., Tian, S., Xu, H., Lin, A. & Gu, L. (2004). Determination of organochlorine pesticides residue in ginseng root by orthogonal array design soxhlet extraction and gas chromatography. *Chromatographia*, 59, 89–93.

Raghavan, S. (2006). *Handbook of Spices, Seasonings, and Flavorings*. CRC Press.

Rahman, S. & Husen, A. (2024). Diversity and geographic distribution of spice and condiment plant species. In *Medicinal Spice and Condiment Crops*. CRC Press.

Ramos, M., Jiménez, A. & Garrigós, M. C. (2019). Il-based advanced techniques for the extraction of value-added compounds from natural sources and food by-products. *TrAC Trends in Analytical Chemistry*, 119, 115616.

Ratiba, B., Wedad, A., Mohamed, O. & Abd El-Baky, H. (2006). Effect of cardamom, thyme and clove powder on the composition and quality of white soft cheese made from goat's milk. *Assiut Journal of Agricultural Science*, 37, 139–157.

Reverchon, E. & Marrone, C. (1997). Supercritical extraction of clove bud essential oil: Isolation and mathematical modeling. *Chemical Engineering Science*, 52, 3421–3428.

Riar, C. S. & Panesar, P. S. (2024). Bioactive compounds and nutraceuticals: Classification, potential sources, and application status. In *Bioactive Compounds and Nutraceuticals from Dairy, Marine, and Nonconventional Sources*. Apple Academic Press.

Riaz, U., Iqbal, S., Sohail, M. I., Samreen, T., Ashraf, M., Akmal, F., Siddiqui, A., Ahmad, I., Naveed, M. & Khan, N. I. (2021). A comprehensive review on emerging importance and economical potential of medicinal and aromatic plants (MAPs) in current scenario. *Pakistan Journal of Agricultural Research*, 34(2), 381–392.

Ribes, S., Fuentes, A. & Barat, J. M. (2019). Effect of oregano (*Origanum vulgare* L. ssp. hirtum) and clove (Eugenia spp.) nanoemulsions on *Zygosaccharomyces bailii* survival in salad dressings. *Food Chemistry*, 295, 630–636.

Rubió, L., Motilva, M.-J. & Romero, M.-P. (2013). Recent advances in biologically active compounds in herbs and spices: A review of the most effective antioxidant and anti-inflammatory active principles. *Critical Reviews in Food Science and Nutrition*, 53, 943–953.

Ryu, B., Kim, H. M., Lee, J. S., Lee, C. K., Sezirahiga, J., Woo, J.-H., Choi, J.-H. & Jang, D. S. (2016). New flavonol glucuronides from the flower buds of *Syzygium aromaticum* (Clove). *Journal of Agricultural and Food Chemistry*, 64, 3048–3053.

Sadowska-Bartosz, I. & Bartosz, G. (2016). Effect of glycation inhibitors on aging and age-related diseases. *Mechanisms of Ageing and Development*, 160, 1–18.

Sarrami, N., Pemberton, M., Thornhill, M. & Theaker, E. (2002). Adverse reactions associated with the use of eugenol in dentistry. *British Dental Journal*, 193, 257–259.

Shahidi, F. & Ambigaipalan, P. (2015). Phenolics and polyphenolics in foods, beverages and spices: Antioxidant activity and health effects–A review. *Journal of Functional Foods*, 18, 820–897.

Shahzadiand, A. & Sonmez, H. (2023). Bioactives from clove oil for antibacterial wound dressings for the treatment and management of wounds in type 2 diabetes mellitus. In *Natural Products and their Bioactives in Antidiabetic Drug Discovery*, Wiley Online Library, 195–210.

Shan, B., Cai, Y. Z., Sun, M. & Corke, H. (2005). Antioxidant capacity of 26 spice extracts and characterization of their phenolic constituents. *Journal of Agricultural and Food Chemistry*, 53, 7749–7759.

Sharma, S., Barkauskaite, S., Duffy, B., Jaiswal, A. K. & Jaiswal, S. (2020). Characterization and antimicrobial activity of biodegradable active packaging enriched with clove and thyme essential oil for food packaging application. *Foods*, 9, 1117.

Simon, J. E., Decker, E. A., Ferruzzi, M. G., Giusti, M. M., Mejia, C. D., Goldschmidt, M. & Talcott, S. T. (2017). Establishing standards on colors from natural sources. *Journal of Food Science*, 82, 2539–2553.

Sindhi, V., Gupta, V., Sharma, K., Bhatnagar, S., Kumari, R. & Dhaka, N. (2013). Potential applications of antioxidants–A review. *Journal of Pharmacy Research*, 7, 828–835.

Singh, P. A., Bajwa, N. & Baldi, A. (2021). Possible role of traditional systems of medicine to manage COVID-19: A review. *Israel Journal of Plant Sciences*, 68(1–2), 3–28.

Singh, P. A., Bajwa, N., Chinnam, S., Chandan, A. & Baldi, A. (2022). An overview of some important deliberations to promote medicinal plants cultivation. *Journal of Applied Research on Medicinal and Aromatic Plants*, 31, 100400.

Singh, P. A., Sood, A. & Baldi, A. (2021a). Determining constraints in medicinal plants adoption: A model geospatial study in the Indian state of Punjab. *Journal of Applied Research on Medicinal and Aromatic Plants*, 25, 100342.

Singh, R., Barden, A., Mori, T. & Beilin, L. (2001). Advanced glycation end-products: A review. *Diabetologia*, 44, 129–146.

Singletary, K. (2014a). Clove: Overview of potential health benefits. *Nutrition Today*, 49, 207–224.

Singletary, K. (2014b). Culinary nutrition. *Nutrition Today*, 49, 207Y224.

Sofia, P. K., Prasad, R., Vijay, V. K. & Srivastava, A. K. (2007). Evaluation of antibacterial activity of Indian spices against common foodborne pathogens. *International Journal of Food Science & Technology*, 42, 910–915.

Srinivasan, K. (2005). Role of spices beyond food flavoring: Nutraceuticals with multiple health effects. *Food Reviews International*, 21, 167–188.

Sritabutra, D., Soonwera, M., Waltanachanobon, S. & Poungjai, S. (2011). Evaluation of herbal essential oil as repellents against Aedes aegypti (L.) and Anopheles dirus Peyton & Harrion. *Asian Pacific Journal of Tropical Biomedicine*, 1, S124–S128.

Starowicz, M. & Zieliński, H. (2019). Inhibition of advanced glycation end-product formation by high antioxidant-leveled spices commonly used in European cuisine. *Antioxidants*, 8, 100.

Sultana, B., Anwar, F., Mushtaq, M., Aslam, M. & Ijaz, S. (2014). *In vitro* antimutagenic, antioxidant activities and total phenolics of clove (*Syzygium aromaticum L.*) seed extracts. *Pakistan Journal of Pharmaceutical Sciences*, 27, 42–49.

Tekin, K., Akalın, M. K. & Şeker, M. G. (2015). Ultrasound bath-assisted extraction of essential oils from clove using central composite design. *Industrial Crops and Products*, 77, 954–960.

Thomas, A., Mazigo, H. D., Manjurano, A., Morona, D. & Kweka, E. J. (2017). Evaluation of active ingredients and larvicidal activity of clove and cinnamon essential oils against Anopheles gambiae (*sensu lato*). *Parasites & Vectors*, 10, 1–7.

Topal, F. (2019). Anticholinergic and antidiabetic effects of isoeugenol from clove (Eugenia caryophylata) oil. *International Journal of Food Properties*, 22, 583–592.

Tungare, K., Bhori, M., Patil, S. A., Iyer, P., Jagtap, N., Kadam, A. & Vijayakuma, S. (2024). Clove Buds (*Syzygium aromaticum*). In *Medicinal Spice and Condiment Crops*. CRC Press.

Uddin, M. A., Shahinuzzaman, M., Rana, M. S. & Yaakob, Z. (2017). Study of chemical composition and medicinal properties of volatile oil from clove buds (Eugenia caryophyllus). *International Journal of Pharmaceutical Sciences and Research*, 8, 895.

Uribarri, J., Del Castillo, M. D., De La Maza, M. P., Filip, R., Gugliucci, A., Luevano-Contreras, C., Macías-Cervantes, M. H., Bastos, D. H. M., Medrano, A. & Menini, T. (2015). Dietary advanced glycation end products and their role in health and disease. *Advances in Nutrition*, 6, 461–473.

Veerapur, V., Prabhakar, K., Thippeswamy, B., Bansal, P., Srinivasan, K. & Unnikrishnan, M. (2012). Antidiabetic effect of Ficus racemosa Linn. stem bark in high-fat diet and low-dose streptozotocin-induced type 2 diabetic rats: A mechanistic study. *Food Chemistry*, 132, 186–193.

Vijayasteltar, L., Nair, G. G., Maliakel, B., Kuttan, R. & Krishnakumar, I. (2016). Safety assessment of a standardized polyphenolic extract of clove buds: Subchronic toxicity and mutagenicity studies. *Toxicology Reports*, 3, 439–449.

Viuda-Martos, M., Ruiz-Navajas, Y., Fernández-López, J. & Pérez-Álvarez, J. (2010). Spices as functional foods. *Critical Reviews in Food Science and Nutrition*, 51, 13–28.

Wang, Y.-F., Jia, J.-X., Tian, Y.-Q., Shu, X., Ren, X.-J., Guan, Y. & Yan, Z.-Y. (2018). Antifungal effects of clove oil microcapsule on meat products. *LWT*, 89, 604–609.

Xiang, B., Zhou, X., Qin, D., Li, C. & Xi, J. (2022). Infrared assisted extraction of bioactive compounds from plant materials: Current research and future prospect. *Food Chemistry*, 371, 131192.

Xiao, J., Chen, G. & Li, N. (2018). Ionic liquid solutions as a green tool for the extraction and isolation of natural products. *Molecules*, 23, 1765.

Xu, J.-G., Liu, T., Hu, Q.-P. & Cao, X.-M. (2016). Chemical composition, antibacterial properties and mechanism of action of essential oil from clove buds against *Staphylococcus aureus*. *Molecules*, 21, 1194.

Yang, J., Kuang, X., Li, B., Zhou, B., Li, J., Cui, B. & Ma, M. (2012). Study on release mechanism of inhibitory components from cinnamon and clove powders. *Journal of Food Safety*, 32, 189–197.

Yousefzadi, M., Riahi-Madvar, A., Hadian, J., Rezaee, F., Rafiee, R. & Biniaz, M. (2014). Toxicity of essential oil of Satureja khuzistanica: In vitro cytotoxicity and anti-microbial activity. *Journal of Immunotoxicology*, 11, 50–55.

Yu, H., Wu, K., Wu, T., Zhong, S., Chai, X. & Li, Z. (2009). Studies on supercritical CO2 extraction-molecular distillation of essential oil from clove (*Eugenia caryophyllata* Thunb.) bud. *Chemistry and Industry of Forest Products*, 29, 74–78.

Zhang, H., Peng, X., Li, X., Wu, J. & Guo, X. (2017a). The application of clove extract protects Chinese-style sausages against oxidation and quality deterioration. *Korean Journal for Food Science of Animal Resources*, 37, 114.

Zhang, L.-L., Zhang, L.-F., Xu, J.-G. & Hu, Q.-P. (2017b). Comparison study on antioxidant, DNA damage protective and antibacterial activities of eugenol and isoeugenol against several foodborne pathogens. *Food & Nutrition Research*, 61, 1353356.

Zhang, Q. H., Schneidmiller, R., Hoover, D., Zhou, G., Margaryan, A. & Bryant, P. (2014). Essential oils as spatial repellents for the brown marmorated stink bug, Halyomorpha halys (*Hemiptera Pentatomidae). Journal of Applied Entomology*, 138, 490–499.

2 Botanical Description

Nurliani Bermawie, Sri Wahyuni,
Adi Setiadi, and Sitti Fatimah Syahid

2.1 INTRODUCTION

The cloves of commerce are the unopened aromatic flower buds of the clove tree, scientifically known as *Syzygium aromaticum* (L. Merrill & Perry). These buds, typically brown and dried, are among the most valuable spices from the Myrtaceae family. Their distinctive aroma and flavor have made them a valued commodity for centuries (Singh et al., 2021, 2024). The high demand for cloves was a significant factor that attracted European colonizers to the regions where they were cultivated. This had a profound effect on the development of international trade.

The high interest in cloves has made cloves no longer confined to the spice islands of Maluku but spread over tropical countries. The major clove-producing countries are Indonesia, India, Malaysia, Sri Lanka, Madagascar, and Tanzania, especially Zanzibar (Kamatou et al., 2012). On a global scale, the annual consumption of cloves reflects their widespread use and demand. The clove production in Indonesia in 2022 reached 133,064 metric tons (FAO), accounting for nearly 73% of the world's production. Being the largest clove producer, Indonesia is also the largest consumer in the world.

Apart from the dried flower buds, leaves and flower stalks or twigs are also commercially valuable. The leaves, flower stalks, or twigs are sources of essential oil. The essential oil of cloves has a variety of advantageous biological functions. Therefore, cloves are highly prized not just for their culinary usage but also for their applications in several other purposes, including dentistry, perfumery, and cosmetics for their pleasing scent, as well as in food flavoring and preservation (Ulanowska & Olas, 2021; Kaufman, 2015; Nejad et al., 2017; Hassan et al., 2020), as an antioxidant (Cortes-Rojas et al., 2014), as a source for kretek—traditional Indonesian cigarettes (Hasibuan et al., 2022), and in agriculture industries (Özgen & Yücel, 2024). Cloves used in the kretek industry accounted for 95% of Indonesia's national production (Hasibuan et al., 2022).

There are several clove varieties grown in the clove-producing countries. The genetic composition influences the different varieties. The morphology of clove can vary among different genotypes or varieties, although they share many common characteristics. Different clove varieties can exhibit variations in the parts of the clove plant (buds, stems, leaves) and might show differences in the quality of clove in several parameters, such as essential oil content, chemical components, and their functional properties. A detailed description of the morphological characteristics of clove plant parts and some notable differences is provided.

 DOI: 10.1201/9781003521785-2

2.2 BOTANICAL CLASSIFICATION

Cloves have been known in the world for centuries. The word "clou," which means nail in French, is where the name "clou" originates because the clove bud's initial appearance resembles a fingernail. According to another report, it comes from the Latin word clavus, which means nail (because of its form). In addition, clove is known by a variety of names in other languages, including clove in English, chiro di Garofano in Italian, qurnafl in Arabic, Nelke in German, ding Xiang in Mandarin Chinese, laung in Hindi, clavo in Spanish, and clou de girofle in French. "Cengkih" is a common local name in Indonesia believed to have originated from Chinese.

Clove was initially assigned to the genus *Eugenia.* Although most is found in tropical Asia, Eugenia is a large genus spread over the New and Old Worlds. In order to manage it, the genus Eugenia was divided into two genera, *Eugenia* and *Syzygium*, to make the genus easier to handle. The generic name *syzygios* (paired) is derived from the Greek because of the leaves and twigs that grow simultaneously in several species. Schmid (1972) offered strong evidence in the form of floral anatomy and also supported the division of the genus *Eugenia* into two genera. A molecular study using internal transcribed spacer (ITS) sequences data on a few clove varieties belonging to the genus *Syzygium* and *Eugenia* of the North Maluku, Sundari et al. (2019) found that the genus *Syzygium* is consistently distinct from Eugenia and thus supports the separation of *Syzygium* and *Eugenia.*

The genus *Syzygium* consists of 500 species, primarily fruit plants, primarily from tropical Asia (Purseglove et al., 1981). Taxonomic challenges arise when determining the correct species. The taxonomic classification of clove is as follows:

Kingdom: Plantae
- Subkingdom: Virideplantae
  - Infrakingdom: Streptophyta
    - Superdivision: Embryophyta
      - Division: Tracheophyta
        - Sub division: Spermatophyta
          - Class: Magnoliopsida
            - Superorder: Rosanae
              - Order: Myrtales
                - Family: Myrtaceae
                  - Genus: *Syzygium* P. Br. Ex. Gaertn.
                    - Species: *Syzygium aromaticum* (L.) Merriel and Perry

(www.itis.gov/servlet/SingleRpt/SingleRpt?search_topic=TSN&search_value=506167#null. May 31, 2024)

The common synonyms are *Caryophyllata aromatica* L. and *Syzygium aromaticum* Thunb. *Eugenia aromatica* Kuntze, *E. caryophyllata* Thunb., and *E. caryophyllus* (Sprengel) Bullock & Harrison.

2.3 ORIGIN AND HISTORY OF CLOVE DISTRIBUTION

The clove tree is indigenous to the Maluku islands of the Indonesian archipelago. The Maluku Islands include Ternate, Tidore, Motir, Makian, and Bacan (Purseglove et al., 1981), and now they are part of the North Maluku province (Purseglove et al., 1981; Milind & Deepa, 2011; Alfian et al., 2019; Mahulette et al., 2019a, 2019b, 2022), and the center of origin of cloves possess a relatively high genetic diversity. Thus, significant variation in the description of clove plant parts may also represent clove variability in the other parts of the world.

Since the early 18th century, the French, British, and Dutch have distributed clove from Maluku, Indonesia, to other parts of the world, such as Brazil, Haiti, India, Kenya, Madagascar, Malaysia, Mauritius, Mexico, Seychelles, Sri Lanka, and Tanzania. It began in 1511 when Portuguese explorers arrived in Maluku in search of a straight path to the Spice Islands. The Portuguese monopolized the clove trade by controlling production and transport routes from the Maluku Islands. However, in the 17th century, the Portuguese lost control of the clove trade to the Dutch East India Company.

The French and British began to break the Dutch monopoly. In the 1770s, the French smuggled clove seedlings out of the Maluku Islands suggested from Halmahera (Lubis, 1954)—now part of the North Maluku province—and introduced them to their colonies, such as Mauritius, Bourbon (Reunion), Seychelles, and the other French colonies. In that period, Bourbon (Reunion) became a commercial clove producer. In the early 19th century (1805), clove cultivation spread to Zanzibar and Pemba in East Africa, where they became significant producers. In 1827, cloves were introduced to Madagascar.

The British also introduced clove cultivation to their colonies. In 1800, Christopher Smith, a botanist of the East Indian Company (EEC), introduced clove to Malaysia (Penang) (Ridley, 1912). In the early 18th century, Penang became an important clove producer. In addition to Penang, some Maluku clove plants were brought to Kew Garden, Calcutta, and Madras, India.

The first report on the distribution of cloves outside Maluku to other parts of Indonesia occurred during British colonialization. The British settlement brought clove from Maluku and planted it in Bencoolen (Bengkulu), Sumatra, in 1798 (another report mentioned in 1823) (Purseglove et al., 1981). However, it was reported that the cloves were eradicated when Bengkulu was handed over to the Dutch. However, Bengkulu still has old clove populations, which are considered to be over 100 years old, especially on Enggano Island. From Bengkulu, cloves were then taken to other areas in Sumatra, including Lampung in 1875, West Sumatra in 1878 (Alwi, 1953), North Sumatra (Tapanuli), and the West Coast of Aceh, maybe including Simeulue island and then to the East coast of Aceh, such as to Sabang island. In the Simeulue and Sabang islands, old clove plants, considered descendants of cloves brought in during the colonial period, are still there.

Missionaries spread clove to North Sulawesi in 1870, bringing clove seeds directly from the Maluku Islands. Missionaries' involvement in distributing clove from Maluku to other parts of Indonesia was also reported in East Nusa Tenggara, such as to Alor, Flores, Ende, East Manggarai, and Timor Island. Though the botanical description of cloves grown in East Nusa Tenggara showed variation, their morphological description was more-or-less similar to those found in the Maluku islands.

In 1932, the Dutch East Indies government brought back clove seeds to Indonesia from Zanzibar, and they were planted in the *Culturtuuin* (now Cimanggu Bogor), Sumatera (Bengkulu and Bukit tinggi), and Laha (Ambon) (Lubis, 1954). From this area, Zanzibar clove seeds were distributed again to the Pasir Madang, Bogor, West Java, and Selokaton, Central Java. In 1940–1941, the Dutch East Indies government also brought clove seeds directly from Maluku grown in the *Culturtuuin* Cimanggu garden, known as Siputih, Ambon, and Sikotok.

Large-scale clove planting in Indonesia was carried out in a self-sufficiency program from the 1970s until the 1980s. The Zanzibar clove trees later became the sources of clove seeds for replanting programs in several provinces in Indonesia. Thousands of hectares were planted with cloves in many provinces using seeds collected from the Zanzibar clove trees grown in the *Cultuurtuin* Bogor. Most of its progeny were distributed to various regions in Indonesia vary morphologically, which may be attributed to the predominantly outcrossing nature of the clove breeding system (Bermawie & Pool, 1991).

2.4 CLOVE BOTANICAL TYPES

Clove genetic resources are very important for clove production and cultivation sustainability. Genetic resources are needed to face various challenges and obstacles in clove cultivation both now and in the future. Climate change has many consequences on the sustainability of clove production, as it can affect flowering patterns, yield, pests, and disease attacks.

Even though there is not as wide genetic variability as it should have been, there are still many types of cloves in the center of origin of the Maluku islands that survived the extermination. A report indicated the presence of an old clove plantation in the mountains at Bukubualawa (local name as Cengkeh Hill) in Halmahera, North Maluku, which survived the extermination by the Dutch.

The clove's center of origin includes Ternate, Tidore, Motir, Bacan, Halmahera, and Makian in the North Maluku province and Ambon, Seram, Saparua, Nusa Laut, Haruku, Buru, and Ambalau in the southern part of the province. Substantial clove genetic diversity is thought to have survived in these islands.

The clove genetic diversity can be divided into botanical or local varieties. Pool et al. (1986) described cloves' genetic resources in three categories: the popular cultivated cloves grown mostly outside Maluku, Indigenous Maluku populations, and wild cloves (forest cloves). The popular cultivated varieties are Zanzibar, Sikotok, Siputih, and Ambon. The forest (wild types) clove and Raja clove (considered a natural hybrid between the cultivated and the forest cloves) are increasingly popular, especially after they have a niche market and provide a source of income for farmers. Detailed botanical descriptions of the cultivated varieties, indigenous Maluku populations, and forest clove are described.

2.4.1 Plant Habit and Botanical Description

The plant habit of clove refers to its overall growth form and structure. Clove trees are characterized by a unique morphology and growth pattern shaped by their genetic composition and environmental condition (Saranya & Ravi, 2018). Clove is a fascinating

FIGURE 2.1 Canopy shape and main trunk branching pattern.
a. Cylindrical, b. Pyramidal, c. Rounded, d. Divide, e. Single, f. Horne-shape

evergreen tree with dense foliage that can reach impressive heights up to 30 m depending on the age and the suitability of the growing conditions. The clove tree can live up to over 100 years and still bear productive cloves. The oldest reported clove tree ever found in the world that survived the Dutch extermination was AFO 1, grown in the Gamalama mountain, on the Ternate island with 36.60 m plant height and trunk circumference of 4.26 m or 1.98 m in diameter, with an approximate age of 350 years. AFO clove trees could produce flowers up to 600 kg for each harvesting season in the 1990s (Andi, 2013). It is estimated that the AFO clove tree existed long before Chinese and Arab traders came to Ternate. The clove tree has varying canopy/crown shapes from pyramidal, round, to cylindrical types, with scarce to very dense leaves as represented in Figure 2.1. The crown size could reach up to 15 m in diameter, depending on the plant's age.

2.5 ROOTS

The root system of the forest clove plant is a taproot consisting of primary roots followed by lateral roots (primary root branches) and secondary, tertiary, and quartet-like roots, which were then overgrown with root hairs. The tap root is shaped like a spear (fusiform), which is very strong, so it can keep the tree upright (Al Muhdhar et al., 2018). At less than 1 year old, clove seedlings have a straight tap root with a root length of around 40–45 cm, and these roots then branch out into 30–35 cm branch roots. During the first year, a mass of fibrous roots spreads out from the tap roots to a depth of about 25 cm and a 36–50 cm radius. The part of the root that grows at the top is covered with many root hairs, which take nutrients from the soil to the plants (Purseglove et al., 1981).

Most clove roots remain at a depth of 0–25 cm. Root density and length generally decrease with increasing soil depth, and the root density is less at a deeper soil layer (50–100 cm depth). Clove roots to a depth of 100 cm are close to the tree's base (0.5 canopy radius) and decrease with increasing distance from the base. The average root length density at 0.5 times the canopy radius is the highest compared to more distant from the canopy radius. The composition of root size with a diameter of < 1 mm comprised 72% of the total root length, whereas roots with a diameter of 1–3 mm and > 3 mm comprised 25 and 3% of the total root length, respectively (Purbopuspito & Van Rees, 2002).

2.6 TRUNK

The clove tree can have a straight and large trunk with hardwood, enabling it to survive for centuries. The trunk has a round shape (teres) and a rough brown or greyish surface. Plant height could reach 16 m and the trunk circumference could reach 200 cm depending on the plant's age and growing condition as mentioned in Table 2.1.

The branching patterns of clove trees also vary. After reaching a certain height, below 1 m or above 3 m, depending on the genetic composition, the trunk begins to fork, forming a few to many branches. The branches have varying growth orientations, from horizontal to semi-erect or multiple branches forming angles to the main trunk range from 45° to 90° (Figure 2.1). The branches are often lengthy and full of little twigs.

2.7 LEAF

The clove leaves are simple, arranged oppositely, with varying shapes, including elliptic, obovate, and lanceolate. They are very narrowly attenuated at the base and pointed upwards as represented in Figure 2.2.

FIGURE 2.2 Characteristics of the clove leaf.

a. Greenish yellow shoot, and elliptic leaf shape, b. reddish pink shoot and obovate leaf shape, c. red shoot and elliptic leaf shape, d. variation in mature leaf shape

TABLE 2.1
Characteristics of clove plant varieties.

S. No.	Clove type	Plant ages (years)	Plant height (m)	Trunk circumference (cm)	Canopy diameter (cm)	Branching pattern	Canopy shape	References
1	Zanzibar Karo	20	NA	NA	NA	Fork	Conical	Ministry of Agriculture (2008, 2010, 2018, 2019a, 2019b, 2019c, 2020a, 2020b, 2023)
2	Zanzibar Gororntalo	40	NA	100.86–131.26	7.57–9.41	Fork	Cylindrical	Ministry of Agriculture (2008, 2010, 2018, 2019a, 2019b, 2019c, 2020a, 2020b, 2023)
3	Siantan Agribun	30–46	14.54–16.24	85.87–127.77	NA	Single	Narrow Cylindrical-conical	Ministry of Agriculture (2010, 2013a, 2013b)
4	Zanzibar Peling	36–46	13.93–15.93	131.82–133.64	5.68–7.5	Single	Narrow cylindrical-wide conical	Ministry of Agriculture (2023)
5	AFO	NA	NA	NA	NA	NA	NA	Minister of Agriculture (2010)
6	Tuni Bursel	30–40	12–15	88–180	NA	Multiple	Cylindrical	Ministry of Agriculture (2008, 2010, 2018, 2019a, 2019b, 2019c, 2020a, 2020b, 2023)

Broad and comprehensive leaf sizes are generally found in Siputih and Siputih cloves, while Zanzibar cloves have narrower sizes and an elliptical leaf shape (Bermawie, 2014). The undersides of the leaves are usually paler than the top. The leaves are leathery and have a shiny texture on the upper (adaxial) surface, with a smooth to wavy margin. Leaves are glabrous, with numerous oil glands on the lower surface. They emit a characteristic aromatic scent when crushed.

The clove leaf varies from short to long and from slim to broad, depending on the variety as mentioned in Table 2.2

The size of the clove leaf varies from 6–13.5 cm long and 2.5–5 cm wide, depending on the genotypes (de Gusman & Siemonsma, 1999). Data from leaf measurements of cultivated varieties grown outside Maluku vary from 7.85–14.50 cm in length and around 3–6 cm in width, with the largest leaf size found in Semarang, Central Jawa.

New leaves (shoots) are light green, bright pink, or reddish brown, turning light to glossy dark green when mature. The leaf petiole of the clove is slightly swelling/thickening at the base, green to red, either entirely or partially encircling the base of the petiole.

There are significant variations in leaf characteristics of the varieties grown outside the Maluku and within the Maluku regions (Maluku and North Maluku). The breeding system of clove may influence these variations. The variation in the leaf characteristics of clove within the Maluku and the North Maluku is more comprehensive, especially when the forest clove is included; regarding leaf shape, there is less variation in clove grown in the North Maluku and Maluku compared to the clove-grown outside Maluku.

There are varying degrees of green intensity in mature leaf and shoot color. Most clove varieties have dark green mature leaves, except Siputih, Ambon, Forest clove, Raja, and Tuni. Shoot color varies from yellowish green to dark red. Red shoot color is often associated with Zanzibar, Afo, and their descendants and may be one of the characteristics used to identify clove varieties.

2.8 FLOWERS

The clove tree is monoecious; its flowers are hermaphrodite and predominantly cross-pollinating. Depending on growing conditions, flowering typically begins after about 4–6 years. Generally, the clove tree reaches a total production capacity of around 50 years.

Clove inflorescences are typically terminal or axillary cymes, where each flower cluster arises from the end of a branch (flos terminals) or the axil of a leaf (axillary). The inflorescence has varying lengths from short to long peduncles, making a variation in the proportion of flower/peduncles ratio.

Each inflorescence grows into several stalks, each stalk consisting of three flowers. The number of flowers per inflorescence varies depending on the genetic and growing conditions and can reach more than 50 in suitable conditions. The flowers are small to large, and the pedicels can be very short to medium.

Individual clove flower buds measure about 1.5–3 cm in length and 0.3–0.6 cm in width. Each flower consists of four serrated calyx lobes, which are fused, forming a tube that opens at the top into four lobes. There are variations in the shape of calyx tubes, from cylindrical to parallel and conical as represented in Figure 2.3.

TABLE 2.2
Leaf characteristics of clove varieties from several origins.

No.	Clove type/variety	Leaf length	Leaf width	Petiole length	Leaf shape	Petiole base color	Leaf color	Shoot color	References
Outside Maluku									
1	Zanzibar Cimanggu, Bogor, West Jawa province	7.95–10.35	3.65–6.10	2.54	Elliptic	Red at encircling the base	Dark green	Dark red	Bermawie (1992)
2	Sikotok Cimanggu, Bogor, West Jawa province	7.85–9.15	3.05–3.55	2.65	Obovate	Red at the upper part	Dark green	Brownish red	Bermawie (1992)
3	Siputih Cimanggu, Bogor, West Jawa province	10.13–12.86	3.34–5.01	2.31	Obovate	Red at the upper part	Yellowish green	Yellowish green	Bermawie (1992)
4	Ambon Cimanggum Bogor, West Jawa province	8–9.85	3.15–4.20	2.20	Elliptic	Red at the upper part	Dark green	Reddish light green	Bermawie (1992)
5	Zanzibar Karo, North Sumatera province	10.35–14.50	4.0–5.8	2.12–2.36	Slim oval, slightly rounded	Greenish red	Dark green	Red	Minister of Agriculture (2010)
6	Zanzibar Gorontalo, Sulawesi province	10.31–13.11	3.78–4.90	2.59–3.49	Narrow Elliptic	Greenish red	Dark green	Yellowish green	Minister of Agriculture (2020)
7	Siantan Agribun, Riau islands province	10.52–14.44	3.97–5.53	2.2–3.1	Obovate	NA	Glossy dark green	Reddish green	Bermawie et al. (2021)
8	Zanzibar Sukabumi, West Jawa province	9.75–12.45,	3.36–4.7	1.9–2.34,	Elliptic	Red	Dark green	Pink	Randriani and Syafaruddin (2011)
9	Siputih Sukabumi, West Jawa province	11.36–12.6	4.47–5.09	2.31–2.71	NA	NA	Green	Light green	Randriani and Syafaruddin (2011)

TABLE 2.2 *(Continued)*
Leaf characteristics of clove varieties from several origins.

No.	Clove type/variety	Leaf length	Leaf width	Petiole length	Leaf shape	Petiole base color	Leaf color	Shoot color	References
10	Zanzibar Peling, South Sulawesi	11.07–13.13	4.41–4.99	2.24–2.60	Elliptic-obovate	NA	Glossy dark green	Reddish green-red	Minister of Agriculture (2023)
11	Semarang, Central Jawa	8.34–12.79	3.15–8.34	1.4–3.11	Elliptic	Red fully encircling the base	Green	Green–reddish green	Susilowati et al. (2024).
Maluku									
1	Raja clove, Central Maluku	11.15–13.39	3.87–5.8	1.5–2	NA	NA	Green–dark green	Strong yellow green Reddish-orange	Mahulette et al. (2022), Wattimena et al. (2023)
2	Forest clove, Maluku	15.13–20.67	6.99–9.85	1.88–2.09	NA	NA	Deep yellowish green	Light/pale preen	Mahulette et al. (2022)
3	Tuni Bursel, Central Sulawesi	10.4–16.0	4.1–6.2	1.7–2.7	Slim oval–slightly rounded	Bright brownish red	Dark green	Light reddish-green	Ministry of Agriculture (2023)
4	Raja Clove, Ambon Island	11.0–16.5	4.4–7.5	1.5–2.2	Elliptic to elliptic-oblong	Brownish green	Dark green	Light reddish-green	Alfian et al. (2019)
5	Tuni clove, Tehoru, Central Maluku	9.91–11.38	3.58–4.19	NA	Elliptic	NA	Yellowish–green	NA	Namakule et al. (2024)
North Maluku									
1	Afo, Ternate	8.7–12.3	3.6–4.6	2.0–2.3	Elliptic	Green	Dark green	Yellowish red	Ministry of Agriculture (2010)

(Continued)

TABLE 2.2 ***(Continued)***
Leaf characteristics of clove varieties from several origins.

No.	Clove type/variety	Leaf length	Leaf width	Petiole length	Leaf shape	Petiole base color	Leaf color	Shoot color	References
3	Ambon panjang, Ternate	NA	NA	NA	Elliptic	Red	Yellowish green	Light yellowish green	Soenarsih et al. (2021)
4	Cengkeh Bogor, Ternate	NA	NA	NA	Elliptic	Red	Dark green	Light yellowish green	Soenarsih et al. (2021)
5	Zanzibar Pangkal Kuning, Ternate	NA	NA	NA	Elliptic	Red	Dark green	Yellowish green	Soenarsih et al. (2021)
6	Cengkeh Rica, Ternate	NA	NA	NA	Elliptic	Red	Dark green	Light pink	Soenarsih et al. (2021)
7	Cengkeh Zanzibar, Ternate	NA	NA	NA	Elliptic	Red	Dark green	Light yellowish green	Soenarsih et al. (2021)
8	Cengkeh Pink, Ternate	NA	NA	NA	Elliptic	Red	Dark green	Light yellowish green	Soenarsih et al. (2021)
9	Cengkeh Posi- Posi, Ternate	NA	NA	NA	Elliptic	Light pink	Dark green	Light pink	Soenarsih et al. (2021)
10	Ambon gemuk, Ternate	NA	NA	NA	Elliptic	Red	Dark green	Red	Soenarsih et al. (2021)
11	Cengkeh Bendera	NA	NA	NA	Elliptic	Red	Dark green	Yellowish green	Soenarsih et al. (2021)
12	Cengkeh Putih, Ternate	NA	NA	NA	Elliptic	Red	Dark green	Light yellowish green	Soenarsih et al. (2021)
13	Local Halmahera, North Maluku	11.2–12.5	4.6–6.4	NA	Oval	NA	Dark green	NA	Cahyaningrum et al. (2021)

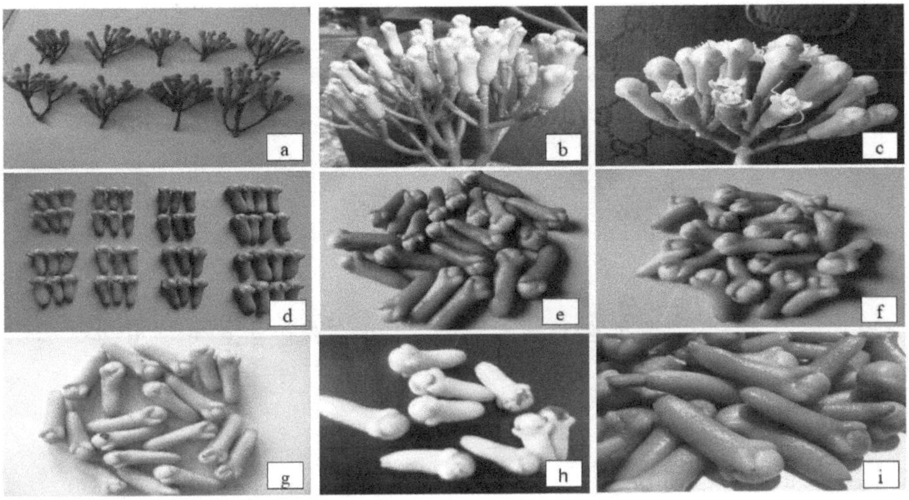

FIGURE 2.3 Characteristics of clove inflorescences and flower bud.

(a) Variation in peduncle lengths, (b) long peduncle, (c) short peduncle, (d) variation in flower bud color and shapes, (e) red flower bud, medium, tapering, (f) reddish green, short, large cone, (g) green, slim conical, (h) beige/pale green, short, conical, (i) red, long, narrow conical

The calyx tube is light green in young buds and turns pink or red when they reach full size with various degrees of red intensity as mentioned in Table 2.3.

There are four petals on the top of the calyx tube. They are initially tightly enclosed within the calyx and remain inconspicuous. The flower contains numerous stamens located at the base, encircling the pistil.

The filament of the stamens bears oil glands beneath the epidermis, and the connective bears a large oil gland at the apex. The ovary is located at the base and inside contains the ovules. The style is slender, extending from the ovary, with a stigma at the tip. The ovary is inferior, located below the other floral parts. The clove heads/petal caps are usually spherical or conical, white or beige; some have red marks when fully developed, depending on the variety.

The inflorescence is harvested when the buds have reached their full size before they open, forming the head of the dried clove. The flowers are aromatic due to essential oil glands in the floral components such as sepals, petals, ovaries, and stamens. The quality of cloves is determined by the chemical compound content, the size of the flower buds, and whether the flowers have bloomed. A fully mature bud before the flower opens is considered a good-quality clove (Razafimamonjison et al., 2013). However, Alfikri et al. (2020) showed that the clove flowers at the flowering stage (petals already fallen off but stamens remaining intact) contain the highest essential oil.

Variation in the inflorescence and flower bud characteristics is found between varieties outside and inside Maluku. However, there is not sufficient data on the flower characteristics in the North Maluku varieties.

Table 2.3

Characteristics of clove inflorescence and flower bud.

No	Clove variety	Peduncle type	Bud size	Bud length (cm)	Bud width (cm)	Bud weight (g)	Number of flower buds/ inflorescence	Bud shape	Flower head/ petal cap shape	Petal color	Mature bud color	References
Outside Maluku												
1	Zanzibar Karo, West Sumatra	Long	NA	NA	NA	NA	12–30	Narrow conical	Conical	Beige with red stripes	Bright red	Ministry of Agriculture (2008)
2	Zanzibar Sukabumi, West Java	Medium	NA	1.45–1.69	0.34–0.42	NA	10.78–22.52	Narrow conical	Conical	Beige– with red stripes	Red	Randriani and Syafaruddin (2011)
3	Zanzibar Gorontalo, North Sulawesi	Long	Medium	1.70–2.00	0.55–0.61	0.3–0.46	19.41–27.79	Broad conical	Conical	Beige	Reddish beige	Ministry of Agriculture (2023)
4	Zanzibar Semarang, Central Java	NA	Small–Large	1.73–2.01	0.37–0.49	NA	8–32	Narrow to broad conical	Conical–rounded	Beige with red stripes	Reddish to bright red	Susilawati et al. (2024)
5	Zanzibar Peling, Central Sulawesi	Medium–long	NA	1.9–2.08	0.46–0.52	0.35–0.45	15.68–22.36	Broad conical	Conical–rounded	Beige with red stripes	Reddish beige-red	Ministry of Agriculture (2023)
6	Siantan Agribun. Riau Province	Medium–long	large	1.92–2.04	0.43–0.55	0.39–0.43	15.64–19.44	Narrow–broad conical	Rounded	Beige	Reddish beige	Bermawie et al. (2021)
7	Siputih Sukabumi, West Java	Short	NA	1.53–1.77	0.41–0.49	NA	17.55–20.29	Broad oblong waisted	Rounded	Greenish beige	Green	Randriani and Syafaruddin (2011)

TABLE 2.2 *(Continued)*

Leaf characteristics of clove varieties from several origins.

Maluku												
1	Tuni Bursel, Maluku	Short	Medium–large	1.6–2.2	0.41–0.50	0.36–0.60	7–30	Broad conical	Rounded	Beige with red stripe	Light reddish beige	Ministry of Agriculture (2023)
2	Small Tuni, Maluku	NA	NA	1.09–1.69	0.29–0.32	0.45–0.48	NA	NA	NA	NA	Pale greenish yellow	Namakule et al. (2024)
3	Large Tuni, Maluku	NA	NA	2.20–3.13	0.42–0.58	0.35–0.37	NA	NA	NA	NA	Yellow	Wattimena et al. (2023)
4	Halmahera, North Maluku	Medium	Small	NA	NA	0.27–0.30	11–21	NA	Rounded	NA	Reddish yellow	Cahyaningrum (2021)
5	Raja clove, Ambon Island	NA	Large	2.1–3.3	0.5–0.6	0.4–0.5	NA	Broad conical	Rounded	Light green	Yellowish green	Alfian et al. (2019)
6	Forest clove Ambon, Maluku	NA	Large	NA	NA	NA	NA	Broad conical	Rounded	NA	NA	Mahulette et al. (2022)
North Maluku												
1.	Local Ternate 1, North Maluku	NA	NA	NA	NA	NA	NA	NA	Oblong, ovoid, breech, elliptic, and cylindrical	NA	Red, reddish green, light pink, yellowish green	Soenarsih et al. (2021)
2.	Local Ternate 2, North Maluku	NA	NA	NA	NA	NA	NA	NA	Sphericle, conical, and cylindrical	NA	Red, light pink, yellowish, green, reddish green	Suparman et al. (2017)
3	Afo Ternate, North Maluku	Long	Small	NA	0.35–0.43	0.27–0.3	18–27	NA	Conical	Beige with red stripe	Reddish beige	Ministry of Agriculture (2010)
4	Afo Halmahera, North Maluku	Long	Small	NA	0.27–0.30	NA	18–27	Slender slightly conical	Round taper	NA	Reddish yellow	Cahyaningrum et al. (2021)
5	Local Halmahera, North Maluku	Medium	Small	NA	0.27–0.30	NA	11–21	Slender slightly conical	Round taper	NA	Reddish beige	Cahyaningrum et al. (2021)

The size of the flower bud (weight, length, width, and head) varies depending on the clove varieties. The Zanzibar clove grown in Indonesia usually has small to medium buds, differing from those grown in other countries, such as Madagascar and Zanzibar, which are reported to have large bud sizes (Razafimamonjison et al., 2013) but no data on their actual size. Most of the Zanzibar cloves grown in Indonesia have small to medium bud sizes (Table 2.3). Large size buds are found in Tuni, Forest, and Raja.

2.9 FRUITS

Clove fruits may result from self-pollination, open pollination with the same variety, or another way. Clove fruit consists of skin, seeds, and an umbilical cord. The clove fruit's shape was ellipsoidal to ovoid, with a dark red color (de Gusman & Siemonsma, 1999). Indonesian clove has variations in fruit size and shape. The shape of the fruit of clove varieties of several regions in Indonesia is oblong, obovate, and conical shape, narrow or relatively wide in diameter as mentioned in Table 2.4.

TABLE 2.4
Fruit characteristics of clove fruit from several origins.

No,	Clove variety	Fruit weight (g)	Fruit length (cm)	Fruit width (cm)	Fruit shape	References
Outside Maluku						
1	Zanzibar Karo. North Sumatra	NA	NA	NA	Long conical	Ministry of Agriculture (2008)
2	Zanzibar Gorontalo, North Sulawesi	0.82–1.22	2.67–2.99	0.96–1.11	Long conical	Ministry of Agriculture (2023)
3	Siantan Agribun, Riau Province	2.5–3.46	1.48–3.82	0.7–2.0	Oblong	Bermawie et al. (2021)
4	Zanzibar Peling, Central Sulawesi	3.57–3.88	2.46–3.36	1.44–1.52	Cylindrical-Oval	Ministry of Agriculture (2023)
5	Aceh, West Sumatra	NA	2.33–2.87	1.22–1.41	Obovate-oblong	Susilowati et al. (2023)
6	Cianjur, Bogor, West Java	NA	2.39–3.17	1.18–1.45	Obovate-oblong	Susilowati et al. (2023)
7	Semarang, Karang anyar, Central Java	NA	1.85–3.09	1.06–1.61	Obovate-oblong	Susilowati et al. (2023)
8	Ende, Alor, East Manggarai, East Nusa Tenggara	NA	2.42–3.29	1.04–1.78	Obovate	Susilowati et al. (2023)
9	East and Central Sulawesi	NA	1.79–3.20	1.25–1.96	Obovate	Susilowati et al. (2023)
10	Zanzibar Sukabumi, West Java	NA	2.2–2.66	1.13–1.35	Long conical	Randriani and Syafaruddin (2011)
11	Siputih Sukabumi, West Java	NA	2.13–2.59	1.13–1.35	Long conical	Randriani and Syafaruddin (2011)

TABLE 2.4 *(Continued)*
Fruit characteristics of clove fruit from several origins.

No,	Clove variety	Fruit weight (g)	Fruit length (cm)	Fruit width (cm)	Fruit shape	References
Maluku						
1	Tuni Bursel, Maluku	2.6–3.9	2.8–3.5	1.1–1.5	Oblong	Ministry of Agriculture (2019a, 2019b, 2019c)
2	Local Tuni, Maluku	2.07–2.27	2.58–2.78	1.25–1.45	Conical	Namakule et al. (2024)
3	Raja I, Ambon Island	2.04–2.57	3.20–3.70	0.9–1.5	Conical	Alfian et al. (2019)
4	Raja II, Maluku	4.11–7.23	3.21–3.86	1.39–1.69	Conical	Wattimena et al. (2023)
North Maluku						
1	Afo, Ternate	3.2–3.5	NA	NA	Long conical	Ministry of Agriculture (2010)
2	Local Halmahera, North Maluku	2.9–3.3	NA	NA	Long conical	Cahyaningrum et al. (2021)

FIGURE 2.4 Young and mature clove fruit shape and colour.

Young fruit is green in color; it will develop into bright red and finally grey-purple when mature. Long and big fruit sizes belong to several clove varieties such as Tuni Bursel, Zanzibar Pelling, and Siantan Agribun; on the other hand, shorter and smaller fruit sizes belong to the Zanzibar Karo variety and several clove varieties from Sulawesi, Central Jawa, East Jawa, and East Manggarai, as represented in Figure 2.4 (Susilowati et al., 2023).

The fruit sizes from several regions in Indonesia vary. The fruit size of Zanzibar varieties grown outside Maluku ranges from very small, found in Zanzibar Gorontalo from 0.8–1.22 g to 3.6–3.9 g in Zanzibar Peling. Larger sizes of fruits were obtained from Cianjur West Jawa, Semarang Central Jawa, and East Nusa Tenggara. Varieties in Maluku, especially Raja II, have the most significant fruit size, in contrast to the Local Tuni. Two samples of clove varieties (Afo and Local Halmahera) were recorded for fruit characters, and slight variation was found. Fruits are green when young and gradually turn to bright red and dark purple when they mature. The shape varied from oblong, to obovate, to conical. They contain a single seed. Fruits mature approximately 9 months after flower initiation and are physiologically mature when the exocarp turns reddish-purple, popularly called the 'mother of clove'.

2.10 SEEDS

Clove seeds may derive from self-pollination, open pollination within the same variety, or cross-pollination with different varieties. If the parents are already heterozygotes, the progenies could produce more variation due to the segregation of characters from the parents. Therefore, new clove variants may be developed in each region. It is a blessing to deviate from the original variety, thus enriching the genetic resources of cloves. However, it also causes a loss of the actual genetic composition of the original variety. These new variants can be seen in the variation in the Zanzibar clove, which differs from the original Zanzibar variety.

Clove seed size varies from small to large, with varying weight, length, and width as represented in Figure 2.5

Seed size in cloves is often considered a determinant of seed viability, even though this is not always true. Genetic factors often influence seed size; small seeds have the same opportunity as large seeds to germinate and develop well. The characteristics of clove seeds and their origins is mentioned in **Table 2.5.**

FIGURE 2.5 Clove seed shape and colour.

TABLE 2.5
Characteristics of clove seeds from several origins.

No	Clove variety	Seed weight (g)	Seed length (cm)	Seed width (cm)	Seed shape	Seed color	References
1	Sukabumi Zanzibar, West Java	NA	1.72–1.96	0.78–0.98	NA	NA	Randriani and Syafaruddin (2011)
2	Siputih, West Java	NA	1.77–2.01	0.79–0.91	NA	NA	Randriani and Syafaruddin (2011)
3	Zanzibar Gorontalo, North Sulawesi	0.82–1.22	1.72–2.08	0.65–0.87	Long conical	Brownish green	Ministry of Agriculture No. (2023)
4	Siantan Agribun, Riau Province	0.7–1.1	1.6–1.88	0.73–1.15	Oblong	Reddish green	Bermawie et al. (2022)
5	Aceh, West Sumatra	NA	1.62–2.29	0.82–0.95	Conical (short–long)	Greenish red–red.	Susilawati et al. (2023)
6	Cianjur, Bogor, West Java	NA	1.71–2.23	0.79–1.01	Oblong	Greenish red	
7	Semarang, Karang anyar, Central Jawa	NA	1.21–2.46	0.58–1.06	Conical (short-long)	Greenish red–red.	
8	Ende, Alor, East Manggarai, East Nusa Tenggara	NA	1.67–2.40	0.72–1.01	Conical (short–long)	Greenish red–red.	
9	East and Central Sulawesi	NA	1.26–2.20	0.79–0.99	Oblong–silindris	Greenish red–red.	
Maluku							
1	Tuni Bursel, Maluku	1.8–2.8	1.9–2.5	0,8–1,3	Long elliptical	Reddish brown	Ministry of Agriculture (2023)
2	Tuni, Maluku	0.83	1.54–1.90	0.34–0.84	Conical	Light Purple	Namakule et al. (2024)
3	Raja, Ambon Island	0.36–0.86	3.2–3.7	0.7–0.9	NA	Pink	Alfian et al. (2019)
Maluku utara							
1	Halmahera Local superior cloves, North Maluku	1.5–2.0	2.2–2.4	NA	Long conical	Beige	Cahyaningrum et al. (2021)
3	Afo Ternate, North Maluku	2.1–2.3	2.50–2.99	1.23–1.32	Long conical	Brownish red	Ministry of Agriculture (2010)

Clove varieties that grow outside Maluku have almost uniform seed sizes; even though they are different varieties, the shape and color of the seeds are more diverse. Meanwhile, in Maluku, Tuni Bursel cloves have a larger seed size than Raja cloves and have red seed colors but differ in the intensity of the red color. Cloves in North Maluku are almost uniform between Afo and local Halmagera, both in size and shape, while the color of the seeds is different. The seed color of the local Halmahera is beige, while Afo is brownish red.

2.11 CONCLUSION

Clove is still a valuable crop commodity. The clove plant's botanical characteristics provide comprehensive insights into its root system, stems, leaves, flowers, fruits, and seeds. Understanding these botanical features is essential for various scientific fields, from agriculture and horticulture to pharmacology and culinary arts, underscoring the clove's multifaceted value.

REFERENCES

Alfian, A., Mahulette, A. S., Zainal, M., Hardin, and Bahrun, A. H. (2019). Morphological character of raja clove (*Syzygium aromaticum* L. Merr & Perry.) native from Ambon Island. *IOP Conf. Series: Earth and Environmental Science.* 343(2019):012150. IOP Publishing. http://doi.org/10.1088/1755-1315/343/1/012150.

Alfikri, F. N., Pujiati, R., Wibisono, M. G., and Hardiyanto, E. B. (2020). Yield, quality, and antioxidant activity of clove (*Syzygium aromaticum* L.) bud oil at the different phenological stages in young and mature trees. *Scientifica.* 2020:Article ID 9701701.1–8. https://doi.org/10.1155/2020/9701701.

Al Muhdhar, M. H. I., Rohman, F., Tamalene, M. N., Nadra, W. S., and Daud, A. (2018). *Diversity of Local Superior Spice and Food Plants.* First edition. Malang University, 160p. ISBN: 978–602-470-067-6.

Alwi. (1953). *Cengkeh.* Teknik Pertanian II, pp. 392–403.

Andi. (2013). *Cengkeh tertua di dunia ada di Ternate (The Old Clove Plant Is in Ternate).* https://news.republika.co.id/berita/msddf4/cengkeh-tertua-di-dunia-ada-di-ternate#google_vignette_May 31 2024.

Bermawie, N. (1992). Clove. Research development of spices and medicinal crops. *Special Edition. Indonesian Spice and Medicinal Crops Research Institute.* 7(2):1–15.

Bermawie, N. (2014). Botany. In: Ferry Karwur and Haryoni Semangun (eds.) *Clove: History, Cultivation and Industry.* PT Gramedia, pp. 11–44.

Bermawie, N., and Pool, P. A. (1991). The use of isozyme to determine the status of Zanzibar type clove (*Syzygium aromaticum* L.). *Journal Pemberitaan Penelitian Tanaman Industri.* 17(1):1–9.

Bermawie, N., Wahyuni, S., and Ginting, R. (2021). Selection of highly productive clove trees in the Anambas population. *IOP Conf. Series: Earth and Environmental Science.* 762(2021):012041.

Cahyaningrum, H., Lala, F., Polakitan, A., and Wahab, A. (2021). Morphological characteristic of local clove varieties in East Halmahera. *IOP Conf. Series: Earth and Environmental Science.* 803:012020. http://doi.org/10.1088/1755-1315/803/1/012020.

Cortes-Rojas, D. F., de Souza, C. R. F., and Oliveira, W. P. (2014). Clove (*Syzigum aromaticum*): A precious spice. *Asian Pacific Journal of Tropical Biomedicine.* 4(2):90–96.

De Gusman, C. C., and Siemonsma, J. S. (1999). Plant resources of South-East Asia no. 13. In: *Spices—Syzygium aromaticum.* Backhuys Publisher, pp. 211–218.

Hasibuan, A. I., Syaukat, Y., and Falatehan, A. F. (2022). Analysis of factors affecting the import of Indonesian cloves. *Jurnal Riset Bisnis.* 6(1):144–160.

Hassan, H. A., Genaidy, M. M., Kamel, M. S., and Abdelwahab, S. F. (2020). Synergistic antifungal activity of mixtures of clove, cumin and caraway essential oils and their major active components. *Journal of Herbal Medicine.* 24.

Kamatou, G. P. P., Vermaak, I., and Viljoen, A. M. (2012). Eugenol—from the remote Maluku Islands to the international marketplace: A review of a remarkable and versatile molecule. *Molecules.* 17(6):6953–6981.

Kaufman, T. S. (2015). The multiple faces of Eugenol. A versatile starting material and building block for organic and bio-organic synthesis and a convenient precursor toward bio-based fine chemicals. *Journal of the Brazilian Chemical Society.* 26:1055–1085.

Lubis, R. (1954). *Cengkeh and Kretek.* Percetakan Makassar, 77 p.

Mahulette, A. S., Alfian, A., Suyadi, S., Supriyanto, S., Situmorang, J., Matatula, A. J., Kilkoda, A. K., Nendissa, J. I., and Wattimena, A. Y. (2022). Type and morphological character of local clove (*Syzygium aromaticum*) from Maluku, Indonesia. *Biodiversitas.* 23(3):1301–1309.

Mahulette, A. S., Hariyadi, H., Yahya, S., Wachjar, A., and Alfian, A. (2019b). Morphological traits of Maluku native forest clove (*Syzygium aromaticum* L. Merr & Perry.). *Journal of Tropical Crop Science.* 6(2):105–111.

Mahulette, A. S., Hariyadi, H., Yahya, S., Wachjar, A., and Marzuki, I. (2019a). Morpho-agronomical diversity of forest clove in Molucas, Indonesia. *Hayati Journal of Bioscience.* 26(4):156–162.

Milind, P., and Deepa, K. (2011). Clove: A champion spice. *International Journal of Research in Ayurveda and Pharmacy.* 2(1): 47–54.

Ministry of Agriculture. (2008). *Decree of the Minister of Agriculture of the Republic of Indonesia Regarding the Release of Cloves of the Composite Variety Zanzibar Karo No. 339/Kpts/SR120/3/2008.* Ministry of Agriculture.

Ministry of Agriculture. (2010). *Decree of the Minister of Agriculture of the Republic of Indonesia Regarding the Release of AFO Clove Populations as High-Yielding Varieties No. 3680/Kpts/SR120/11/2010.* Ministry of Agriculture.

Ministry of Agriculture. (2013a). *Decree of the Minister of Agriculture of the Republic of Indonesia Regarding the Release of Gorontalo Clove Populations as High-Yielding Varieties No. 198/Kpts/SR120/1/2013.* Ministry of Agriculture.

Ministry of Agriculture. (2013b). *Decree of the Minister of Agriculture of the Republic of Indonesia Regarding the Release of the Population of Tuni Bursel Cloves as a High-Yielding Variety No. 4964/Kpts/SR120/12/2013.* Ministry of Agriculture.

Ministry of Agriculture. (2019a). *Decree of the Minister of Agriculture of the Republic of Indonesia Concerning the Release of the Establishment of Local Superior Clove Seed Source Gardens in Simeuleu, Aceh Province No. 85/Kpts/KB.020/6/2019.* Ministry of Agriculture.

Ministry of Agriculture. (2019b). *Decree of the Minister of Agriculture of the Republic of Indonesia Concerning the Release of the Establishment of Local Superior Clove Seed Source Gardens in Donggala, Province of Central Sulawesi Tengah No. 09/Kpts/KB.020/2/2019.* Ministry of Agriculture.

Ministry of Agriculture. (2019c). *Decree of the Minister of Agriculture of the Republic of Indonesia Concerning the Release of the Establishment of Local Superior Clove Seed Source Gardens in Natuna, Riau Island Province No. 14/Kpts/KB. 020/2/2019.* Ministry of Agriculture.

Ministry of Agriculture. (2020a). *Decree of the Minister of Agriculture of the Republic of Indonesia Regarding the Release of Siantan Agribun as Clove Populations as High-Yielding Varieties No. 41/Kpts/KB. 010/3/2020.* Ministry of Agriculture.

Ministry of Agriculture. (2020b). *Decree of the Minister of Agriculture of the Republic of Indonesia Concerning the Release of the Establishment of Local Superior Clove Seed Source Gardens in Pesisir Selatan Regency, West Sumatra Province No. 09/Kpts/KB. 020/2/2020.* Ministry of Agriculture.

Ministry of Agriculture. (2023a). *Decree of the Minister of Agriculture of the Republic of Indonesia Concerning the Release of the Establishment of Local Superior Clove Seed Source Gardens in Sabang, Aceh Province No.58/Kpts/KB.020/05/2023.* Ministry of Agriculture.

Ministry of Agriculture. (2023b). *Decree of the Minister of Agriculture of the Republic of Indonesia Regarding the Release of Zanzibar Peling Clove Populations as High-Yielding Varieties No. 48/Kpts/KB.020/05/2023*. Ministry of Agriculture.

Namakule, M. Y., Mahulette, A. S., and Matatula, A. J. (2024). Morphological diversity of tuni clove (*Syzygium aromaticum* L.) in Tehoru, Central Maluku, Maluku Province. *Jurnal Agro Wiralodra*. 7(1):1–7.

Nejad, M. S., Özgüneş H., and Başaran, N. (2017). Pharmacological and toxicological properties of eugenol. *Turkish Journal of Pharmaceutical Sciences*. 14(2):201–206.

Özgen, A., and Yücel, İ. (2024). Nature's gift: *Syzygium aromaticum. Health and Technology Journal (HTechJ)*. 2(1):28–33. https://doi.org/10.53713/htechj.v2i1.139

Pool, P. A., Eden-Green, S. J., and Muhammad, M. T. (1986). Variation in clove (*Syzyhium aromaticum* L.) germplasm in the Moluccan islands. *Euphytica*. 35:149–159.

Purbopuspito, J., and Van Rees, K. C. J. (2002). Root distribution at various distances from clove trees growing in Indonesia. *Plant and Soil*. 239:313–320.

Purseglove, J. W., Brown, G., Green, C. L., and Robbins, S. R. J. (1981). *Spices*. Vol. 2. Longman Scientific and Tecnial Co. and John Wiley and Sons, Inc., 457.

Randriani, E., and Syafaruddin, S. (2011). Uji Kekerabatan Aksesi Cengkeh di Kebun Percobaan Sukapura. *Buletin Plasma Nutfah*. 17(1).

Razafimamonjison, G., Jahiel, M., Ramanoelina, P., Fawbush, F., and Danthu, P. (2013). Effect of phenological stages on yield and composition of essential oil of *Syzygium aromaticum* buds from Madagascar. *International Journal of Basic and Applied Sciences*. 2(4):312–318.

Ridley, H. N. (1912). *Clove*. Macmillan and Co.

Saranya, D., and Ravi, R. (2018). *Syzygium aromaticum*: A subtropical tree scrutinized for the genetic differentiation among the altitudes of Nilgiris. *Annals of Plant Sciences*. 7(4):2134–2138.

Schmid, R. (1972). A resolution of the Eugenia–Syzygium controversy (Myrtaceae). *American Journal of Botany*. 59:423–436.

Singh, P. A., Bajwa, N., and Baldi, A. (2021). Possible role of traditional systems of medicine to manage COVID-19: A review. *Israel Journal of Plant Sciences*. 68(1–2):3–28.

Singh, P. A., Dash, S., Choudhury, A., and Bajwa, N. (2024). Factors affecting long-term availability of medicinal plants in India. *Journal of Crop Science and Biotechnology*. 27(2):145–173.

Soenarsih, S., Wahyudiono, E., and Mandea, A. R. (2021). Diversity and relationship of clove (*Syzyhium aromaticum* L.) in Ternate Island. *Cannarium* (Jurnal Ilmu-Ilmu Pertanian). 19(2):65–84.

Sundari, S., Nurhasanah, N., Mas'ud, A., Amin, M., Arumingtias, E. L., and Azrianingsih, R. (2019). Update phylogenetic information of the local varieties of cloves (*Syzygium aromaticum*) from North Maluku, Indonesia based on ITS sequences data. *Biodiversitas*. 20(6):1510–1515.

Suparman, S., Nurhasanah, J., and Papuangan, N. (2017). Population mapping and types of local varieties of clove plants (*Syzygium aromaticum* L.) in Ternate Island District. *Proceedings of the National Seminar on the Application of Science and Technology*. 239–244.

Susilowati, M., Wahyuni, S., Setiadi, A., and Bermawie, N. (2023). Fruit and seed morphology variation of clove (*Syzygium aromaticum* L.) from various regions in Indonesia. *Proceedings of the 9th International Symposium on Innovative Bioproduction Indonesia on Biotechnology and Bioengineering 2022. AIP Conf. Proc. 2972*, 030012–1–030012–9. https://doi.org/10.1063/5.0183279

Susilowati, M., Wahyuni, S., Setiadi, A., and Bermawie, N. (2024). Yield and morphological characteristics of cloves (*Syzygium aromaticum* L.) from Semarang Plantation, Indonesia. *IOP Conference Series Earth and Environmental Science*. 1377:012097. https://doi.org/10.1088/1755-1315/1377/1/012097

Ulanowska, M., and Olas, B. (2021). Biological properties and prospects for the application of Eugenol: A review. *International Journal of Molecular Sciences*. 22(7):3671. http://doi.org/10.3390/ijms22073671. PMID: 33916044; PMCID: PMC8036490.

Wattimena, A. Y., Mahulette, A. S., Makaruku, M. H., and Lating, M. F. A. (2023). Morphological character of clove "Raja" (*Syzygium aromaticum* L.), Endemic to Maluku, Indonesia. *Journal of Tropical Crop Science*. 10(1): 71–78.

3 Agrarian Conditions

Ireng Darwati, Rudi Suryadi, Muchamad Yusron, Devi Rusmin, Octivia Trisilawati, Nurliani Bermawie, and R. Vitri Garvita

3.1 INTRODUCTION

Clove (*Syzygium aromaticum* L. Meer) is an aromatic, a member of the Myrtaceae family and is originally from Indonesia, namely the Moluccas Islands, which is one of its local origins (Alfian et al., 2019). Clove is a valuable spice crop used as a food preservative and culinary spice. Flower buds contain 80–90% of eugenol, a chemical responsible for the clove's distinctive aroma, which has antiseptic, anesthetic, and antimicrobial functions (Shaikh et al., 2018).

Most cloves are cultivated primarily to satisfy their domestic needs, as they are a crucial ingredient in the production of cigarettes. Less than 10% of the production is allocated to spices, medicinal compounds, food preservatives, and other purposes. Cloves are important for Indonesian farmers because most (99.18%) are on farmers' plantations. The Indonesia clove planting area is 574,539 ha, with low clove productivity, about 336–424 kg/ha (Directorate General of Estate Crops-Indonesia, 2023). The highest clove production centers in Indonesia included ten provinces: Center of Sulawesi, North Sulawesi, South Sulawesi, Maluku, East Java, Center of Java, Southeast Sulawesi, West Java, North Maluku, and Aceh. Cloves also have a vital role in Indonesia's economy, specifically by significantly contributing to the country's national income and foreign exchange through the taxation of cigarettes. In 2022, the export value of cloves will reach US$ 2,285,060 (Directorate General of Estate Crops-Indonesia, 2023). Enhancing the country's clove production with the growing demand from emerging clove cigarette makers is imperative. The clove cigarette industry predominantly utilizes the nation's clove harvest, and any remaining need is met by procuring additional resources.

Sustainable clove production and improving farmers' income depend on adherence to good agricultural practices (GAP). Implementing these practices involves monitoring the entire process from cultivation to consumption, including soil management, water use efficiency, and natural resource conservation (Singh et al., 2022; Singh et al., 2021; Singh & Baldi, 2018). Effective agricultural practices require integrating and executing production systems tailored for commercial agricultural product cultivation in a manner that prioritizes economic viability alongside social and environmental sustainability factors. Although there is a common underlying concept behind excellent agricultural practices, different countries have specific definitions with accompanying guidelines to improve yields while addressing broader socio-economic concerns. Additionally, opinions vary regarding the importance of

 DOI: 10.1201/9781003521785-3

GAP in addressing multifaceted environmental, economic, and social issues while enhancing production through established rules and available data application without compromising long-term sustainability goals or overall profitability objectives. Good agricultural practice is a quality assurance system that guarantees safe product production by regulating the production process and maintaining crop standards. It emphasizes ideal weather conditions, soil health, nutrient levels, parent trees as sources for planting material such as seeds and shoots, generative and vegetative plant propagation techniques, seed and Scion management, fertilizer use, and overall farm maintenance, including sanitation and cleanliness.

Clove cultivation is influenced by climate and soil and the altitude and topography of its growing area. Cloves are usually cultivated in highland areas, taking advantage of relatively low temperatures and sufficient rainfall. Understanding the specific environmental factors that influence the growth and quality of clove plants is critical for farmers to make decisions regarding the placement and management of their plantations. Moreover, it is essential to consider the specific type and properties of soil to maintain soil fertility and reduce environmental impact. Cloves thrive in well-drained soils with high organic matter content and pH neutrality. Warm temperatures, high humidity, and well-distributed rainfall characterize the ideal climate for clove cultivation. These conditions are crucial for the proper development of clove plants and the formation of their aromatic buds.

Implementing GAP is essential for enhancing the overall quality of clove production. It ensures the security of the products and promotes sustainable agricultural methods that prioritize preserving the environment and long-term economic viability. By adhering to the principles of GAP, farmers can enhance their cultivation techniques, minimize harmful pesticides, and boost the yield of their clove farms. In the following sections, we will thoroughly analyze these agricultural conditions, exploring their importance in successfully cultivating cloves and implementing sustainable farming practices.

3.2 AGRO-CLIMATE

Some critical agroclimatic factors are necessary for effective clove growing. The initiation of clove flowering requires a fluorogenic hormone, the production of which is influenced by climatic conditions. For clove plants to flourish, particular environmental factors are needed, such as a suitable soil type, climate, and water availability (Umasugi et al., 2020). Clove trees are susceptible to extremes in temperature and irregular precipitation. Thus, they must have ideal temperatures and rainfall patterns. Climatic factors, including temperature, sunshine, and rainfall, affect crops' overall growth and development when they act in concert rather than separately.

Tropical areas are characterized by high rainfall. Rainfall will directly affect humidity, air temperature, and the intensity of solar radiation. Optimal rainfall for clove growth is 1,500–3,500 mm/year spread evenly throughout the year, providing the plant's water needs and creating a humid environment. Continuous heavy rains inhibit the formation of flower primordials, causing low production. Apart from that, high rainfall can reduce the quality of clove flowers. High rainfall, especially on land with poor aeration and soil drainage, causes plant death due to flooded roots. Low to moderate rainfall (63–187 mm) will increase production. Low rainfall causes reduced soil water availability, high air temperatures, and low humidity. The critical rainfall limit for clove plants is 80 mm/month (Wahid et al., 1985). Drought can cause plant death, particularly in young plants. On the other hand, high rainfall, especially on land with poor aeration and soil drainage, causes plant death owing to flooded roots.

Field experience shows that cloves flower profusely after a dry period of 2–4 months, followed by sufficient rainfall during flower primordial formation. On the island of Java, primordial flowers appear in November, December, and January; in Sumatra in October, November, and December; in North Sulawesi in December, January, and February; and in the Maluku Islands in May, June, and July. It takes 6 months from the release of the primordial flower until it is ready to harvest. Flower failure usually occurs in the highlands or areas with high rainfall. The initiation of clove flowering requires a fluorogenic hormone, the production of which is influenced by climatic conditions.

The optimal air temperature for clove plant growth ranges from 22–32°C. Air temperatures that are too high (≥ 34°C) harm the growth of clove plants. High air temperatures cause large amounts of evaporation; therefore, plants become stressed with insufficient soil water. On the other hand, low air temperatures, especially in the highlands, cause better vegetative growth but lower production.

Air temperatures (± 32°C) due to low rainfall can encourage the release of flower primordials. Meanwhile, low air temperatures (<19°C) can cause flowers that have appeared to turn into leaves (flower failure). Air temperature can also affect the quality of clove flowers. In Zanzibar, with an air temperature of 21–29°C and an average annual rainfall of 1,650 mm/year evenly distributed, produces the best quality cloves for kretek cigarettes.

Relative humidity (RH) is the ratio between the water vapor in the air and the amount of water vapor that can be held at the same temperature and pressure. It affects the plant transpiration rate and indirectly affects water absorption and nutrients. Thus, within certain limits, the RH affects the growth of clove plants. The optimal relative humidity for clove plant growth ranges from 60–80%.

The intensity of solar radiation and the influence of sunlight on clove flowering are often associated with rainfall. Sunlight intensity between 61–66% combined with low to moderate rainfall (2–177 mm) will increase clove production. The intensity of sunlight on average 1–4 months before the formation of flower primordials is the most determining factor in the amount of production.

Particularly in Southeast Asia and the Spice Islands, cloves are a crop of great economic and cultural significance. They are highly valued spices made from the dried flower buds of the clove tree (Umasugi et al., 2020; Athaillah et al., 2021). Numerous agroclimatic conditions significantly impact clove cultivation and are critical in determining the growth and productivity of this valuable crop. The agroclimate of the clove tree can be seen in Table 3.1.

Land suitability and analysis of limiting factors are essential for sustainable land management for clove plants. Sloping land can be developed for cloves, overcoming the severe dangers of erosion and slopes with a reforestation system and helping preserve sloping land by making terracing. Fertilization can improve soil conditions for clove cultivation (Yusuf et al., 2023).

TABLE 3.1
Agroclimatic of clove tree.

Agroclimatic	Value	Reference
Elevation (m. asl)	200–600	Wahid et al. (1985)
Slope (%)	< 8	Ritung et al. (2011)
Rainfall (mm/years)	1,500–2,500	Wahid et al. (1985)
Dry month (month)	< 2	Wahid et al. (1985)
Temperature (ºC)	25–28°C	Ritung et al. (2011)
Relative Humidity (%)	60–80%	Ritung et al. (2011)

3.3 PLANT MATERIAL

Planting materials, including clove, are crucial for successful crop cultivation. Seeds are planting materials or plant parts used to reproduce and breed plants. Seeds can be generative (true seeds) or seeds resulting from vegetative propagation.

The availability of high-quality clove seeds is essential for successful agricultural practices. Therefore, the seeds must be derived from superior varieties that have received distribution recommendations and guaranteed superiority. However, breeding or crop improvement efforts for cloves' superior varieties have been limited. As clove is a perennial crop, breeding efforts will take a long time. Consequently, there are few reports on clove plant breeding, and information regarding recommended varieties for clove cultivation needs to be included.

In most clove-producing countries, quality seeds are generally obtained by selecting high-yielding and regular-bearing trees (termed mother trees) from existing cultivated/farmer populations. Mother trees should be selected from homogeneous environmental conditions so that the superior characteristics are genetically controlled and passed on to their offspring.

3.3.1 Selection of High-Yielding Population

There are several criteria for a cultivated clove population to be selected as high-yielding. In Indonesia, the criteria include acreage, plant age, the number of trees in a population, environmental conditions, cultural practices, minimum yield per tree, and healthy population (Ministry of Agriculture, 2023a, 2023b). An acreage of a yielding population should be at least 0.5 ha, consist of a minimum of 50 mature trees, have average yield of 30 kg fresh flower bud/tree, have a homogenous and favorable growing environment for clove, cultural practices should follow the standard operational procedures, and the trees within the population should be free from the main pests and diseases.

3.3.2 Selection of Mother Trees

Clove mother trees are selected from the best performers in a high-yielding population and grown in homogenous and favorable environments. This allows the superior trees to exhibit their superior phenotype and genetic characteristics.

Clove trees are predominantly outcrossing or, more precisely, a mixed mating crop, so seeds derived from a mother tree could be from selfing, self-pollination, or outcrossing (Bermawie & Pool, 1991). Genetic variation occurs in the population developed from seeds of outcrossed plants. High genetic variability provides a high chance of finding new variants that have important agronomic characteristics). Conversely, self-pollinating derived seeds in clove often result in growth depressions (Bermawie & Pool, 1991).

The distance between trees should be considered when selecting mother trees. The mother tree chosen should not be too close to the phenotypically similar trees to prevent self-pollination, which can cause inbreeding depression and poor adaptability of the progeny. The progenies will usually be well adapted if the seeds collected are from climatic conditions similar to the climate in which they will be planted. The selected mother tree should have superior characteristics such as high yield, regular bearing, and high quality, determined by genetic and environmental factors and interactions between genetics and the environment.

A large number of mother trees should be selected to prevent mating between the same variety, which can lead to inbreeding depression. In forest trees, 30 of the best trees are recommended for seed sources (www.cifor-icraf.org/publications/pdf/books/Agroforestry-primer-06.pdf/10 June 2024). In clove, about 5–10% of the best-performing plants in the clove population are selected, depending on the number of clove plants. Thirteen to twenty selected clove mother trees have been recommended as superior mother trees.

However, finding clove populations with large numbers of trees takes work. The clove mother trees selected should be collected only from the best of these trees. If the number of trees in a high-yielding population is less than the requirement, then several farmers whose land is adjacent to each other can join and be grouped into one high-yielding population. Each farmer gathers and mixes seeds from the chosen mother trees for distribution.

Crosses between diverse plants will preserve genetic diversity and prevent inbreeding depression, resulting in stunted growth. Plants located far from other plants should not be chosen because they could produce seeds by inbreeding (self-pollination), often resulting in growth depressions in progeny. Besides that, adjacent mother trees with similar phenotypes assumed to have the same genetic background should not be selected to prevent self-pollination, which can result in growth depression in the progeny.

In Indonesia, the criteria for selecting clove mother trees are regulated (Ministry of Agriculture, 2023a, 2008) to provide seed quality guarantees for farmers. The criteria for selecting clove mother trees in Indonesia are the same as those for selecting the population, added with the minimum plant age and yield of the mother trees to be selected (Ministry of Agriculture, 2023a, 2023b). The minimum age should not be less than 10 years, and the yield/tree should be more than the average yield of the population. The yield of some of the selected trees is 20% higher than the average yield of the population.

Several clove-superior varieties have been recommended for cultivation in Indonesia, selected primarily based on yield performance as mentioned in Table 3.2.

TABLE 3.2
Recommended clove varieties grown in Indonesia.

Country	Cultivated varieties	Clove characteristics	Leaf	Mature bud size and color	References
Indonesia	Zanzibar Karo	High EO and high eugenol	Narrow oval, slightly rounded	Small size, bright red	Ministry of Agriculture (2020)
	AFO	High EO and eugenol	Elliptic	Small-large size, reddish beige–red	Ministry of Agriculture 2016)
	Zanzibar Gorontalo	High EO and high eugenol	Narrow Elliptic	Small–large, reddish beige	Ministry of Agriculture (2013a, 2013b)
	Tuni Bursel	High EO and high eugenol	Slim oval–slightly rounded	Large size, light reddish beige	Ministry of Agriculture (2016)
	Siantan Agribun	Low-medium EO, high eugeno	Obovate	Large, reddish beige	(Bermawie et al., 2021)
	Zanzibar Peling	Low–medium EO, medium–high eugenol	Elliptic–obovate	Large, reddish beige–red	Ministry of Agriculture (2023a)

Besides that, popular clove varieties commonly cultivated are Zanzibar, Sikotok, Siputih, and Ambon. In clove-producing countries, several clove varieties are also known in trade: Zanzibar, Penang, Amboyna, Ceylon, and Malabar. Seeds, as propagating materials, are collected from regular-bearing mother trees. Besides these varieties, local clove varieties have also been developed in various countries. Local varieties may be developed due to a need for superior varieties. Most farmers collected seeds from their clove plantations, which are selected based on the farmer's preferences.

In Tanzania, the Zanzibar clove variety originated from Indonesia and is well known for its high-quality clove. It has a large bud size and a high oil content, particularly eugenol. The leaves are broader than other varieties, and the aroma is pronounced. In Madagascar, the clove has broad leaves, with half of the calyx tube turning red when the flower bud reaches its full size/mature, and it has medium-sized flowers. The leaf is green, broad lanceolate, with red shoots. The crown is conical and has a rosette branching pattern with a conical canopy shape (Danthu et al., 2014), similar to the Indonesian Sikotok variety.

3.3.3 Quality Seeds

Apart from the mother trees, seed quality is also an important factor in crop production as good seed interacts with the good environment around it and determines the performance of the crop or plant. Seeds must be collected from healthy seed sources.

Seeds can be procured from traditional/local varieties or from improved and high-yielding varieties. To increase the likelihood of collecting high-quality seeds, clove mother trees should be selected from populations of good-quality trees (Ministry of Agriculture, 2020). When many high-quality trees surround a clove mother tree, its progeny will exhibit desirable characteristics. This method has been practiced for many years to select clove-superior mother trees and obtain high-quality clove seeds. In forest trees, it is recommended that poor-quality trees in the population should be removed when possible (www.cifor-icraf.org/publications/pdf/books/Agroforestry-primer-06.pdf/accessed June 10, 2024) because if a mother tree is chosen from a population surrounded with poor quality trees, the seeds collected will likely produce poor-quality progenies.

Seed quality includes four components:

(1) Genetic quality is related to the correctness of the variety, both phenotypically (physically) and genetically inherited characteristics from the parent trees.
(2) Physiological quality related to the viability of seeds if they are grown (germinated), both under favorable (optimum) and less-favorable (suboptimum) conditions.
(3) Physical quality is related to the physical condition of the seeds, which concerns color, shape, size, weight, surface texture, level of physical damage, cleanliness, and uniformity.
(4) Health quality is related to the presence or absence of pest or disease attacks on the seeds and the level of attack that occurs.

Quality seeds are labeled with a high level of purity and growth capacity. Genetic quality can be assured if the seeds are collected from the recommended mother trees. In this way, true clove seeds come from open pollination, where only the mother plant is known, while the pollen source for the formation of the fruit/seed is unknown. Therefore, such seeds are referred to as "illegitimate" seeds. As a result, the mother tree's phenotypic performance exclusively reflects the maternal traits of the seeds that were gathered. Since the male parent tree is unknown, it is impossible to assess its phenotypic traits properly. However, since the surrounding trees are most likely the source of pollen that pollinated the mother tree's flowers, an accurate assessment of the male parent tree's traits may be established.

3.4 PLANT PROPAGATION TECHNIQUES

Clove plant propagation techniques involve generative and vegetative methods, each with advantages and applications. The propagation of clove plants is generally done by generative or seed propagation, as it is more accessible and practical. Seeds should be taken from clove-selected mother trees that have been recommended as seed sources, a mature tree with a minimum age of 10 years old, which has a high yield and regular bearing. This method is often used for large-scale cultivation because it can produce plants in large quantities. However, generative propagation has several drawbacks: (1) genetic variability between new progeny due to the cross-pollinated nature of cloves, which can cause inconsistent crop quality and yields (selecting the right seeds from mother trees that are high-yielding, stable, and disease-resistant is very important to ensure the best results in growth and production), (2) longer production time, and (3) rapid decrease in seed viability.

Vegetative propagation is a standard method for reproducing new plants, especially for outbreeding crops such as clove. It is crucial to obtain true-to-type progenies. This method involves growing new plants from plant parts of mother trees, such as stems and shoots.

Vegetative propagation methods, such as top grafting, approach grafting, cuttings, budding, and marcotting. The plant parts used are derived from healthy, disease-free branches and then planted in appropriate media until they establish strong growth. This technique preserves the genetic makeup of the parent trees, ensuring uniform quality and yield. It is crucial for maintaining a supply of healthy, vigorous clove plants, thereby enhancing the productivity and sustainability of clove cultivation. Selecting and employing the correct propagation method is anticipated to improve genetic quality and ensure stable production.

Clove flowers form at the tips of shoots, known as terminal flowers. Following harvest, new vegetative shoots emerge from the leaf axils, eventually developing into new flowers as they enter the generative phase. The flower grows to a maximum supported by more than 3 leaves under the flower (Figure 3.4). This flowering cycle leads to irregular bearing, where a high-yield season is often followed by a low-yield season, causing fluctuations in crop production (Martínez-Fuentes et al., 2013).

Plant propagation affects plant performance. Trees derived from vegetative propagation differ in their performance from generative propagation. Vegetatively propagated plants will have the same characteristics as their parents but less vigorous, making it easier to harvest and implement better GAP, such as fertilizing, pests, and plant disease protection or pruning shoots on the main stem to decrease internal tree humidity that which cause them to be easily attacked by disease (Protacio, 2013).

3.4.1 Generative Propagation

The seeds of cloves belong to the recalcitrant seed group, where the seed viability rapidly decreases after harvest. Clove seeds that have been separated from their fruit flesh and stored at a temperature of 28°C lose 10% of their moisture content within 4 days (Anilkumar et al., 2000). The problems in clove seed storage include the inability to withstand low temperatures (< 35°C) and germination during storage. Proper handling of the seeds, starting from harvesting, sorting, packaging, and storage, is crucial to maintaining seed quality until they are used as planting material (Hasanah & Rusmin, 1997).

One way to obtain high-quality clove seeds is to determine the right harvesting time. Physiologically mature Zanzibar clove seeds are obtained when the fruit turns red based on vigor index (73.75%), growth rate (1.19% per day), seed dry weight (26.7 g), and germination capacity (100%) (Pelealu et al., 2019). Previous research reported that the ideal seed harvesting time is 90 days after flowering, with the criteria for dark (purple) colored fruits (Hasanah & Rusmin, 1997).

The seed size and the fruit's position on the mother trees also determine the quality of clove seeds. Large seeds weighing ≥ 0.9 g from fruits exposed to morning sunlight have the best viability and vigorous growth, with a germination rate of 92% and a vigor index of 50%. Conversely, small seeds weighing < 0.9 g from fruits not exposed to sunlight result in low germination (36.70%) and a vigor index of 18.3%. Large seeds exposed to sunlight have a higher energy reserve in the form of starch than those not exposed. Fruits exposed to sunlight affect the hormonal balance, which can increase germination capacity (Melati & Rusmin, 2023).

The proper storage conditions for clove seeds are crucial in maintaining their high quality during storage. Clove seeds have a high moisture content at harvest, which determines the seeds' ability to maintain viability during storage (Hasanah & Rusmin, 1997). Clove seeds peeled and stored in polyethylene bags at 20°C and 30°C can maintain their viability for 6–9 months, even though they sprout during storage (Anilkumar et al., 2000). Another study demonstrates that storing clove seeds using ginger flour-based hydrogel, combined with 150 mg L^{-1} of vitamin C and 200 mg L^{-1} of Paclobutrazol as a storage medium, can maintain a germination rate of 98% after 8 weeks of storage. The hydrogel with ginger flour mixture can maintain moisture and suppress fungal attacks during storage. The application of vitamin C aims to suppress oxidation, while Paclobutrazol is used to suppress sprouting (Ediningsih et al., 2018).

After peeling, clove seeds could be sowed individually or collectively in a seedbed for germination. Geminating individual seeds can be done by planting clove seeds directly in polybags. The 15 × 20 cm polybag contains a mixture of soil and cow manure/humus or vermicompost in a ratio of 2:1 (v/v) as represented in Figure 3.1.

This mixture gives an ideal nutrient content for seedling with best root length and dry weight (Thankamani et al., 1996). Other research results show that a mixture of sand + soil + cocopeat/vermicompost (1:1:1) + Trichoderma produces the best growth of clove seeds, including plant height (11.40 cm), internode length (3.80 cm), number of leaves (9.73), root length (12.21 cm) and survival rate (96.17%). This media can be used as a choice of media for seedlings, according to the availability of

FIGURE 3.1 Seedling method: collective seedling (left) and individual seedling (right).

materials at the nursery location. By using good-quality seeds, the growth percentage is almost 100%, and sowing individual clove seeds is more efficient.

Collective sowing is done by planting clove seeds in seedbeds or a wooden bed (Figure 3.1). The seedbed should be sloping and close to a water source to make watering more manageable and allow seeds to germinate and grow well. The requirements for a nursery are sun intensity of around 25%, air temperature of 22–30°C, and humidity (RH) of more than 80%. The bed height is 15 cm to provide a medium for seed roots to grow straight. Seeds are sown standing up at a distance of 2 cm × 2 cm. The prospective roots face downwards, with the top surface of the seed almost flush with the planting medium.

The seedbed is placed in a shady place and not exposed to rainwater. Watering must be done sufficiently every 2 days. Fungicide must be applied once a week to prevent the growth of fungus on the seeds and planting medium. After 2–3 weeks, the seedlings are transferred to a polybag with 5–6 cm long shoots and roots. The seedlings can be transferred to the field at 8 to 30 months. The polybags sizes used are 15 cm × 20 cm × 0.4 mm for seedlings up to 12 months old and 20 cm × 30 cm × 0.4 mm for seedlings 13–30 months old (Ministry of Agriculture, 2023a). Polybags are filled with a mixture of soil and cow manure/humus in a ratio of 2:1 (v/v). The seed is planted by making holes in the media ±10 cm deep. The seedling must be shaded at about 50%. Maintenance is carried out, including watering, fertilizing, controlling weeds, and controlling plant pests.

3.4.2 Vegetative Propagation

The success rate of vegetative propagation of cloves (cuttings and budding) is still low, while marcotting, although the success rate is higher, cannot be done because it can damage the clove tree. Clove cuttings from soft, somewhat stiff stems taken from etiolate or non-etiolate shoots treated with hormones produce cuttings without roots (Rema & Krishnamoorthy, 1994). Shoots dipped in 4,000 mg L^{-1} IBA solution result in a low percentage of rooted cuttings (Chezhiyan et al., 1996). Therefore, further research is needed to improve rooted clove cuttings.

Clove marcotting's success rate was 64%, but 45% of marcotted clove plants died during transplanting to polybag (Pool et al., 1991). Propagation by marcotting is not recommended because it will damage the clove mother plant.

Approach grafting using 6-month-old rootstocks of the same or different species in the Myrtaceae family is compatible and grows vigorously. The success rate for vegetative propagation using approach grafting is relatively high, namely 87% (Rema et al., 1997). Both Scion and Rootstock should be taken from the recommended variety.

There are two approach grafting techniques, namely the spliced approach grafting and the tongue approach grafting. In spliced grafting, the bark of the two rootstock plants peeled off to a length of ±2 cm, and the two cuts are tied together with a plastic rope while the plant still has roots and crown and is actively growing (Figure 3.2.a). Another technique is tongue approach grafting, in which the bark of the plant stem is cut off to a length of ±2 cm, and a gap is cut in the stem of the clove plant, which acts as the upper stem. The stem is cut at the same height as the upper stem for plants used as rootstocks, and a gap is made. One side of the stem is slit open to a length of ±2 cm, and the other side is sharpened so that it resembles a tongue.

FIGURE 3.2 Spliced approach grafting *Syzygium syzygoides* (rootstock) and *Syzygium aromaticum* (scion): (a) tongue grafting; (b) approach grafting).

The sharpened end of the stem is then inserted into the upper stem gap and tied with a plastic rope (Figure 3.2 b).

Epicotyl grafting combines two different plant species from the same Myrtaceae family to form a new plant. The scion is derived from a good variety already produced and not infested with pests. The rootstock is a clove plant or another Myrtaceae family with a 4–8 week-old epicotyl. Dipping the scion in a solution of 125 mg L^{-1}–130 mg L^{-1} ascorbic acid and a solution of 2 mg L^{-1} NAA + 10 mg L^{-1} BAP + 1 mg L^{-1} GA3 and grafting with a 4-week-old rootstock resulted in a survival rate of 56.67%–75% (Suryadi et al., 2020; Darwati et al., 2023). The preparation for epicotyl grafting goes through several stages, as follows:

3.4.2.1 Rootstock Preparation

Clove seeds, derived from superior or local varieties, are initially planted in a nursery. After 30 days, the seedlings are transplanted into 15 × 20 cm polybags, filled with suitable soil media. Between 4–8 weeks after the seed germinates, these seedlings reach the optimal stage for transplantation into the field.

3.4.2.2 Scion Preparation

Scion preparation involves carefully selecting scion from healthy clove mother trees of the desired variety, free from pests and diseases. The scion is between 5 and 7 cm long, with flushing buds. Scion collection is recommended between 8:00 and 11:00 am. Prior to grafting, it is advisable to soak the scion in a vitamin C solution with a concentration ranging from 125 mg L^{-1} to 130 mg L^{-1} for 2 hours.

3.4.2.3 Epicotyl Grafting Technique

The seedling for rootstock is cut to a height of ±5 cm above the ground level, and a silicone tube is inserted. The cut surface of the rootstock is split right down the middle with a depth of ±1 cm. The base of the stem for scion is cut to resemble the letter V and then inserted into the rootstock. The silicone hose is pulled up until it covers the connection area, then covered with tape so that water does not enter the connection area. Before grafting, the rootstock planting medium is doused with 2% fungicide to prevent fungal growth, carried in a plastic bag, and stored in a wooden bed. The connection results are stored in a room containing lit shelves with a temperature set between 24–25°C and humidity between 75–80%.

3.4.2.4 Acclimatization and Maintenance of Grafting

Before acclimatization, newly developed shoots on the rootstock are removed, and acclimatization can be carried out in a glass house for 2 months after grafting, with a light intensity of 25%. At the time of acclimatization, the top part of the plastic cover is cut at both ends and left for 2 weeks to prevent stress due to changes in temperature and humidity. After 2 weeks, the plastic cover is totally removed. Watering and weed control are carried out as necessary. Four months after grafting, the silicon tube is removed as represented in Figure 3.3.

FIGURE 3.3 Epicotyl grafting technique on clove plants: four-week-old seedling, decapitated rootstock shoot, splitting the middle of the stem and making a V-shaped scion base, scion insert into the rootstock and joint with silicon tube, covered with a transparent plastic bag, put on the rack in the room temperature 24–25 °C and humidity 75–80%, acclimatization in the greenhouse, five-month-old grafting plant.

3.5 CULTIVATION MANAGEMENT

Cultivation management is a fundamental requirement of agricultural practice, including strategic planning, implementation, and optimization of techniques used to increase growth and production. Effective cultivation management is essential to increasing agricultural productivity, efficient resource use, and sustainability of agricultural ecosystems. The growth and yield of clove plants are significantly influenced by the availability of nutrients in the soil, which play a crucial role in their development and productivity. Soil nutrient management is vital in clove cultivation, ensuring optimal plant growth and yield (Office, 2006).

3.5.1 Land Preparation for Planting

Land preparation plays a crucial role in clove cultivation, as it determines the growth and productivity of the plant itself. The main aspects that must be checked include site selection, plowing the soil, planting holes, and fertilizing.

3.5.1.1 Site Selection

The ideal sites for clove cultivation are eastern and northeastern hill slopes and well-drained valleys. It is very important to select an ideal site with loose, fertile, and well-drained soil at an altitude of 200–600 m asl. Cultivating the soil by plowing or hoeing should improve its structure and fertility.

3.5.1.2 Land Clearing

Land clearing for clove cultivation involves several steps to prepare the land for planting and ensure optimal growth conditions for clove trees. Any existing vegetation from the site, including trees, shrubs, weeds, and grasses, should be removed. Land clearing can be done manually with machetes or using machinery such as bulldozers or excavators for larger areas. If trees are cleared, stumps should be removed to prevent interference and cultivation activities. Stump grinders and excavators can be used for this purpose. Before planting, the soil around the planting hole was tilled or plowed to loosen and improve the aeration. Soil amendments, such as compost or organic matter, can be added to enhance soil fertility and structure.

3.5.1.3 Preparing the Planting Holes

Planting holes is an essential part of preparing the soil for planting cloves. The planting hole serves as a place for the clove seedling and provides an optimal environment for the plant's root growth. Planting holes should measure 75 × 75 × 75 cm, but in soil with a higher clay content, the planting hole size is 100 × 100 × 100 cm. Choosing the right planting hole influences the growth and productivity of the clove plants. Planting distance of 8 × 8 m, 8 × 10 m, or 10 × 10 m provides the clove plants with an optimal place to grow. Once the planting hole has been dug, it should rest for one month before the clove seedlings are planted to give the soil time to settle and make the soil structure more stable. Making the right planting hole has several advantages, including sufficient space for the roots of the clove plant to grow, improved aeration of the soil so that the plant roots can get enough oxygen, facilitated planting of clove seedlings, and reduced the risk of clove plants collapsing during growing.

3.5.2 Fertilizer

Basic fertilization is important for preparing the soil for planting cloves. Basic fertilization aims to provide the clove plants with the nutrients they need for optimal plant growth in the early stages of growth. Basic fertilizer consists of 5–10 kg manure and 150 g dolomite per hole. Base fertilization offers several benefits, including providing the nutrients clove plants need in the early stages of growth, increasing clove plants' growth, and increasing clove production.

Clove growth and productivity are highly influenced by soil characteristics such as porosity, organic matter content, and nutrient composition (Athaillah et al., 2021). The nutrients that clove plants need include nitrogen, phosphorus, and potassium. Nitrogen is crucial in forming chlorophyll, is essential for photosynthesis, and contributes to the plant's overall vegetative growth (Johnson & Mirza, 2020). Phosphorus, on the other hand, is essential for root development, flowering, and fruit formation, while sulfur is necessary for synthesizing various amino acids and proteins (Johnson & Mirza, 2020). The soil's availability and balance of these macronutrients can significantly affect clove crops' growth, yield, and quality. Proper fertilization practices that consider the specific needs of clove plants can help optimize their growth and productivity (Athaillah et al., 2021; Widaningrum & Aji, 2023).

Managing macro- and micro-nutrients has been identified as crucial in suppressing irregular bearing (Ruhnayat et al., 2022). Applying appropriate levels of nitrogen, phosphorus, potassium, and essential micronutrients can significantly enhance the productivity and quality of cloves (Laban et al., 2020; Athaillah et al., 2021). A lack of nutrients in the soil can lead to various problems, such as slow clove production, a large number of non-producing plants, and lower clove yields (Athaillah et al., 2021).

Irregular bearing might occur due to the imbalance between clove plants' vegetative and reproductive growth, particularly after a large harvest. During heavy flowering, the plant's nutrient reserves become depleted as they are diverted toward reproductive development. This nutrient depletion can lead to declining subsequent harvests as the plant struggles to recover and rebuild its nutrient reserves.

The ideal fertilizer dosage for clove plants should consist of organic and inorganic sources (Athaillah et al., 2021; Umasugi et al., 2020). The organic matter provides essential nutrients and improves soil structure, allowing for better root development and nutrient absorption. Table 3.3 shows the fertilizer dosage based on the clove development stage. Fertilizer applications should be tailored to the specific nutrient needs of the crop at different stages of development to promote healthy growth and maximize yields. Hence, growers can optimize nutrient utilization, minimize nutrient losses, and achieve better crop yields and quality.

3.5.3 Plant Maintenance

Clove plants from 1–5 years old are in a critical period. About 10–30% of the plants die or need to be replanted for various reasons, e.g., pests, diseases, drought, inability

TABLE 3.3
Fertilizer dosage based on the development stage of the clove tree.

Tree developmental stage	Organic fertilizer (kg/tree/year)	Inorganic fertilizer (g/tree/year)			References
		Urea	SP 36	KCl	
Seedling stage < 2 years	0.50	0.06–0.12	0.058–0.102	0.035–0.075	(Karwur & Semangun, 2014)
Vegetative stage 2–5 years	5–10	0.25–0.60	0.192–0.512	0.10–0.20	(Karwur & Semangun, 2014)
Early productive stage 5–10 years	5–10	0.90–3.90	0.512–1.024	0.60–2.00	(Karwur & Semangun, 2014)
Productive stage 11–30 Years	5–10	4.30–8.80	1.024–2.048	2.10–5.10	(Karwur & Semangun, 2014)
Older productive stage > 30 years	5–10	9.40	2.432	5.40	(Karwur & Semangun, 2014)

Note: Urea, Sp36, KCl contains 46% N; 36% P_2O_5 and 60% K_20, respectively.

to compete with weeds, or other reasons. It is best to replant during the rainy season to avoid plants dying due to a lack of water. Intensive maintenance is required to prevent many clove plants from dying. Loosening the soil around the roots of clove plants aims to improve the physical soil so that the area becomes better and clears weeds, which are competitors in absorbing water and nutrients (Das & Yaduraju, 1999). Pruning is done on water branches because they are not productive, on dead branches so that new shoots grow, and on double stems so that when they are big they do not split or break easily (Bhagya et al., 2024). In the dry season, it is necessary to water inside the plant canopy, especially young clove plants. Mulch must be applied to the plant canopy to avoid rapid water loss in the soil due to evapotranspiration. The mulch can utilize harvested intercropping plants or weeds around the clove plants.

Loosen the soil around the plant at the roots with a shallow hoe (±10 cm) at least twice a year, at the beginning and end of the rainy season and in preparation for fertilization. Weeds must be removed with a hoe/fork or sprayed with herbicides down to the roots.

In the early stages of growth, clove plants need sufficient shade with a light intensity of ±50%. This is because young clove plants cannot withstand exposure to sunlight, and their roots have not developed well, so they cannot absorb sufficient amounts of water to compensate for evapotranspiration. Embroidery is carried out on dead and poorly growing clove plants with seedlings of the same variety, relatively the same age, and good vigor to obtain relatively uniform plant growth. The height of the shade needs to be adjusted according to the height of the clove plants. There are two types of shade, namely artificial shade and natural shade. Artificial shade is maintained for a maximum of two dry periods after planting. Natural shade trees are designed to withstand the sun's heat and protect the plant from the wind and heavy rain. The planted shade trees are usually Theoprocia, Flumingia, and Congesta. If the shade level is too high, it must be pruned back immediately at the beginning of the rainy season. After the clove plant has reached an age of 5 years, the natural shade (shade trees) is completely removed, as the plant is already resistant to all external influences.

At the early stage of clove growth, the plant must be watered during the dry season as necessary. Watering is less necessary for mature plants, except in arid climatic conditions. Mulching is applied before the dry season. The aim is to retain soil moisture around the plants and create better conditions for root growth.

Fertilization aims to improve plant growth and increase clove production after harvest. Based on the root distribution pattern, clove plants are fertilized under and within the canopy. Fertilization is applied twice a year, namely at the beginning and the end of the rainy season. Fertilization can be done with organic fertilizer (manure or compost) and inorganic fertilizer, either alone or in the form of compound fertilizer in granular or tablet form. Inorganic fertilizer in granular form (Urea, SP-36, KCl) is applied twice a year, at the beginning and end of the rainy season. The highest root density occurs closest to the tree's base at a distance of 0.5 times the canopy radius (Grandgirard et al., 2002). Fertilizer is placed closer to the tree's base so trees grow flat to increase P uptake. Trees growing on sloping land should receive fertilizer placed next to the base of the tree in an upslope position rather than a downslope position because it has a higher root density. Fertilizer applications are closer to trees, which will minimize the impact of competition with other intercropping species grown on plantations.

Clove plants that have been harvested must be fertilized to restore the nutrients lost. Clove plants in West Nusa Tenggara Province have many dry branches because the plants are not fertilized after harvesting the clove flowers (Suniantara et al., 2020).

3.6 HARVESTING

Clove flower harvesting must be well managed to maintain the quantity and quality of clove flowers. Management determines the right harvest time to reduce yield losses due to clove flowers falling off during harvesting, transporting, or handling of harvest. Generally, clove trees from generative and vegetative propagation begin to flower at 5–7 and 4–5 years old, respectively and reach a peak of flowering at 40 years of age. Under good maintenance, cloves will remain productive for more than 70 years. The flowering period varies from September to October or December to January, depending on the growing area. The clove flower initiation until ready for harvest takes 4–6 months, when the bud heads are wholly formed, and the color of the flower petals changes to reddish-yellow (Hussain et al., 2017). Harvesting can be performed every 10–14 days for 3–4 months by picking the clove flower buds in clusters just above the last leaf node as represented in Figure 3.4. If the flowers are not harvested, they become fruit, elongated, dark red berries (Nurdjannah & Bermawie, 2012). The flowers should be picked above the leaves to support the initiation of new shoots from the leaf axils (Darwati, 2017).

Clove harvesting techniques can be done from the inside and outside of the crown. Clusters of flower buds were hand-picked by climbing a clove tree with a bamboo ladder as represented in Figure 3.5. Next, the clove flowers and bunches were separated by threshing. In general, the threshing of clove flowers is still conventional, using human resources with an average capacity of 30 kg/day.

Clove flower buds are removed from the peduncles and dried under the sun using bamboo math at 32–34°C (daytime) for 4–5 days. Good-quality dried clove flower buds are characterized by a complex, crunchy texture, dark brown color, and water

FIGURE 3.4 Flower harvesting method: (a) clove flowers are picked (red arrow) between flowers and leaves; (b) new shoots emerge at the location where the flower was picked; (c) flowers appear (red arrow) on new shoots that are old enough; (d) number of leaves for maximum development of flowering is more than three; (e) flower initiation; (f) inflorescence.

FIGURE 3.5 (a) Planting cloves on sloping land; (b) terminal flower of clove; (c) harvesting using a bamboo ladder.

content of < 12%. Clove plants aged 15–20 years produce 3–4 kg dried clove flowers. The yield value from fresh and dry clove flowers was 33.33% (Lutfhi & Kurniati, 2018). Store dried buds for 4 months to increase eugenol levels from 56.69% to 60.95%–64.64% (Hastuti et al., 2017). Dried clove flower buds can be packaged using burlap sacks and can be stored for 1–2 years. Some traders use dryers to speed up drying and maintain the quality of dried clove flower buds (Simbolon et al., 2022).

3.7 CONCLUSION

In conclusion, it is essential to implement good agricultural practices when cultivating cloves to ensure optimal growth and yield. Good agricultural practices (GAPs) for clove cultivation encompass a range of practices aimed at ensuring sustainable and high-quality production while minimizing environmental impact and ensuring the safety of the final product. By following these GAPs, clove growers can optimize production, improve product quality, and promote environmental sustainability in their farming practices.

REFERENCES

Alfian, A., A. S. Mahulette, M. Zainal, Hardin, and A. H. Bahrun. 2019. Morphological character of raja clove (*Syzygium aromaticum* L. Merr & Perry.) native from Ambon Island. *IOP Conference Series: Earth and Environmental Science* 343: 012150. http://doi.org/10.1088/1755–1315/343/1/012150.

Anilkumar, C., P. N. Krishnan, and N. Salim. 2000. Seed viability of *Syzygillm aromatic* lim (L.) Merill & Perry during storage. *Centennial Conference on Spices and Aromatic Plants* (September). https://eurekamag.com/research/003/556/003556131.php?srsltid=AfmBOorqTu943LHiQM-GJyWsMUm5a6So-CaqzM3zx8qGoMVqb_kqpbHe

Athaillah, T., Raidayani, O. Sanjaya, and Bagio. 2021. The SWOT analysis for the development of clove plants (*Syzygium Aromaticum*) in Teupah Barat Sub-District, Simeulue District. *Proceedings of the 2nd International Conference on Science, Technology, and Modern Society*. https://doi.org/10.2991/assehr.k.210909.042.

Bermawie, N., and P. A. Pool. 1991. The use of isozyme to determine the status of Zanzibar type clove (*Syzygium aromaticum* L.). *Journal Pemberitaan Penelitian Tanaman Industri* 17, no. 1: 1–9.

Bermawie, N., S. Wahyuni, and R. Ginting. 2021. Selection of highly productive clove trees in the Anambas population. *IOP Conf. Series: Earth and Environmental Science* 762: 012041.

Bhagya, H. P., A. J. Dinakar, M. Shamsudheen, G. L. Veena, M. Babli, K. Vanitha, G. N. Manjesh, K. Manjunatha, and C. Aswathy. 2024. Canopy management in cashew through training and pruning technique. *AgriGate: An International Multidisciplinary e-Magazine* 4, no. 1: 24–33.

Chezhiyan, N., M. Ananthan, and P. G. B. Vedamuthu. 1996. Vegetative propagation studies of clove (*Syzygium aromaticum* (L.) Merr. & Perry). *Indian Cocoa, Arecanut & Spices Journal* 20: 50–54.

Danthu, E. Penot, K. M. Ranoarison, J. C. Rakotondravelo, I. Michel-Dounias, M. Tiollier, T. Michels, F. Normand, G. Razafimamonjison, F. Fawbush, and M. Janiel. 2014. The clove tree of Madagascar: A success story with an unpredictable future. *Bois et Forêts des Tropiques* 320, no. 2.

Darwati, I. 2017. Suppression of fluctuations clove (*Syzygium aromaticum*) production with physiology mechanism. *Perspektif* 16, no. 2: 93–101. http://doi.org/10.21082/psp.v16n2.2017.

Darwati, I., R. Suryadi, N. Bermawi, O. Trisilawti, D. Rusmin, S. Wahyuni, Siswanto, and R. V. Garvita. 2023. Influence of rootstock age and gibberellin on the survival rate of clove (*Syzygium aromaticum* L. Merr. & Perr.) grafting. *Turkish Journal of Botany* 47: 556–566. http://doi.org/10.55730/1300-008X.2784.

Das, T. K., and N. T. Yaduraju. 1999. Effect of weed competition on growth, nutrient uptake and yield of wheat as affected by irrigation and fertilizer. *Journal of Agricultural Science*, 133: 45–51.

Directorate General of Estate Crops-Indonesia. 2023. *Statistic of Estate Crops*. http://repo.iain-tulungagung.ac.id/5510/5/BAB 2.pdf (accessed May 20, 2024).

Ediningsih, R. A. Permadi, A. Kartikawati, I. Darwati, D. Rusmin, Erizal, S. N. Samsiyah, and Zainudin. 2018. *Teknologi Penyimpanan dan Invigorasi yang Dapat Memperpanjang Daya Simpan Benih Cengkeh*. Laporan Teknis Penelitian Tahun Anggaran 2018, Balai Penelitian Tanaman Rempah Dan Obat, 217–231.

Grandgirard, J., D. Poinsot, L. Krespi, J. P. Nénon, and A. M. Cortesero. 2002. Costs of secondary parasitism in the facultative hyperparasitoid Pachycrepoideus dubius: Does host size matter? *Entomologia Experimentalis et Applicata* 103, no. 3: 239–248. https://doi.org/10.1023/A

Hasanah and Rusmin. 1997. *Aspek Perbenihan Cengkeh*. Monograf Cengkeh. Balai Penelitian Tanaman Rempah dan Obat 210 p.

Hastuti, L. T., E. Saepudin, A. H. Cahyana, D. U. C. Rahayu, V. W. Murni, and J. Haib. 2017. The influence of sun drying process and prolonged storage on composition of essential oil from clove buds *(Syzygium aromaticum). International Symposium on Current Progress in Mathematics and Sciences AIP Conf. Proc.* 1862, 030092–1–030092–6. http://doi.org/10.1063/1.4991196

Hussain, S., R. Rahman, A. Mushtaq, and A. El Z. Belaskri. 2017. Clove: A review of a precious species with multiple uses. *International Journal of Chemical and Biochemical Sciences* 11: 129–133.

Johnson, V. J., and A. Mirza. 2020. Role of macro and micronutrients in the growth and development of plants. *Excellent Publishers* 9, no. 11: 576–587. https://doi.org/10.20546/ijcmas.2020.911.071

Karwur F. F., and H. Semangun. 2014. *Cengkeh Sejarah, Budidaya dan Industri*. PT. Gramedia. Cetakan I: 45–72.

Laban, S., M. S. Solle, A. Ahmad, and M. Jayadi. 2020. Land suitability evaluation for clove plants in Bacan Island. *IOP Conf. Series: Earth and Environmental Science* 486:012067. https://doi.org/10.1088/1755-1315/486/1/012067

Lutfhi, M., and A. Kurniati. 2018. Pengelolaan Panen Bunga Cengkih (*Syzygium aromaticum* L.) di Kebun Branggah Banaran, Blitar, Jawa Timur. *Buletin Agrohorti* 6, no. 2: 188–197.

Martínez-Fuentes, A., C. Mesejo, N. Muñoz-Fambuena, C. Reig, M. C. González-Mas, D. J. Iglesias, E. Primo-Millo, and M. Agusti. 2013. Fruit load restricts the flowering promotion effect of paclobutrazol in alternate bearing Citrus spp. *Scientia Horticulturae* 151, February: 122–127. http://doi.org/10.1016/j.scienta.2012.12.014.

Melati, M., and D. Rusmin. 2023. Effect of seed size and pod position of clove (*Syzigium aromaticum*) to seed viability. *IOP Conf. Series: Earth and Environmental Science* 1253: 012035. http://doi:10.1088/1755-1315/1253/1/012035

Ministry of Agriculture. 2008. *Decree of the Minister of Agriculture of the Republic of Indonesia Regarding the Release of Cloves of the Composite Variety Zanzibar Karo No. 339/Kpts/SR120/3/2008*. Ministry of Agriculture.

Ministry of Agriculture. 2010. *Decree of the Minister of Agriculture of the Republic of Indonesia Regarding the Release of AFO Clove Populations as High-Yielding Varieties No. 3680/Kpts/SR120/11/2010*. Ministry of Agriculture.

Ministry of Agriculture. 2013a. *Decree of the Minister of Agriculture of the Republic of Indonesia Regarding the Release of Gorontalo Clove Populations as High-Yielding Varieties No. 198/Kpts/SR120/1/2013*. Ministry of Agriculture.

Ministry of Agriculture. 2013b. *Decree of the Minister of Agriculture of the Republic of Indonesia Regarding the Release of the Population of Tuni Bursel Cloves as a High-Yielding Variety No. 4964/Kpts/SR120/12/2013*. Ministry of Agriculture.

Ministry of Agriculture. 2016. *Decree of the Minister of Agriculture of the Republic of Indonesia Regarding Amendments to the Attachment to the Decree of the Minister of Agriculture No. 315/Kpts/KB.020/10/2015 Regarding Guidelines for the Production, Certification, Distribution and Supervision of Clove Seeds (Syzygium aromaticum (L.)*. Merr & Perr. Ministry of Agriculture.

Ministry of Agriculture. 2020. *Decree of the Minister of Agriculture of the Republic of Indonesia Regarding the Release of Siantan Agribun as Clove Populations as High-Yielding Varieties No. 41/Kpts/KB. 010/3/2020*. Ministry of Agriculture.

Ministry of Agriculture. 2023a. *Decree of the Minister of Agriculture of the Republic of Indonesia Regarding the Release of Zanzibar Peling Clove Populations as High-Yielding Varieties No. 48/Kpts/KB.020/05/2023*. Ministry of Agriculture.

Ministry of Agriculture. 2023b. *Decree of the Minister of Agriculture of the Republic of Indonesia Regarding Guidelines for the Production, Certification, Distribution and Supervision of Clove Seeds (Syzygium aromaticum (L.)*. Merr & Perr. Ministry of Agriculture.

Nurdjannah, N., and N. Bermawie. 2012. *Handbook of Herbs and Spices*. Woodhead Publs.

Office, C H. 2006. *CABI Is a Trading Name of CAB International*. http://endcap.eu/wp-content/uploads/2015/06/Chubb-editor-2006Stereotypic-Animal-Behaviour.pdf

Pelealu, R. V. H. R., E. Widajati, and F. C. Suwarno. 2019. Effect of physiological maturity and germinating media on Zanzibar clove seeds viability. *Buletin Penelitian Tanaman Rempah dan Obat* 30, no. 2: 81–89.

Pool, P. A., N. Bermawie, and U. Daras. 1991. Factors affecting marcotting in clove (Syzygium aromaticum). *Industrial Crops Research Journal* 4, no. 1: 12–16 ref. 8.

Protacio, C. M. 2013. Towards a good agricultural practice (GAP)-compliant mango production system in the Philippines. *Acta Horticulturae* 992: 69–74. http://doi.org/10.17660/actahortic.2013.992.7.

Rema, J., and B. Krishnamoorthy. 1994. Vegetative propagation of clove, Eugenia caryophyllus (Sprengel) B&H. *Tropical Agriculture* 71, no. 2: 144–146.

Rema, J., B. Krishnamoorthy, and P. A. Mathew. 1997. Vegetative propagation of major tree spices a review. *Journal of Spices and Aromatic Crops* 6, no. 2: 87–105.

Ritung, S., K. Nugroho, A. Mulyani, and E. Suryani. 2011. *Petunjuk teknis evaluasi lahan untuk komoditas pertanian*. Balai Besar Penelitian dan Pengembangan Sumberdaya Lahan Pertanian.

Ruhnayat, A., S. Setiawan, and O. Rostiana. 2022. Management of macro and micro nutrients to suppress flower yield fluctuations in productive clove. *IOP Conf. Series: Earth and Environmental Science* 974: 012063. https://doi.org/10.1088/1755-1315/974/1/012063.

Shaikh, N., S. D. Ramteke, A. V. Mane, U. B. Pethe, and V. Urkude. 2018. Studies on effect of different propagation media on growth and survival of clove (*Syzygium aromaticum* L. Meer) seedling. *International Journal of Agriculture Science* 10, no. 19: 7328–7330.

Simbolon, E., J. Alfredo, B. S. Kaligis, M. F. Pomalingo, I. D. Piri, and S. V. C. Kawuwung. 2022. Desain dan Pabrikasi Mesin Pengering Cengkeh Berbahan Bakar Limbah Kelapa Untuk Mempercepat Proses Penjemuran Cengkeh. *Jurnal Keteknikan Pertanian Tropis dan Biosistem* 10, no. 1: 21–28. https://doi.org/10.21776/ub.jkptb.2022.010.01.03

Singh, P. A., and A. Baldi. 2018. Good agricultural practices: A prequisite approach for enhancing the quality of Indian herbal medicines. *Biomedical Journal of Scientific & Technical Research* 2, no. 4: 1–4.

Singh, P. A., N. Bajwa, S. Chinnam, A. Chandan, and A. Baldi. 2022. An overview of some important deliberations to promote medicinal plants cultivation. *Journal of Applied Research on Medicinal and Aromatic Plants* 31: 100400.

Singh, P. A., A. Sood, and A. Baldi. 2021. Determining constraints in medicinal plants adoption: A model geospatial study in the Indian state of Punjab. *Journal of Applied Research on Medicinal and Aromatic Plants* 25: 100342.

Suniantara, D. K., I. K. Budastra, and T. Sjah. 2020. Feasibility of clove farming in Gangga District North Lombok Regency. *Proceedings of the 2nd Annual Conference on Education and Social Science (ACCESS 2020) Advances in Social Science, Education and Humanities Research*, 556. Atlantis Press.

Suryadi, R., E. R. Pribadi, and I. Darwati. 2020. The effect of ascorbic acid on clove (*Syzigium aromaticum*) grafting. *IOP Conf. Series: Earth and Environmental Science* 418: 012026. http://doi.org/10.1088/1755-1315/418/1/012026.

Thankamani, C. K., K. Sivaraman, and K. Kandiannan. 1996. Response of clove (*Syzygium aromaticum* (L) Merr & Perry) seedlings and black pepper (*Piper nigrum* L) cuttings to propagating media under nursery conditions. *Journal of Spice and Aromatic Crops* 5, no. 2: 99–104.

Umasugi, B., S. Prijono, Soemarno, and Ariffin. 2020. Potential for land and lengas available to water needs on the clove land in Ternate Island. *Proceedings of the 5th International Conference on Food, Agriculture and Natural Resources*, March. https://doi.org/10.2991/aer.k.200325.010.

Wahid, P., I. Las, and I. Dwiwarni. 1985. *Peta Kesesuaian iklim dan lahan untuk tanaman cengkeh*. Balai Penelitian Tanaman Rempah dan Obat.

Widaningrum, I., and J. M. M. Aji. 2023. Efisiensi Alokatif Penggunaan Faktor Produksi pada Usahatani Cengkeh di Desa Sawahan Kecamatan Sawahan Kabupaten Nganjuk. *Agro Info Galuh* 10, no. 2: 1235–1235. https://doi.org/10.25157/jimag.v10i2.9857

Yusuf, W. A., R. Neswati, and M. Nathan. 2023. Analysis of land suitability of clove (*Syzygium aromaticum* L.) in the humid tropics of South Sulawesi. *IOP Conference Series: Earth and Environmental Science* 1230, no. 1: 1–10. http://doi.org/10.1088/1755-1315/1230/1/012056.

4 Post-harvest Management

Subhajit Hazra, Preet Amol Singh, Sneha Kumari, Ritika Sindhwani, Neha Bajwa, and Shiva Tushir

4.1 INTRODUCTION

Post-harvest management encompasses the procedures for handling, storing, and transporting agricultural commodities after harvest (Singh et al., 2022; Singh et al., 2024; Singh et al., 2021; GrainPro, 2024). This protocol is essential for minimizing losses and ensuring the safety and quality of agricultural products throughout the value chain (Mkojera & Chove, 2021; FAO-WHO, 2021). Inadequate post-harvest management can lead to significant issues such as food insecurity and hunger. According to the World Hunger Organization, approximately 333 million people were affected by food insecurity in 2016 (Hunger-Notes, 2018). The situation worsened by 2022, with around 402 million Asians, 43 million Latin Americans and Caribbeans, 282 million Africans, and 3 million Oceanic people experiencing hunger (ACDI/VOCA, 2023). Therefore, the effective post-harvest management of crops, particularly those with medicinal value like cloves (*Syzygium aromaticum*), is crucial for the sustainability of medical products derived from them.

Clove trees take approximately 15 years to reach their full-bearing stage. The flowering stage begins in the fourth year after planting, typically between October and December, with the flowers harvested when they transition from olive green to pink (Sarma et al., 2014). During harvesting, handpicking is employed to collect the flower buds from the branches, ensuring no damage occurs to the branches to prevent future yield reduction (KAU, 2024). The process flowchart for the post-harvest management of clove is illustrated in Figure 4.1. Generally, each clove tree produces about 3.5 to 7.0 kilograms of cloves, depending on the tree's age, size, and health.

4.2 PRIMARY PROCESSING

4.2.1 Pre-treatment of Clove

Clove buds are removed from the clove tree stalks by gripping a cluster in one hand, pushing it against the palm of the other hand, and gently twisting until the buds detach. To prevent contamination, both the processor's hands and the clove processing area must be kept meticulously clean. The stems and buds are then separated and dried individually. The stems can be utilized for oil extraction via distillation (EPGP, 2024). Various pre-treatment methods can be employed to enhance the yield and concentration of

DOI: 10.1201/9781003521785-4

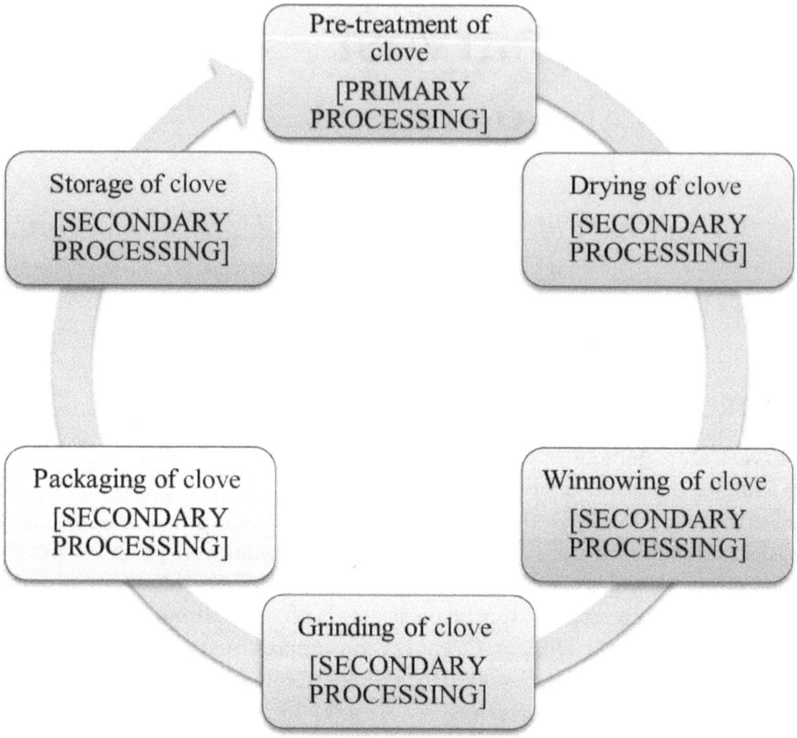

FIGURE 4.1 Process flow in the post-harvest management of clove.

chemical compounds (CBI, 2020). For instance, a study by Alashti et al. demonstrated that pre-treatment with cold plasma altered the color parameters of cloves and significantly increased the total phenolic content (TPC) when pre-treated with dielectric barrier discharge plasma for 10 minutes (Alashti et al., 2024). Such procedures can be adopted by researchers and industries involved in traditional medicine for clove extraction.

4.3 SECONDARY PROCESSING

4.3.1 Drying of Clove

Clove buds are dried to remove excess moisture, reducing storage rot and allowing the buds to become compact and enter a dormant stage. The drying and curing process is complete when the clove changes from pale russet to darker brown, typically taking around 4–5 days, indicated by the buds snapping easily (Saxena, 2024). Microwave-assisted drying (MAD) has emerged as a preferred method due to its speed and high efficiency (Kusuma et al., 2023). In the USA, market standards require the final moisture content of dried cloves to be less than 8%, extraneous matter to be less than 1%, and moldy buds to be less than 1%. According to the International Food Standards, 2022 (document code: CXS 344–2021), the physical and chemical requirements of dried clove floral parts are detailed in Table 4.1 and Table 4.2, respectively (Codex-Alimentarius-Commission, 2022).

TABLE 4.1
Physical requirements of dried floral parts of *Syzygium aromaticum*.

Product	Style	Mammalian and/or other excreta* (w/w)/mg/Kg	Live insect count/ 100g	Mold visible (% w/w)	Insect defiled/ Infested (% w/w)	Extraneous matter (% w/w)	Foreign matter (% w/w)	Defects (headless/ mother clove[2]/ khoker[3])	Reference
Dried Clove	Whole	10	0	1	1	1	1	5/6/5	(Codex-Alimentarius-Commission, 2022)
	Ground	N/A	0	N/A	1	1	1	N/A	

N/A, Not available; * such as reptiles and birds; 1 clove consisting of only the receptacle and sepals, which has lost the dome-shaped head; 2 fruit of the clove tree in the form of an ovoid brown berry surmounted by four incurved sepals; 3 clove having undergone fermentation as a result of incomplete drying, as evidenced by its pale brown colour, whitish mealy appearance, and often wrinkled surface.

TABLE 4.2
Chemical requirements of dried floral parts of *Syzygium aromaticum*.

Product	Style	Total ash (w/w) *dry basis]*	Acid insoluble ash (% w/w) *[Dry basis]*	Moisture content (% w/w)	Volatile oils (ml/100 g) *[dry basis]*	Crude Fiber (% w/w)	Reference
Dried Clove	Whole	7	0.5	12	17	13	(Codex-Alimentarius-Commission, 2022)
	Ground	7	0.5	10	14	13	

4.3.2 Winnowing and Grinding of Clove

Winnowing is employed to remove dust and other foreign matter from clove buds. Traditionally, this process involves using baskets to eliminate dust. However, modern small cleaning machines equipped with blowers are now available to efficiently remove dirt and dust. Grinding cloves adds value by increasing the extractive yield of their essential oil (EO) (Saxena, 2024). For instance, a study by Ben Hassine et al. observed that unground clove buds (particle size between 1.5 and 2 cm) had an EO extraction yield (EY) of 7.1 ± 0.8%, significantly lower than ground cloves, which had an EY of 14.3 ± 0.6% (Ben Hassine et al., 2021). Similarly, Kapadiya and Desai (2017) reported an EO extractive yield of 11.6% for ground clove buds of *S. aromaticum* (Kapadiya & Desai, 2017). This increase in yield is attributed to the grinding process, which reduces the particle size of the cloves, thereby increasing the contact area between the solid medium and the extractive agent.

However, grinding cloves is generally not recommended because it makes the spice more susceptible to spoilage and destabilizes its flavor and aromatic compounds, which can quickly dissipate from the ground product. Additionally, grinding reduces the storage shelf life of the spice, causing it to degrade much faster compared to whole spices. This also raises concerns for consumers regarding the quality of ground spices, as it becomes easier for unscrupulous processors to adulterate the product with cheaper materials. Consequently, most consumers prefer to purchase whole spices (Saxena, 2024).

4.3.3 Packaging of Clove

Various materials are commonly used for the transport and storage of cloves, including polythene bags of various sizes, jute fabric bags with a storage capacity of up to 50–65 kg, bales, boxes, and bast mats or gunny bags with a storage capacity of up to 75 kg (TIS, 2024). Regardless of the material used, the bags must be properly sealed to prevent moisture entry, typically using sealing machines. Additionally, labeling should adhere to the General Standard for the Labelling of Pre-packaged Foods (document code: CXS 1–1985). Labels should include the product name (common and trade names), product indication, variety or cultivar, country of origin, country of harvest, region of harvest, and year of harvest. For non-retail customers, labeling must follow the guidelines set by the General Standard for the Labelling of Non-Retail Containers in Foods (document code: CXS 346–2021) (Codex-Alimentarius-Commission, 2022).

4.3.4 Storage of Clove

Dried cloves should be stored in moisture-proof containers and kept away from direct sunlight. It is crucial that the cloves are fully dry before storage, as any residual moisture can cause the cloves to rot. Therefore, it is essential to regularly check stored cloves for signs of dampness and, if necessary, to re-dry them to a moisture content of 10% and a relative humidity of less than 75%. The optimal temperature range for storing and transporting cloves is between 5°C and 25°C, with a shelf life of up to 24 months. Additionally, the storage room should be clean, dry, cool, and free from pests, which can be managed by installing mosquito netting on the windows. Furthermore, substances with strong odors, such as smelly foods, detergents, and paints, should not be stored in the same room, as they can compromise the aroma and flavor of the cloves (BMT, 2024).

4.4 STANDARDS OF PROCESSED CLOVE

Clove (Syzygium aromaticum) is a highly valued spice known for its aromatic and medicinal properties. The term "processed clove" refers to clove buds that have undergone cleaning, drying, and sorting to meet specific quality standards before reaching consumers. The processing of cloves is crucial as it enhances their shelf life, ensures safety, and maintains the desired flavor and medicinal qualities. Processed cloves are used extensively in culinary applications, pharmaceuticals, and aromatherapy, highlighting their global significance (Codex Alimentarius Commission, 2022; FAO, 2024).

The Codex Alimentarius Commission (CAC) to the spice industry is the establishment of the Codex Standard for Spices and Culinary Herbs (CXS 328–2024). This standard provides a comprehensive framework for the quality and safety of spices, including cloves. The Codex Standard for Spices and Culinary Herbs outlines the essential quality factors, permissible levels of contaminants, hygiene requirements, and labeling standards for spices to ensure their safety and quality in international trade (Codex Alimentarius Commission, 2022). The quality factors detailed in the Codex standard for cloves include the spice's physical and chemical characteristics. These parameters ensure that the cloves are free from visible mold, insect infestation, and other defects. The standard specifies moisture content limits, which are crucial in preventing microbial growth and spoilage. According to the Codex standards, the maximum permissible moisture content for cloves is 12%, ensuring they remain dry enough to avoid mold and other spoilage issues. Additionally, it addresses the permissible levels of extraneous matter, such as plant debris and foreign materials, ensuring that the cloves meet high purity standards (FAO, 2021).

Contaminant limits are another critical aspect of the Codex standard for cloves. The standard sets maximum residue limits (MRLs) for pesticides, mycotoxins, heavy metals, and other harmful substances that could pose a risk to consumer health. For example, the MRL for the pesticide carbendazim in cloves is set at 0.1 mg/kg, and for the mycotoxin aflatoxin, the limit is 2 μg/kg. These limits are established based on scientific risk assessments conducted by expert committees, such as the Joint FAO/WHO Expert Committee on Food Additives (JECFA). By adhering to these limits, producers and exporters of cloves can ensure that their products are safe for consumption and comply with international safety standards (JECFA, 2024).

Hygiene requirements specified in the Codex standard emphasize the importance of good manufacturing practices (GMP) and good hygienic practices (GHP) throughout the production and processing of cloves. This includes guidelines for the harvesting, drying, storage, and packaging of cloves to prevent contamination and ensure their safety. For instance, cloves must be harvested by hand or by using tools that do not cause contamination, and they must be dried on clean surfaces under hygienic conditions. Adherence to these practices helps in maintaining the cloves' quality and extending their shelf life (Codex Alimentarius Commission, 2024).

Labeling standards outlined by the Codex include essential information that must be provided on the packaging of cloves. This includes the product name, country of origin, net weight, and any additives used. Additionally, the labeling must provide information on any potential allergens and the proper storage conditions to ensure consumer safety. These labeling requirements not only help consumers make informed choices but also facilitate transparency and traceability in the supply chain. For example, labels must indicate if the cloves contain any substances that could cause allergic reactions, and instructions for safe storage should be included to maintain product quality (Codex Alimentarius Commission, 2022).

In terms of forms and formulations, the Codex Alimentarius Commission allows cloves to be traded in several forms, including whole cloves, ground cloves, clove oil, and clove extracts. Each form must adhere to the specified quality and safety standards. For instance, clove oil must contain at least 85% eugenol, the active ingredient responsible for its aroma and medicinal properties. There are also guidelines on the permissible limits for certain active compounds in clove extracts used in food products to ensure they do not exceed safe levels for human consumption.

4.4.1 Regional Standards of Clove

4.4.1.1 Africa

4.4.1.1.1 Tanzania

The Tanzanian Bureau of Standards (TBS) is the primary regulatory body overseeing the quality and safety of processed cloves in Tanzania. According to TBS regulations, the quality of cloves is assessed based on several parameters, including moisture content, extraneous matter, and the presence of pests or mold. The moisture content of cloves is crucial as it affects their shelf life and susceptibility to microbial contamination. TBS standards specify that the moisture content of processed cloves should not exceed 12% (TBS, 2021). Additionally, cloves must be free from visible mold, insect infestation, and any foreign matter that could compromise their quality. The acceptable extraneous matter in cloves is set at a maximum of 0.5% by weight.

In terms of safety requirements, the TBS has set limits on pesticide residues, mycotoxins, and heavy metals. These contaminants pose significant health risks if present in high concentrations. The TBS enforces maximum residue limits (MRLs) for various pesticides commonly used in clove cultivation. For example, the MRL for DDT (dichlorodiphenyltrichloroethane) in cloves is set at 0.05 mg/kg, aligning with CAC standards (Codex Alimentarius Commission, 2021; Codex Alimentarius Commission, 2022). Moreover, the presence of mycotoxins, such as aflatoxins, must be below 2 µg/kg, ensuring the safety of the cloves for consumption. Heavy metals like lead and cadmium are restricted to maximum levels of 2 mg/kg and 1 mg/kg, respectively. Cloves in Tanzania can be consumed in various forms, including whole cloves, ground cloves, clove oil, and clove extracts. The TBS allows these formulations, provided they meet the set quality and safety standards. Clove oil, in particular, must have a eugenol content of not less than 80% to ensure its efficacy and quality (TBS, 2021).

4.4.1.1.2 Madagascar

In Madagascar, the National Office for Standards (Office National des Normes) sets the standards for cloves. Malagasy standards emphasize the importance of maintaining high quality to compete in the international market. Madagascar is one of the world's leading exporters of cloves, and compliance with these standards is critical for market access and consumer safety.

Quality parameters specified in Malagasy standards include moisture content, foreign matter, and essential oil content. The moisture content in cloves should be maintained at or below 13% to prevent spoilage and maintain product quality during storage and transport (ONN, 2020). The standards also stipulate that cloves must be free from extraneous matter such as dust, dirt, and other foreign substances. The presence of essential oils, which contribute to the cloves' flavor and aroma, is another critical quality parameter. Malagasy standards require that cloves contain a minimum of 15% essential oil content, measured through steam distillation methods. To ensure safety, Malagasy standards also address pesticide residues and microbial contamination. Similar to Tanzanian regulations, the permissible levels of pesticide residues must align with international benchmarks (ONN, 2020).

4.4.1.1.3 Comoros

In Comoros, the national standards for cloves are followed by the Common Market for Eastern and Southern Africa (COMESA). Quality parameters include the absence of mold, pests, and foreign matter, with a permissible moisture content of up to 14% (FAO, 2023). This slightly higher threshold accommodates the local climatic conditions while still ensuring that the cloves remain safe and marketable. The COMESA also sets strict guidelines for the handling and processing of cloves to maintain their quality from harvest to export. These guidelines include best practices for drying, storing, and packaging cloves to prevent contamination and degradation. For example, cloves must be dried on clean, elevated platforms to avoid contact with soil and other contaminants (COMESA, 2015). They should be stored in well-ventilated, dry conditions to prevent mold growth and pest infestation. Safety requirements in Comoros include monitoring for pesticide residues and heavy metals. The MRLs for pesticides are aligned with those of international standards, such as those set by the Codex Alimentarius. The MRL for chlorpyrifos in cloves is set at 0.5 mg/kg. Additionally, the COMESA requires regular testing for heavy metals, such as lead and cadmium, to ensure that these do not exceed safe levels for human consumption. Comoros allows the export of whole cloves, ground cloves, and clove oil. The essential oil content in clove oil must be at least 80% eugenol (FAO, 2023). There are no specific restrictions on the consumption of cloves, but products must be properly labeled with usage instructions and potential health warnings.

4.4.1.2 Asia

4.4.1.2.1 India

India, one of the largest producers and exporters of cloves, adheres to strict standards set by the Food Safety and Standards Authority of India (FSSAI). These standards are in place to ensure the quality and safety of cloves, whether they are sold domestically or exported (FSSAI, 2016).

4.4.1.2.1.1 Permissible Limits (FSSAI, 2022)

- Moisture content: maximum of 12% by weight. This limit is crucial to prevent microbial growth and spoilage.
- Volatile oil content: minimum of 15 ml/100 g on a dry weight basis.
- Extraneous matter: includes both organic and inorganic substances. The limit for extraneous matter is 1% by weight, ensuring the purity of the cloves.
- Defects: the proportion of headless cloves (cloves without the flower bud) should not exceed 2%, and insect-damaged cloves are also limited to 2%.

4.4.1.2.1.2 Adulteration and Contaminants (FSSAI, 2011)

- The FSSAI has strict policies against adulteration. Cloves should be free from added coloring agents, preservatives, or any other substances that could compromise their quality.
- Pesticide residues: maximum residue levels (MRLs) for pesticides such as chlorpyrifos (0.05 mg/kg) and DDT (0.1 mg/kg).
- Heavy metals: the limits for heavy metals are stringent and are restricted to 0.3 mg/kg for lead and 0.1 mg/kg for cadmium.
- Aflatoxins: these mycotoxins are restricted to 2 μg/kg to ensure safety against carcinogenic contaminants.

4.4.1.2.1.3 Packaging and Labeling (FSSAI, 2011)

- Packaging requirements: cloves must be packed in clean, dry, and sound containers. Suitable materials include glass, tin, aluminum, or food-grade plastic. The packaging must prevent contamination and preserve the quality of the cloves.
- Labeling: labels must specify product details, net weight, date of packing, best before date, batch number, and manufacturer details.

4.4.1.2.1.4 AGMARK Certification (Indian Standards, 2010)

- The Agricultural Marketing Advisor to the Government of India oversees the AGMARK certification, which is a quality certification mark for agricultural products.
- Grades: cloves are graded into three categories based on quality parameters such as moisture content, volatile oil content, and extraneous matter. Grade I clove has the highest quality, followed by Grades II and III.

4.4.1.2.1.5 Processing Standards (Barai, 2023)

- Clove processing includes steps like drying, winnowing, and grading. Proper drying techniques are crucial to achieving the desired moisture content and preventing fermentation. Mechanical driers are recommended during the rainy season to ensure uniform drying.
- Grinding: ground cloves must possess at least 16 ml of volatile oil per 100 grams to ensure their quality and efficacy. Ground cloves are more susceptible to quality degradation, so whole cloves are often preferred for longer storage.

4.4.1.2.1.6 Special Considerations (Barai, 2023; FSSAI, 2011)

- Hygiene: processing facilities must adhere to strict hygiene standards to prevent contamination. Personnel involved in processing must maintain personal hygiene and health standards.
- Storage: cloves should be stored in hermetically sealed containers, shielded from direct light. Regular inspection is necessary to check for signs of spoilage or moisture ingress.

4.4.1.2.1.7. Regulatory Bodies (FSSAI, 2011) The FSSAI, in conjunction with the Bureau of Indian Standards (BIS), ensures that these standards are enforced and regularly updated to align with international benchmarks, such as the ISO standards for cloves.

4.4.1.2.2 Indonesia

The Indonesian National Standard (SNI) for cloves, established by the National Standardization Agency of Indonesia (BSN), outlines comprehensive quality and safety criteria to ensure cloves meet both domestic and international standards. The key quality indicators specified in SNI include moisture content, extraneous matter, and essential oil content. The SNI sets the maximum allowable moisture content for cloves at 13%, ensuring the product remains dry enough to prevent spoilage (BSN, 2024). The acceptable level of extraneous matter, such as plant debris and foreign materials, is limited to 1% by weight. Furthermore, the essential oil content, which is crucial for the clove's flavor and aroma, must be at least 15% (BSN, 2024). Safety criteria under SNI include permissible limits for pesticide residues, mycotoxins, and heavy metals.

4.4.1.2.3 Sri Lanka

Sri Lankan standards for clove processing are established by the Sri Lanka Standards Institution (SLSI). These standards focus on maintaining high quality and safety levels to ensure that cloves meet both domestic and international market requirements. Quality parameters in Sri Lankan standards include moisture content, foreign matter, and essential oil content. The moisture content in cloves should not exceed 12%, which helps in preventing microbial growth and maintaining product quality during storage and transport (SLSI, 2024). The permissible level of foreign matter, including dust, dirt, and other contaminants, is set at 0.5% by weight. The essential oil content, which is critical for the clove's aromatic properties, must be at least 15% (EDB, 2024; SLSI, 2024).

4.4.1.2.4 China

The National Food Safety Standards (GB) for cloves in China are overseen by the National Health Commission (NHC) and the State Administration for Market Regulation (SAMR). These standards ensure that cloves meet stringent quality and safety criteria for both domestic consumption and export (Liu, 2024). Quality controls under the GB standards include moisture content, extraneous matter, and essential oil content. The maximum permissible moisture content for cloves is 12%, which helps prevent spoilage. The essential oil content, which is vital for the clove's flavor and aroma, must be at least 15% Chinese regulations for food products, including cloves, are supervised by the State Administration for Quality Supervision, Inspection, and Quarantine (AQSIQ). Labeling and marking are critical for compliance, and noncompliance can lead to consumer complaints and import rejections.

4.4.1.2.4.1 Labeling and Certification Requirements (Cui et al., 2022)

- Language requirements: all products sold in China must be labeled in Chinese.
- China compulsory certification (CCC) mark: Many products must receive the CCC mark before sale. The application for the CCC mark involves submission to the China National Certification and Accreditation Administration (CNCA), sample testing in accredited labs, and factory inspections. Follow-up inspections occur every 12–18 months.

4.4.1.2.4.2 Food Labeling Standards (GB7718–2011) (ITA, 2024)

- Mandatory information: imported foods must explicitly state the country of origin, along with the name and address of the primary distributor registered in China.
- •Pre-packaged food requirements: as per the 2015 Food Safety Law, product labels must include the following details: product name, specifications, production information, and any supplementary data mandated by relevant laws, regulations, and food safety standards.

4.4.1.2.5 Japan

In Japan, the Ministry of Health, Labour, and Welfare (MHLW) sets the regulations for food safety and quality standards for cloves. These regulations are designed to ensure that cloves meet high-quality benchmarks and are safe for consumption. Quality standards under MHLW regulations include moisture content, extraneous matter, and essential oil content. The maximum allowable moisture content for cloves is 12%, which helps prevent microbial growth and spoilage (MHLW, 2024). The permissible level of foreign matter is set at 0.5% by weight. The essential oil content, which is crucial for the clove's aromatic properties, must be at least 15% (MHLW, 2024).

4.4.1.2.5.1 Regulations and Procedural Requirements (JETRO, 2011)

- Plant Protection Act: fresh, unprocessed spices and herbs undergo quarantine procedures to check for pests and contaminants. Items with attached soil are not allowed. Quarantine is managed by regional quarantine stations.
- Food Sanitation Act: spices and herbs must comply with standards for food additives, pesticide residues, and contaminants. Import bans can be imposed if levels exceed permissible limits. The positive list system introduced in 2006 requires all foods to meet specific pesticide residue limits.
- Customs Act: proper labeling and accurate declaration of origin are required. Mislabeling can lead to bans on imports.

4.4.1.2.5.2 Labeling and Sales Requirements (JETRO, 2011)

- Food Sanitation Act: products must be free from harmful substances and meet hygiene standards. Labeling must include food additives, allergens, raw ingredients, and genetic modification information.
- Act on Specified Commercial Transactions: protects consumers in direct commercial transactions like mail orders and telemarketing.

4.4.1.2.5.3 Import and Sales Procedures (JETRO, 2011)

- Plant Inspection: required documents include an application for import inspection and a phytosanitary certificate for fresh spices and herbs.
- Food sanitation inspection: documents required include an import notification form, ingredient table, production flow chart, and analysis results. Non-compliance can result in rejection, destruction, or return of the shipment.
- •Customs clearance: importers must make a declaration to customs and provide necessary documentation, such as invoices, packing lists, and bills of lading. Customs may inspect cargo before granting an import permit.

4.4.1.3 The European Union

The European Food Safety Authority (EFSA) is responsible for ensuring that food products, including cloves, meet stringent safety and quality standards across the European Union (EU). Suppliers entering the EU market must focus on low pesticide residue levels, secure food safety certifications, and conduct regular laboratory tests. Sustainable production techniques and commitment to corporate social responsibility (CSR) standards offer supplementary benefits (European Commission, 2024). The primary suppliers of cloves to Europe include Madagascar, Indonesia, Sri Lanka, and Comoros, with additional competition from Tanzania and Vietnam. To enter the European market, suppliers must meet various compulsory criteria that prioritize consumer health and safety, with an increasing emphasis on sustainability (CBI, 2024).

4.4.1.3.1 *Food Safety Certification and Official Controls*

All cloves imported into the European Union are subject to official food inspections to verify adherence to European food safety standards. The European Commission for Health and Food Safety administers these regulations and oversees their enforcement. Instances of non-compliance are reported through the Rapid Alert System for Food and Feed (RASFF) (RASFF, 2024). Should imports from a particular country consistently fail to comply with EU food safety standards, the frequency of official inspections at the border is increased.

4.4.1.3.2 *Pesticide Residues*

The EU's Maximum Residue Levels (MRLs) regulation sets limits on pesticide residues in food products. Items that surpass these limits are removed from the market. In 2022, a notable issue with cloves was the detection of excessive pesticide levels, specifically 2-chloroethanol, which was found at a concentration of 0.22 mg/kg compared to the MRL of 0.1 mg/kg. Additional problems reported include the presence of chlorpyrifos in organic cloves in 2021 and contaminants such as stones in cloves from Madagascar in 2023 (EFSA, 2024).

4.4.1.3.3 *Contaminant Control*

Food contaminants, which may be inadvertently introduced during production, packaging, transportation, or due to environmental factors, pose significant health risks. To mitigate these risks, the EU has established maximum permissible levels for specific contaminants (Henry-Lamotte-Foods, 2010):

- Microbiological contaminants: while there are no specific limits for cloves, salmonella must be absent.
- Polycyclic aromatic hydrocarbons (PAHs): the allowable limit is 10 μg/kg for benzo(a)pyrene and 50 μg/kg for the total PAH content.
- Metal contaminants: the permissible lead concentration in cloves is 0.1 mg/kg.

4.4.1.3.4 *Labelling Requirements*

Export packages of cloves must be comprehensively labeled with the product name, batch code, net weight, shelf life, recommended storage conditions, lot identification number, and country of origin. Labels on consumer packaging must adhere to the EU Regulation on food information, encompassing nutritional details, origin, and allergen labeling and must also comply with the minimum font size requirements. Organic cloves must also display the inspection body name/code and certification number (European Commission, 2024).

4.4.1.3.5 *Additional Requirements and Certifications*

European buyers often have additional quality and sustainability requirements, including compliance with the European Spice Association's (ESA) minimum quality requirements. Common certifications include Food Safety System Certification (FSSC 22000), International Featured Standards (IFS), and British Retail Consortium Global Standards (BRCGS).

4.4.1.3.6 Quality Requirements for Cloves

Several factors determine the quality of cloves (TIS, 2024):

- Purity and cleanliness: cloves should be devoid of diseases, foreign materials, and any unwanted odors. The ESA Quality Minima Document specifies that foreign objects larger than 2 mm are unacceptable, and the extraneous matter should be below 1% by weight.
- Moisture levels: the maximum permissible moisture content is 12% for whole cloves and 10% for ground cloves. Buyers may prefer lower moisture levels for enhanced quality.
- Granule size: for powdered cloves, a specific mesh size is required, typically ensuring that 95% to 99.5% of the powder passes through a sieve with openings between 600 and 900 microns.
- Aroma and taste: cloves must have a distinct aroma and flavor, primarily determined by their essential oil content, particularly eugenol.
- Essential oil content: the essential oil content is crucial for quality. Whole cloves should contain at least 17 ml per 100 grams, while ground cloves should have at least 14 ml per 100 grams.

4.4.1.3.7 Packaging Standards

Cloves are generally shipped in bulk using multi-wall laminated bags of varying capacities, typically 12.5 kg and 25 kg. Retail packaging ranges from 10 g to 50 g, often in glass or plastic containers or small bags. Transparent packaging is popular for consumer inspection. Packaging must comply with EU standards for labor health and safety, with increasing use of smaller weights such as 10–20 kg to accommodate lifting restrictions in some European countries (TIS, 2024).

4.4.1.3.8 Food Safety Certification

GFSI-recognized certifications, such as FSSC 22000, IFS, and BRCGS, are crucial for accessing the European market. These certifications demonstrate adherence to high food safety standards and are often required by European buyers. Sustainable practices and social responsibility are also increasingly important, with buyers seeking suppliers that comply with codes of conduct and third-party schemes like Rainforest Alliance (EFSA, 2024; TIS, 2024).

4.4.1.3.9 Other European Countries

Other European countries have established specific standards to ensure the safety and quality of cloves. Post-Brexit, the United Kingdom follows regulations set by the Food Standards Agency (FSA), which are similar to EU standards but tailored to the UK market. In general, the standards across these countries specify a maximum permissible moisture content of 12%, and extraneous matter is limited to 1% by weight. For pesticide residues, common limits include cypermethrin at 0.05 mg/kg. The acceptable levels for aflatoxins are typically 2 μg/kg. Heavy metals such as lead and cadmium are generally restricted to 0.3 mg/kg and 0.1 mg/kg, respectively (European Commission, 2024).

In Germany, the Federal Office of Consumer Protection and Food Safety (BVL) enforces similar standards, with specific additional limits for pesticides like malathion at 0.1 mg/kg and cypermethrin at 0.5 mg/kg (TIS, 2024). France's standards, regulated by the French Agency for Food, Environmental and Occupational Health and Safety (ANSES), also align closely, with minor variations in permissible pesticide levels (ANSES, 2024). Italy, through its Ministry of Health, maintains these core standards but specifies a higher limit for aflatoxins at 5 μg/kg. Spain, regulated by the Spanish Agency for Food Safety and Nutrition (AESAN) and the Netherlands, overseen by the Netherlands Food and Consumer Product Safety Authority (NVWA), adhere to these general standards with slight adjustments based on national requirements (European Commission, 2024).

4.4.1.4 North America

4.4.1.4.1 United States

In the United States, the U.S. Food and Drug Administration (FDA) establishes extensive standards for food safety and quality, including those for cloves. These standards are enforced under the Food, Drug, and Cosmetic Act (FD&C Act) to prevent adulteration and protect consumer safety. Cloves and their derivatives must meet rigorous criteria to be classified as generally recognized as safe (GRAS) (CFR, 2023).

4.4.1.4.1.1 FDA Standards for Cloves

- Adulteration policies: derivatives include essential oils from buds, leaves, and stems, oleoresins, and natural extractives. Under the FD&C Act, cloves must be pure and free from adulteration, which includes any substitution or addition that diminishes quality or strength (FDA, 2020).
- Permissible Limits and Specifications: The FDA has specific standards for clove products. For example, the essential oils from cloves must meet the specifications outlined in the "Food Chemicals Codex" (CFR, 2023; FDA, 2020). These include:
  - Phenol content (as Eugenol): the concentration of eugenol, a key phenolic compound, must be measured.
  - Optical rotation: the acceptable range for optical rotation is between –2° and 0°.
  - Refractive index: the refractive index should fall between 1.527 and 1.538 at 20°C.
  - Specific gravity: the specific gravity must be within the range of 1.036 to 1.060.
  - Residual solvents: any residual solvents must be generally recognized as safe (GRAS) or within the tolerance levels specified in Part 173, Subpart C of the FDA regulations (CFR, 2023; FDA, 2020).
- Usage and safety: clove and its derivatives are utilized as flavoring agents in foods at concentrations that comply with good manufacturing practices. The FDA imposes strict limits on contaminants, including heavy metals, pesticides, and mycotoxins. For example, the allowable lead content in

cloves is 0.1 mg/kg, and aflatoxins are restricted to 20 ppb. Pesticide residues, such as DDT and chlorpyrifos, must not exceed 0.05 mg/kg and 0.1 mg/kg, respectively (CFR, 2023; FDA, 2020).

- Labeling requirements: proper labeling is mandated, including product name, net weight, ingredients, manufacturer or distributor details, and necessary nutrition information. For cloves, labels must specify whether they are whole or ground and any processing methods used (CFR, 2023; FDA, 2020).
- Regulatory action guidance: the FDA provides criteria for direct reference seizure requests if cloves contain an average of 5% or more stems by weight (CFR, 2023; FDA, 2020).

4.4.1.4.1.2 Safety Concerns and Use in Veterinary Medicine The FDA's Center for Veterinary Medicine (CVM) has addressed inquiries about using clove oil as an anesthetic for fish. Clove oil, primarily composed of eugenol, also contains isoeugenol and methyl eugenol, compounds with potential anesthetic effects. However, none of these compounds are approved by the FDA for use as anesthetics in fish due to safety concerns. Studies by the National Toxicology Program (NTP) have shown that eugenol is an equivocal carcinogen, while methyl eugenol is carcinogenic to rodents. Consequently, the CVM advises against using clove oil or its components as fish anesthetics due to potential risks to human and animal food safety (FDA, 2002).

4.4.1.4.2 Canada

In Canada, the Canadian Food Inspection Agency (CFIA) regulates the safety and quality of food products, including cloves, under the Safe Food for Canadians Regulations (SFCR). These regulations ensure that cloves meet specific quality and safety standards to protect consumer health (CID, 2024).

4.4.1.4.2.1 CFIA Regulations for Cloves (CFR, 2023; FDR, 2023)

- Quality and safety requirements: the CFIA mandates that cloves meet rigorous standards for various quality parameters. According to the regulations, whole or ground cloves must contain no more than 5% clove stems, 6% total ash, 0.5% ash insoluble in hydrochloric acid, 10% crude fiber, 8% moisture, and at least 13 ml of volatile oil per 100 grams of spice (NHPID, 2004).
- Contaminant control: the CFIA sets maximum levels for contaminants such as heavy metals and pesticides. For instance, the permissible level of lead in cloves is 0.1 mg/kg. Pesticide residues are also tightly controlled, with MRLs for substances like cypermethrin are set at 0.05 mg/kg.
- Microbiological standards: cloves must be free from harmful microorganisms, with no detectable levels of Salmonella, to prevent foodborne illnesses.
- Labeling requirements: cloves sold in Canada must comply with CFIA labeling requirements, including product name, net weight, country of origin, lot number, and the name and address of the manufacturer or importer. Organic cloves must display certification details.
- Preventive control plan (PCP): importers must have a PCP that demonstrates compliance with regulatory requirements for food safety.

4.4.1.4.2.2. T&T Clove Oil T&T clove oil, a by-product of clove oil processing, is widely utilized in self-care products and as an ingredient in pesticides. It is found in fragrances, body lotions, essential oils, massage products, oral care items, and pest control solutions. Under Canada's Chemicals Management Plan, the potential risks associated with chemical substances such as T&T clove oil are thoroughly reviewed and managed. This process begins with a draft chemical risk assessment. T&T clove oil, used in various self-care and pest control products, has undergone safety evaluations to ensure its safe usage (Health Canada, 2021).

Some identified health effects of T&T clove oil include liver hypertrophy (enlarged liver) and alterations in the lymphatic system, such as the presence of blood cells and reduced organ weight. However, it has been determined that T&T clove oil does not pose a significant health risk when used in body lotions, massage products, mouthwash, and as a body fragrance or perfume for individuals over the age of nine. For occupational exposure, employees should discuss relevant laws, safe handling, and storage practices with their employer and occupational health and safety (OHS) officials (Health Canada, 2021).

4.4.1.5 Oceania

In Australia and New Zealand, the regulation of food safety and quality, including cloves, is managed by Food Standards Australia New Zealand (FSANZ). FSANZ establishes and upholds standards to ensure the safety and integrity of the food supply chain, thereby protecting public health and supporting fair trade practices (ACCC, 2021; FSANZ, 2024).

4.4.1.5.1 Safety and Quality Standards (ACCC, 2021; FSANZ, 2024)

- Imported cloves: FSANZ has stringent standards for the importation of cloves to ensure they meet specific safety and quality requirements
- Maximum residue limits (MRLs): FSANZ sets MRLs for pesticides in food products, including cloves. These limits are crucial to ensure that pesticide residues in imported cloves do not exceed safe levels. For example, the MRL for pesticides such as chlorpyrifos in cloves is set at 0.05 mg/kg.
- Contaminant limits: FSANZ also establishes limits for various contaminants, including heavy metals and mycotoxins. For instance, the permissible level of lead in cloves is 0.1 mg/kg, and the limit for aflatoxins is 2 μg/kg.
- Microbiological standards: cloves must meet specific microbiological criteria to ensure they are free from harmful bacteria and pathogens.

4.4.1.5.2 Packaging and Labeling Requirements (ACCC, 2021; FSANZ, 2024)

- Labeling standards: FSANZ mandates that all food products, including cloves, must be properly labeled. Labels must include the product name, country of origin, net weight, and the name and address of the importer or distributor.
- Nutritional information: labels must also provide nutritional information, including energy content, protein, fat, carbohydrates, and sodium.
- Packaging requirements: cloves must be packaged in materials that protect them from contamination and preserve their quality. Packaging should also be designed to prevent moisture ingress, which can lead to spoilage.

4.4.1.5.3 *Import Inspection and Compliance (ACCC, 2021; FSANZ, 2024)*

- Inspection protocols: imported cloves are subject to inspection upon arrival in Australia to ensure they comply with FSANZ standards. This includes checks for pesticide residues, contaminants, and microbiological quality.
- Compliance measures: if imported cloves do not meet FSANZ standards, they may be rejected, destroyed, or required to be re-exported.

4.4.1.5.4 *Quality and Storage (ACCC, 2021; FSANZ, 2024)*

- Oil content: cloves should contain 15.0–20.0% essential oils—primarily eugenol—and 7.0–10.0% fatty oils.
- Storage duration: cloves can be stored for up to 24 months under recommended conditions, which include keeping them in cool, dry places with good ventilation.

4.4.1.5.5 *Transportation and Handling*

- Transport: handling during damp weather should protect cloves from moisture to prevent mold and spoilage.
- Stowage: proper stowage is crucial to prevent shifting during transport. Containers should be stored below deck to avoid exposure to high temperatures, which can cause self-heating and loss of essential oils.

4.4.1.5.6 *Risk Factors and Loss Prevention (ACCC, 2021; FSANZ, 2024)*

- Temperature: the optimal transport temperature for cloves is between 5°C and 25°C. Storage above 25°C should be avoided to prevent loss of essential oils and self-healing.
- Humidity: cloves should be stored with a relative humidity of 60–70% and a water content of 10–12%.
- Ventilation: proper ventilation is essential to disperse water vapor and prevent mold growth.

Table 4.3 gives a comparative list of quality and safety standards for cloves in various regions of the world.

4.5 CONSUMER SAFETY AND QUALITY ASSURANCE

Consumer safety and quality assurance are paramount in the food industry, ensuring that products reaching consumers are safe, nutritious, and of high quality (Figure 4.2). The importance of standards in this process cannot be overstated, as they form the foundation for regulatory frameworks that protect public health globally. Standards for food safety and quality encompass a range of criteria, including permissible levels of contaminants, microbial limits, pesticide residues, and other harmful substances (ISO, 2024). These standards are established based on scientific research and risk assessments conducted by expert committees and regulatory bodies. The establishment of regulatory frameworks by national and international bodies, such as the CAC, FDA, and EFSA, provides a foundation for consumer

TABLE 4.3
Quality and safety standards for cloves in various regions.

Region	Regulatory body	Pesticide residue limits	Heavy metal limits	Mycotoxin limits	Microbiological standards	Additional contaminant controls	Refs.
Africa	**Tanzania**: TBS	DDT: 0.05 mg/kg	Lead: 2 mg/kg, Cadmium: 1 mg/kg	Aflatoxins: 2 μg/kg	No visible mold or insect infestation	Regular testing for heavy metals	(TBS, 2021)
	Madagascar: Office National des Normes	Aligns with international benchmarks	Similar to Tanzania	Similar to Tanzania	No specific limits set	Similar pesticide residue limits in Tanzania	(ONN, 2020)
	Comoros: COMESA	Chlorpyrifos: 0.5 mg/kg	Lead: 2 mg/kg, Cadmium: 1 mg/kg	Aflatoxins: 2 μg/kg	Similar to Tanzania	Detailed drying, storage, and packaging guidelines	(COMESA, 2015)
Asia	**India**: FSSAI	Chlorpyrifos: 0.05 mg/kg, DDT: 0.1 mg/kg	Lead: 0.3 mg/kg, Cadmium: 0.1 mg/kg	Aflatoxins: 2 μg/kg	Free from mold and insect damage	Strict policies against adulteration	(FSSAI, 2016).
	Indonesia: BSN	Similar to international standards	Lead: 2 mg/kg, Cadmium: 1 mg/kg	Aflatoxins: 2 μg/kg	No visible mold or infestation	Essential oil content ≥ 15%	(BSN, 2024).
	Sri Lanka: Sri SLSI	Aligns with international benchmarks	Lead: 0.3 mg/kg, Cadmium: 0.1 mg/kg	Aflatoxins: 2 μg/kg	Similar to other countries	Essential oil content ≥ 15%	(SLSI, 2024).
	China: NHC and SAMR	Similar to international benchmarks	Lead: 0.1 mg/kg	Aflatoxins: 2 μg/kg	Free from Salmonella	CCC mark required for imported goods	(NHC/SAMR, 2024)
	Japan: MHLW	Pesticide residues must not exceed positive list limits	Lead: 0.3 mg/kg, Cadmium: 0.1 mg/kg	Aflatoxins: 2 μg/kg	No harmful microorganisms	Compliance with multiple regulatory acts	(MHLW, 2024)

(Continued)

TABLE 4.3 *(Continued)*
Quality and safety standards for cloves in various regions.

Region	Regulatory body	Pesticide residue limits	Heavy metal limits	Mycotoxin limits	Microbiological standards	Additional contaminant controls	Refs.
Europe	**European Union**: EFSA	Cypermethrin: 0.05 mg/kg	Lead: 0.1 mg/kg	Aflatoxins: 2 µg/kg	No specific limits but Salmonella absent	Compliance with ESA and other quality documents	(CBI, 2020, 2024)
	UK: FSA	DDT: 0.05 mg/kg, Chlorpyrifos: 0.1 mg/kg	Lead: 0.3 mg/kg, Cadmium: 0.1 mg/kg	Aflatoxins: 2 µg/kg	Similar to EU	Post-Brexit standards adapted from EU	(CBI, 2020, 2024)
	Germany: BVL	Malathion: 0.1 mg/kg, Cypermethrin: 0.5 mg/kg	Lead: 0.3 mg/kg, Cadmium: 0.1 mg/kg	Aflatoxins: 2 µg/kg	No specific limits but Salmonella absent	Adheres to both EU and national regulations	(TIS, 2024)
	France: ANSES	Chlorpyrifos: 0.05 mg/kg, Malathion: 0.1 mg/kg	Lead: 0.3 mg/kg, Cadmium: 0.1 mg/kg	Aflatoxins: 2 µg/kg	Similar to EU	High national safety benchmarks	(ANSES, 2024)
	Italy: Ministry of Health	DDT: 0.05 mg/kg	Lead: 0.3 mg/kg, Cadmium: 0.1 mg/kg, Aflatoxins: 5 µg/kg	Aflatoxins: 5 µg/kg	No specific limits but Salmonella absent	High-quality import focus	(European Commission, 2024)
	Spain: AESAN	Cypermethrin: 0.05 mg/kg	Lead: 0.3 mg/kg, Cadmium: 0.1 mg/kg	Aflatoxins: 2 µg/kg	Similar to EU	Compliance with national safety standards	(European Commission, 2024)
	Netherlands: NVWA	Glyphosate: 0.1 mg/kg	Lead: 0.3 mg/kg, Cadmium: 0.1 mg/kg	Aflatoxins: 2 µg/kg	Similar to EU	High quality and safety benchmarks	(European Commission, 2024)
North America	**United States**: FDA	DDT: 0.05 mg/kg, Chlorpyrifos: 0.1 mg/kg	Lead: 0.1 mg/kg	Aflatoxins: 20 ppb	No specific limits but Salmonella absent	Detailed adulteration policies	(CFR, 2023)
	Canada: CFIA	Cypermethrin: 0.05 mg/kg	Lead: 0.1 mg/kg	Aflatoxins: 2 µg/kg	Free from harmful microorganisms	Comprehensive Preventive Control Plan (PCP)	(CID, 2024)
Oceania	**Australia and New Zealand**: FSANZ	Chlorpyrifos: 0.05 mg/kg	Lead: 0.1 mg/kg	Aflatoxins: 2 µg/kg	Free from Salmonella	Stringent import inspection protocols	(ACCC, 2021; FSANZ, 2024)

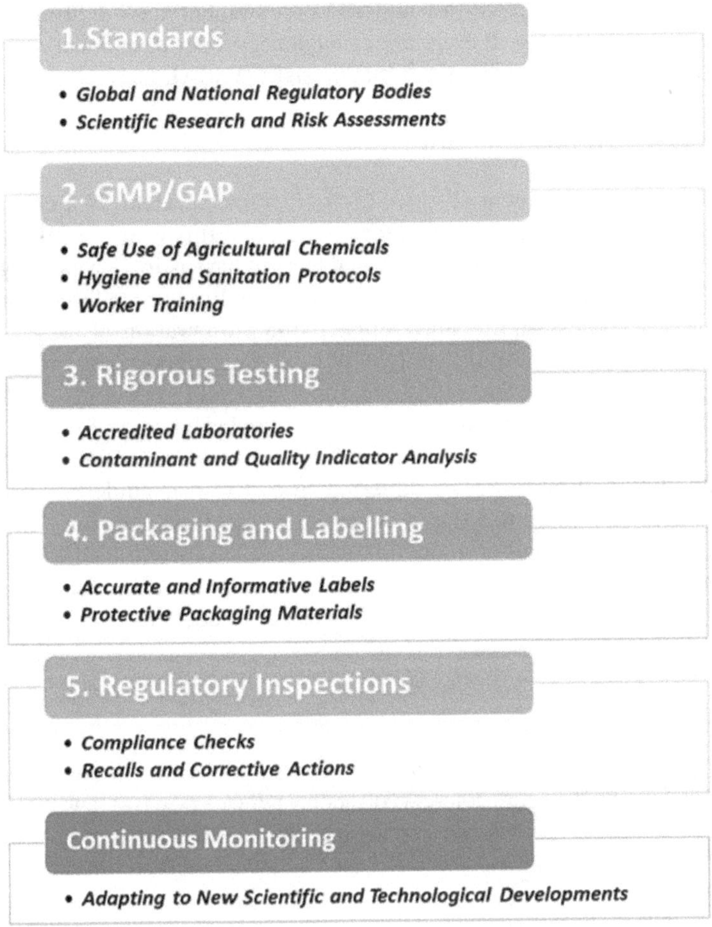

FIGURE 4.2 Consumer safety and quality assurance process.

safety. These frameworks set forth guidelines and mandatory requirements that food products must meet to ensure they are safe for consumption. This includes specifying acceptable levels of contaminants, such as pesticide residues, mycotoxins, and heavy metals, as well as defining microbiological criteria (Codex-Alimentarius-Commission, 2022; CFR, 2023). Quality control measures are integral to maintaining product integrity. These measures involve rigorous testing and monitoring of products at various stages of the supply chain—from production and processing to distribution and retail. Regular laboratory tests check for compliance with established standards, ensuring that products do not contain harmful substances beyond permissible limits. Quality control also includes sensory evaluations to assess the physical and chemical properties of the product, such as moisture content, essential oil levels, and purity. The GMP is a set of principles and procedures that manufacturers must follow to produce safe and high-quality food products. GMP encompasses all aspects of production, from raw material procurement and handling to processing, packaging, and storage. It emphasizes hygiene, proper equipment maintenance, and staff training to minimize contamination risks (FSSAI, 2011).

4.6 CONCLUSIONS

The post-harvest management of clove (*Syzygium aromaticum*) is a critical aspect of ensuring the quality, longevity, and market value of this valuable spice. Effective post-harvest practices encompass timely and careful harvesting, proper drying techniques, efficient storage methods, and meticulous handling procedures. Each step in the post-harvest process significantly impacts the preservation of clove's aromatic and medicinal properties. Additionally, understanding market demands and maintaining high standards in post-harvest handling can lead to better economic returns for producers. This is only possible through the standardization of processed clove, which is also essential for ensuring the quality, safety, and consistency of this globally traded spice. Furthermore, by adhering to established guidelines and regulations, producers can guarantee that their products meet the necessary standards to protect consumer health and promote fair trade practices. Continued efforts in these areas such as implementing good agricultural and post-harvest practices, supported by research and technology, can help in overcoming challenges related to quality control and market competition.

REFERENCES

ACCC. (2021). *Country of Origin Food Labelling.* Gov.au. www.accc.gov.au/system/files/Country%20of%20origin%20food%20labelling.pdf (Accessed 30 June 2024).

ACDI/VOCA. (2023). *World Hunger Education Service.* Worldhunger.org. www.worldhunger.org/wp-content/uploads/2024/01/Understanding-Global-Hunger-Fact-Sheet-2-1.pdf (Accessed 30 June 2024).

Alashti, F. J., Sohbatzadeh, F., Ahmadian, S., Kenari, R. E., & Nazifi, E. (2024). Impact of atmospheric cold plasma pretreatment on morphology, structure, and chemical properties of clove (Syzygium aromaticum). *Lebensmittel-Wissenschaft, 191*(115639), 115639.

ANSES. (2024). *ANSES.* Anses—Agence Nationale de Sécurité Sanitaire de l'alimentation, de l'environnement et Du Travail. www.anses.fr/en (Accessed 30 June 2024).

Barai, K. (2023). *Export Import Guru.* Export Import Guru. https://exportimport.guru/Exim_SEO/Single/Exporting/how-to-import-cloves-into-india-2023 (Accessed 30 June 2024).

Ben Hassine, D., Kammoun El Euch, S., Rahmani, R., Ghazouani, N., Kane, R., Abderrabba, M., & Bouajila, J. (2021). Clove buds essential oil: The impact of grinding on the chemical composition and its biological activities involved in consumer's health security. *BioMed Research International, 2021*, 1–11.

BMT. (2024). *Cloves—Cargo Handbook—The World's Largest Cargo Transport Guidelines Website.* Cargohandbook.com. https://cargohandbook.com/Cloves (Accessed 30 June 2024).

BSN. (2024). *BSN—Badan Standardisasi Nasional—National Standardization Agency of Indonesia—Setting the Standard in Indonesia ISO SNI WTO.* Bsn.Go.Id. www.bsn.go.id/ (Accessed 30 June 2024).

CBI. (2020). *Centre for the Promotion of Imports from Developing Countries—Exporting Cloves to Europe.* Srilankabusiness.com. www.srilankabusiness.com/reports/cbi/spices/cbi-cloves.pdf (Accessed 30 June 2024).

CBI. (2024). *The European Market Potential for Cloves.* Cbi.Eu. www.cbi.eu/market-information/spices-herbs/cloves-0/market-potential (Accessed 30 June 2024).

CFR. (2023). *CFR—Code of Federal Regulations Title 21.* www.accessdata.fda.gov/scripts/cdrh/cfdocs/cfcfr/CFRSearch.cfm?fr=184.1257 (Accessed 30 June 2024).

CID (Canadian Importers Database). (2024). *090720—Cloves, Whole Fruit, Cloves and Stems, Crushed or Ground.* Canada.Ca. https://ised-isde.canada.ca/app/ixb/cid-bdic/productReport.html?hsCode=090720 (Accessed 30 June 2024).

Codex-Alimentarius-Commission. (2021). *Nutrition and Labelling.* Fao.org. www.fao.org/fao-who-codexalimentarius/thematic-areas/nutrition-labelling/en/ (Accessed 30 June 2024).

Codex-Alimentarius-Commission. (2022). *Standard for Dried Floral Parts: Cloves.* Fao.org. www.fao.org/fao-who-codexalimentarius/sh-proxy/en/?lnk=1&url=https%253A%252F%252Fworkspace.fao.org%252Fsites%252Fcodex%252FStandards%252FCXS%2B344-2021%252FCXS_344e.pdf (Accessed 30 June 2024).

Codex-Alimentarius-Commission. (2024). *International Food Standards.* Fao.org. www.fao.org/fao-who-codexalimentarius/committees/cac/about/en/ (Accessed 30 June 2024).

COMESA. (2015). *COMESA.* Comesa.int. www.comesa.int/wp-content/uploads/2020/10/Final-COMESA-SQA-Strategic-Plan-2.pdf (Accessed 30 June 2024)

Cui, J., Yan, R., Astell-Burt, T., Gong, E., Zheng, L., Li, X., Zhang, J., Xiang, L., Ye, L., Hu, Y., Tang, Y., Gao, C., Xiao, L., Jiang, Y., Shao, R., Feng, X., Zhang, J., & Yang, Y. (2022). Types and aspects of front-of-package labeling preferred by parents: Insights for policy making in China. *Nutrients, 14*(4), 800.

EDB. (2024). *Clove from Sri Lanka—Ceylon Cloves—EDB Sri Lanka.* Srilankabusiness.com. www.srilankabusiness.com/spices/about/cloves-sri-lanka.html (Accessed 30 June 2024).

EPGP. (2024). *Clove Processing and Drying.* Inflibnet.Ac.In. https://epgp.inflibnet.ac.in/epgp-data/uploads/epgp_content/food_technology/technology_of_spices_and_condiments/31.cloves__processing_and_drying_/et/2901_et_m31.pdf (Accessed 30 June 2024).

European Commission. (2024). *Health and Food Safety.* European Commission. https://commission.europa.eu/about-european-commission/departments-and-executive-agencies/health-and-food-safety_en (Accessed 30 June 2024).

European Food Safety Authority (EFSA). (2024). Peer review of the pesticide risk assessment of the active substance clove oil. *EFSA Journal, 22*(4), e8671.

FAO. (2021). *Strategic Framework 2022–31.* Fao.org. https://openknowledge.fao.org/server/api/core/bitstreams/29404c26-c71d-4982-a899-77bdb2937eef/content (Accessed 30 June 2024).

FAO. (2023). *In Comoros, FAO Concludes an Evaluation of the National Food Control System.* Fao.org. www.fao.org/food-safety/news/news-details/en/c/1648492/ (Accessed 30 June 2024).

FAO. (2024). *FAO.* FAOHome. www.fao.org/home/en/ (Accessed 30 June 2024)

FAO-WHO. (2021). *FAO/WHO Joint Media Advisory: Codex Alimentarius Commission.* Who.int. www.who.int/news/item/05-11-2021-fao-who-joint-media-advisory-codex-alimentarius-commission (Accessed 30 June 2024).

FDA. (2002). *Status of Clove Oil and Eugenol for Anesthesia of Fish.* Umces.edu. www.umces.edu/sites/default/files/guidance_for_industry_0.pdf (Accessed 30 June 2024).

FDA. (2020, February 13). *CPG sec 525.250 Cloves—Adulteration with Stems.* U.S. Food and Drug Administration; FDA. www.fda.gov/regulatory-information/search-fda-guidance-documents/cpg-sec-525250-cloves-adulteration-stems (Accessed 30 June 2024).

FDR. (2023, November 24). *Food and Drug Regulations.* Justice.Gc.Ca. https://laws-lois.justice.gc.ca/eng/regulations/c.r.c.,_c._870/page-23.html (Accessed 30 June 2024).

FSANZ. (2024). *Allergen Labelling for Food Businesses.* Gov.au. www.foodstandards.gov.au/business/labelling/allergen-labelling (Accessed 30 June 2024).

FSSAI. (2011). *Food Safety and Standards (Packaging And Labelling) Regulations*. Gov.In. www.fssai.gov.in/upload/uploadfiles/files/Packaging_Labelling_Regulations.pdf

FSSAI. (2016). *Food or Health Supplements, Nutraceuticals, Foods for Special Dietary Uses, Foods for Special Medical Purpose, Functional Foods and Novel Food) Regulations*. Gov.In. https://fssai.gov.in/cms/product-standards.php (Accessed 30 June 2024).

FSSAI. (2022). *Import of Clove Stem*. Gov.In. https://fssai.gov.in/upload/advisories/2022/06/62ac4e5e84f6cOrder_Import_Clove_17_06_2022.pdf (Accessed 30 June 2024).

GrainPro. (2024). *Better Post Harvest Management*. Grainpro.com. www.grainpro.com/better-post-harvest-management (Accessed 30 June 2024).

Health Canada. (2021, May 7). *T & T Clove Oil*. Canada.Ca. www.canada.ca/en/health-canada/services/chemicals-product-safety/t-t-clove-oil.html (Accessed 30 June 2024).

Henry-Lamotte-Foods. (2010). *Clove Bud Oil*. Sevron.co.uk. https://safety365.sevron.co.uk/substances/accessSDS/SDS-43506-5787568d74c917.24443250 (Accessed 30 June 2024).

Hunger-Notes. (2018, September 17). *Africa Hunger Facts, Africa Poverty Facts*. World Hunger News. www.worldhunger.org/africa-hunger-poverty-facts-2018/ (Accessed 30 June 2024).

Indian-Standards. (2010). *Spices and Condiments—Cloves, Whole and Ground*. Resource.org. https://law.resource.org/pub/in/bis/S06/is.4404.2010.pdf (Accessed 30 June 2024).

ISO. (2024). *ISO 2254:2004(en) Cloves, Whole and Ground (Powdered)—Specification*. Iso.org. www.iso.org/obp/ui/ (Accessed 30 June 2024).

ITA. (2024). *China—Labeling/Marking Requirements*. Privacyshield.gov. www.privacyshield.gov/ps/article?id=China-Labeling-Marking-Requirements (Accessed 30 June 2024).

JECFA. (2024). *JECFA*. Fao.org. www.fao.org/food-safety/scientific-advice/jecfa/en/ (Accessed 30 June 2024).

JETRO. (2011). *Guidebook for Export to Japan (Food Articles)*. Jetro.Go.Jp. www.jetro.go.jp/ext_images/mexico/mercadeo/2Espice.pdf (Accessed 30 June 2024).

Kapadiya, S., & Desai, M. A. (2017). Isolation of essential oil from buds of Syzygium aromaticum using hydrodistillation: Multi-response optimization and predictive modelling. *IJAST*, *6*(1), 405–418.

KAU. (2024). *Clove Processing*. Celkau.In. http://celkau.in/Agrienterprises/enerprise/11.Processing%20of%20spices/5.%20CLOVE%20PROCESSING.pdf (Accessed 30 June 2024).

Kusuma, H. S., Lantip, G. I. A., Mutiara, X., & Iqbal, M. (2023). Evaluation of mini bibliometric analysis, moisture ratio, drying kinetics, and effective moisture diffusivity in the drying process of clove leaves using microwave-assisted drying. *Applied Food Research*, *3*(1), 100304. https://doi.org/10.1016/j.afres.2023.100304 (Accessed 30 June 2024).

Liu, J. (2024, March 14). *China NHC Released 47 National Food Safety Standards Including GB 2760–2024 and 6 Amendments*. Cirs-group.com. www.cirs-group.com/en/food/china-nhc-released-47-national-food-safety-standards-including-gb-2760-2024-and-6-amendments (Accessed 30 June 2024).

MHLW. (2024). *Food Safety (Ministry of Health, Labour and Welfare)*. Mhlw.go.jp. www.mhlw.go.jp/stf/seisakunitsuite/bunya/kenkou_iryou/shokuhin/index_00006.html (Accessed 30 June 2024).

Mkojera, B. T., & Chove, B. (2021). Assessment of harvesting and post-harvest handling practices on organically grown cloves and black pepper in Tawa Ward, Morogoro. *TAJAS*, *20*(2), 317–325. www.ajol.info/index.php/tjags/article/view/225891 (Accessed 30 June 2024).

NHC/SAMR. (2024). *National Health Commission (NHC) and State Administration for Market Regulation (SAMR)*. Gov.Cn. www.nhc.gov.cn/ (Accessed 30 June 2024).

NHPID. (2004). *Natural Health Products Ingredients Database*. https://webprod.hc-sc.gc.ca/nhpid-bdipsn/search-rechercheReq.do (Accessed 30 June 2024).

ONN. (2020). *Standards for Cloves in Madagascar.* Onn.Mg. www.onn.mg/ (Accessed 30 June 2024).

RASFF. (2024). *Food Safety.* https://food.ec.europa.eu/safety/rasff_en (Accessed 30 June 2024)

Sarma, Y. R., Babu, K. N., & Aziz, S. (2014). Spices and aromatics. In *Encyclopedia of Agriculture and Food Systems* (pp. 211–234). Elsevier.

Saxena, D. C. (2024). *Technology of Spices and Condiments.* Inflibnet.Ac.In. https://ebooks.inflibnet.ac.in/ftp07/chapter/clove-processing-and-drying/ (Accessed 30 June 2024).

Singh, P. A., Bajwa, N., Chinnam, S., Chandan, A., & Baldi, A. (2022). An overview of some important deliberations to promote medicinal plants cultivation. *Journal of Applied Research on Medicinal and Aromatic Plants, 31*, 100400.

Singh, P. A., Dash, S., Choudhury, A., & Bajwa, N. (2024). Factors affecting long-term availability of medicinal plants in India. *Journal of Crop Science and Biotechnology, 27*(2), 145–173.

Singh, P. A., Sood, A., & Baldi, A. (2021). Determining constraints in medicinal plants adoption: A model geospatial study in the Indian state of Punjab. *Journal of Applied Research on Medicinal and Aromatic Plants, 25*, 100342.

SLSI. (2024). *Sri Lanka Standards Institution.* Slsi.Lk. www.slsi.lk/ (Accessed 30 June 2024).

TBS. (2021). *Cloves—Specification.* Tbs.Go.Tz. www.tbs.go.tz/ (Accessed 30 June 2024).

TIS. (2024). *Cloves—Transport Informations Service.* Tis-gdv.de. www.tis-gdv.de/tis_e/ware/gewuerze/nelken/nelken-htm/ (Accessed 30 June 2024).

5 Biochemistry

Fatemeh Jamshidi Alashti and
Farshad Sohbatzadeh

5.1 INTRODUCTION

Clove is derived from dried flower buds of the clove tree, *Syzygium aromaticum*, which is a rich source of mineral and bioactive compounds. Previous findings state that the clove tree possesses diverse macro and micronutrients (Haro-González et al., 2021; Kaur et al., 2019). Overall, clove's primary metabolism possesses macronutrients like potassium, calcium, magnesium, sodium, and phosphorus, and also micronutrients like manganese, zinc, and iron are found in clove (Kaur et al., 2019). Potassium is a vital mineral that helps regulate fluid balance in the body, supports muscle function, and plays a role in nerve transmission (Weaver, 2013). Calcium is crucial for the upkeep of sturdy bones and teeth, while iron is necessary for oxygen transport in the blood and energy generation (Li et al., 2016). Magnesium is essential for maintaining healthy blood pressure levels and is necessary for the function of muscles and nerves. Phosphorus is essential for bone health, energy production, and maintaining proper pH balance in the body. Manganese, one of the primary minerals found in cloves, is necessary for several of the body's metabolic functions. Manganese is necessary for the metabolism of carbohydrates, lipids, and protein as well as for the formation of cartilage, bones, and connective tissues (Cotzias, 1958; O'Neal & Zheng, 2015). Zinc is vital for the production of DNA, healing from wounds, and the immune system. Clove is a good source of trace minerals like copper. Copper is required for the formation of connective tissues, antioxidant defense-related enzymes, and red blood cells (Song et al., 2021).

5.2 THE TECHNIQUES UTILIZED TO EXTRACT BIOACTIVE COMPOUNDS FROM CLOVE

The bioactive compound extraction methods play a critical role in the variation of compositions due to extraction techniques that could affect the quality of compounds. Also, bioactive compounds possess diverse biological activities, and selecting the appropriate extraction technique could enhance their effects, especially on human health (Carpentieri et al., 2021; Ebrahimi & Lante, 2022; Singh et al., 2024). Developing effective extraction methods for bioactive compounds is essential for harnessing their potential health benefits and incorporating them into various products for human consumption. Bioactive compound extraction from plant material is categorized into traditional and emerging technologies. Conventional methods include solvent extraction, maceration, steam distillation,

 DOI: 10.1201/9781003521785-5

Soxhlet extraction, and cold processing. However, traditional extraction possesses negative impacts such as extended extraction times, slow processing speeds, reduced phenolic compound yields, high energy usage, unsuitability for heat-sensitive compounds, and the necessity for large solvent volumes. As a consequence, innovative methods referred to as "green extraction" have been developed. These advanced extraction techniques are becoming increasingly popular due to their effectiveness, sustainability, and capacity to maintain the biological activity of the compounds. The term "green extraction methods" refers to a variety of techniques such as cold atmospheric plasma (CAP), microwave-assisted extraction (MAE), ohmic heating (OH), and ultrasound-assisted extraction (UAE) (Armenta et al., 2019; Picot-Allain et al., 2021; Rodriguez Garcia & Raghavan, 2022). For instance, the impact of microwave-assisted extraction on the yield and composition of essential oils (EOs) from clove buds was investigated (Gonzalez-Rivera et al., 2021). For comparison purposes, the traditional method of hydrodistillation of clove buds using standard heating was carried out. After 90 min of MAE, the EO extraction yield obtained from 5.5 ± 0.8 up to 16 ± 1.5%. The extraction yield of EO through conventional hydrodistillation (120 min) was 7.8 ± 0.8% despite using over 30% more extraction time. So, the MAE method was faster and more environmentally friendly than conventional hydrodistillation. In extracting dry clove buds using the ultrasound scheme, the effectiveness of batch and flow ultrasonic apparatus (20 L in 45 min) was assessed compared to traditional maceration. The strongest radical scavenging activity and maximal phenolic content have been identified with flow UAE operating at 1,350 mL/min. On the other hand, the maceration showed the lowest extraction yield. The effect of dielectric barrier discharge of cold atmospheric plasma on the physicochemical properties of clove was assessed (Alashti et al., 2024). The maceration extraction is considered a control group. The findings indicate that the reactive oxygen-nitrogen species (RONS) produced during the plasma process disturbed the clove's cell wall structure, enhancing the extraction efficiency compared to the maceration extraction. The cold plasma pretreatment increased 10% of the total phenolic compounds rather than the maceration extraction. The cold plasma sample at 10 min exhibited the highest antioxidant activities among all samples. EOs of clove bud were extracted using ohmic heating-assisted hydrodistillation. The extraction yield and chemical characteristics were compared with the traditional hydrodistillation method (Tunç & Koca, 2019). The yield of clove bud EOs obtained by ohmic heating-assisted hydrodistillation and hydrodistillation was 13.18 ± 1.50 and 8.23 ± 0.35 respectively. In the ohmic heating-assisted hydrodistillation method, eugenol yield was higher than the hydrodistillation method, with amounts of 6.95 ± 0.6 and 5.13 ± 0.28, respectively.

5.3 CLOVE BIOACTIVE COMPOUNDS STRUCTURE AND COMPOSITION

5.3.1 Clove Essential Oil Composition

Eugenol comprises about 70 to 90% of the essential oil found in clove. Gas chromatography (GC) and gas chromatography-mass spectrometry (GC-MS) methods were used to study the essential oil of clove leaves, yielding a composition of 23 components that formed approximately 99% of the total oil. The predominant constituents identified were eugenol (76.8%), followed by β-caryophyllene (17.4%), alpha-humulene (2.1%), and eugenol acetate (1.2%). Furthermore, more substances were detected in amounts lower than 0.5% (Jirovetz et al., 2006). GC and GC-MS were employed to investigate the essential oil extracted from the clove's dried flower buds. The main components detected were eugenol (76.23%), β-caryophyllene (11.54%), and eugenol acetate (1.76%) (Xu et al., 2016; Singh et al., 2021). The composition of clove essential oil was determined using GC-MS analysis. The chemical analysis revealed the presence of five compounds, with eugenol being the major component at 74.6%. This was followed by caryophyllene at 3.5% and caryophyllene oxide at 0.5% (Kaur et al., 2019). As a consequence, other key components present in clove essential oil are eugenyl acetate, beta-caryophyllene, and alpha-humulene. Table 5.2 shows the general composition of clove essential oil (Golmakani et al., 2017). The chemical composition of the essential oil extracted from clove buds is shown in Table 5.1.

5.3.1.1 Eugenol

Eugenol ($C_{10}H_{12}O_2$) is a guaiacol substituted with an allyl chain, belonging to the allylbenzene class of chemical compounds with a molecular weight of 164.20 g/mol. Eugenol is a widely recognized and researched substance. It was initially identified as a volatile component of Eugenia caryophyllata in 1929 (Marchese et al., 2017; Tammannavar et al., 2013).

Also, eugenol is a main component of the aromatic oil extract from clove, comprising between 70% and 90% of the extract (Bhuiyan et al., 2010). The characteristic

TABLE 5.1
Chemical composition of essential oil from clove buds.

Compound	Molecular formula	Molecular weight (g/mol)	Peak area
Eugenol	$C_{10}H_{12}O_2$	164.20	87.3
Eugenyl acetate	$C_{12}H_{14}O_3$	206.24	10.4
β-Caryophyllene	$C_{15}H_{24}$	204.35	1.35
α-Humulene	$C_{15}H_{24}$	204.35	0.19
Caryophyllene oxide	$C_{15}H_{24}O$	220.35	0.2
Chavicol	$C_9H_{10}O$	134.17	0.31
Methyl salycilate	$C_8H_8O_3$	152.15	0.08
Benzaldehyde	C_7H_6O	106.12	0.07
Benzyl acetate	$C_9H_{10}O_2$	150.17	0.05
2-Nonanone	$C_9H_{18}O$	142.24	0.04
Benzyl benzoate	$C_{14}H_{12}O_2$	212.24	0.03
Ethyl benzoate	$C_9H_{10}O_2$	150.17	0.01

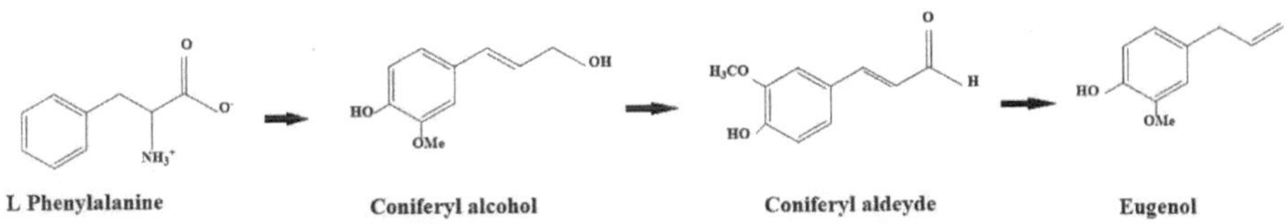

FIGURE 5.1 Biosynthesis of eugenol.

aroma and flavor of clove comes from eugenol in its essential oil. Eugenol has diverse biological activities and is commonly used as an analgesic, antibacterial, and potent antioxidant. It is also a key ingredient in the production of fragrances and soaps (Ulanowska & Olas, 2021; Zari et al., 2021). The primary precursor for eugenol is derived from the phenylpropanoid pathway. Phenylalanine is transformed into cinnamic acid via phenylalanine ammonia-lyase (PAL) and cinnamate 4-hydroxylase. Cinnamic acid is then converted to coumaric acid by the enzyme cinnamic acid 4-hydroxylase. Coumaric acid is converted to piperonylic acid through a series of enzyme-catalyzed reactions. Piperonyl CoA is then converted to eugenol. Overall, the biosynthesis pathway of eugenol in clove involves the phenylpropanoid pathway and several enzyme-catalyzed reactions to convert precursor molecules into eugenol as represented in Figure 5.1. (Ouadi et al., 2022).

5.3.1.2 Eugenyl Acetate

Eugenyl acetate ($C_{12}H_{14}O_3$), a yellowish liquid, is a phenylpropanoid derivative of eugenol. Eugenyl acetate's molecular weight is 206.24 g/mol, and it is a member of phenols and a benzoate ester.

It is found in small amounts in clove bud and leaf EO, as well as cinnamon leaf EO (Golmakani et al., 2017; Guan et al., 2007). There are multiple phases in the clove eugenyl acetate production. The phenylpropanoid path produces eugenyl acetate. Eugenol can then be further modified to eugenol acetate through the esterification of eugenol with acetic acid, catalyzed by an acetyltransferase enzyme (Santolin et al., 2021).

5.3.1.3 *β*-Caryophyllene

β-Caryophyllene ($C_{15}H_{24}$), is a bicyclic sesquiterpene that occurs naturally and is frequently found in essential oils from various spices. It is categorized as a phytocannabinoid and is accepted for use as a food additive and flavoring agent by the United States Food and Drug Administration and European organizations (Sharma et al., 2016). The biosynthesis pathway of *β*-Caryophyllene in clove is farnesyl pyrophosphate (FPP), which is derived from the mevalonic acid pathway. FPP is converted to *β*-Caryophyllene through a series of enzyme-catalyzed cyclization reactions. *β*-Caryophyllene is one of the major components of clove oil, contributing to its characteristic aroma and flavor (Di Sotto et al., 2020; Gertsch et al., 2008; Yang et al., 2016). In recent years, *β*-Caryophyllene has been the focus of significant study. Preclinical studies have shown that *β*-Caryophyllene modulates the nervous system and has beneficial effects on various neurodegenerative and inflammatory conditions (Chang et al., 2013; Machado et al., 2018).

5.3.1.4 α-Humulene

α-Humulene ($C_{15}H_{24}$) or α-caryophyllene is a monocyclic sesquiterpene with a 204.35 g/mol molecular weight. Overall, the biosynthesis pathway of α-humulene in clove follows the general terpene biosynthesis pathway, starting from isoprene building blocks and proceeding by several enzymatic reactions to synthesize α-humulene specifically. This process is catalyzed by the enzyme farnesyl pyrophosphate synthase. FPP undergoes further cyclization and rearrangement reactions catalyzed by enzymes such as sesquiterpene synthase to form various sesquiterpenes, including α-humulene (Guo et al., 2022; Wei et al., 2016). The essential oil compounds of clove are represented in Figure 5.2.

5.3.1.5 Chavicol

The production of chavicol in clove involves the phenylpropanoid pathway to p-coumaric alcohol, which is then transformed to chavicol through deoxygenation. This transformation is facilitated by chalcone synthase (CHS), an important enzyme in the biosynthesis of flavonoids. CHS also plays a role in plant defense mechanisms,

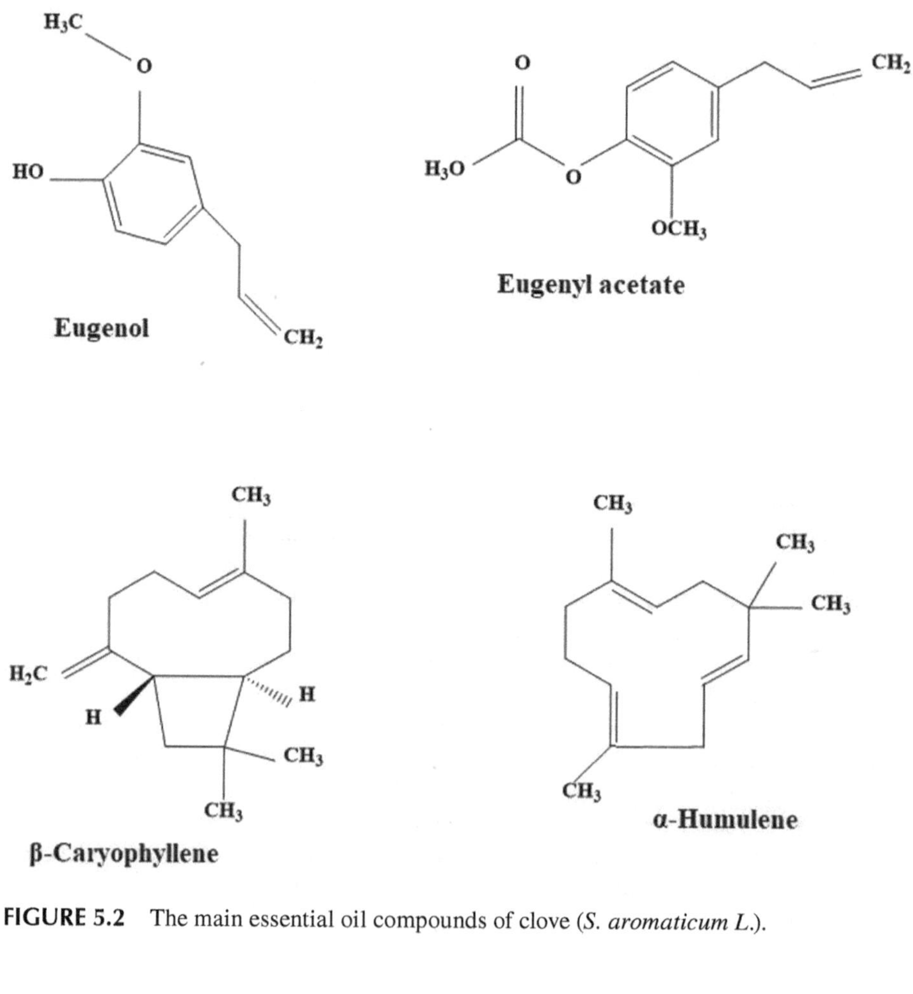

FIGURE 5.2 The main essential oil compounds of clove (*S. aromaticum L.*).

as its expression is triggered under stressful conditions, leading to the accumulation of flavonoid and isoflavonoid phytoalexins. The crystal structure of CHS has been determined, providing a detailed understanding of the cyclization process involved in chalcone synthesis (Vassão et al., 2006).

5.3.1.6 Caryophyllene Oxide

Caryophyllene oxide ($C_{15}H_{24}O$), with a 220.35 g/mol molecular weight, is derived from caryophyllene, as its alkene group has become an epoxide. Additionally, it is an approved food additive often used as a flavoring. It is worth noting that caryophyllene oxide may have negligible cannabinoid activity (Yang et al., 2000).

5.3.2 Flavonoid

Flavonoids are a diverse group of plant compounds that are widely distributed in fruits, vegetables, and spices (Havsteen, 2002). Flavonoids are important pigments in plants that give flowers their color by creating yellow, red, and blue hues in the petals to attract pollinators. They also have various functions in plants, including providing protection from UV rays, aiding in nitrogen fixation, and enhancing flower pigmentation. Flavonoids are the predominant polyphenolic compounds in the human diet and are widely present in various plants. These compounds have been extensively studied for their potential health benefits, including antioxidant, anti-inflammatory, anticancer, and antimicrobial properties. Flavonoids are known for their ability to scavenge free radicals in the body and protect cells from oxidative stress, which has been linked to various chronic diseases such as cancer and neurodegenerative disorders (Batra & Sharma, 2013; Hazafa et al., 2020; Jucá et al., 2020; Masuoka et al., 2012; Silva et al., 2002; Soto-Cabrera et al., 2016). They can also modulate the immune system and help fight off infections (Panche et al., 2016; Rice-Evans, 2001). The most common polyphenols found in plants are identified as flavonoids, and they are relatives of the benzopyran derivative group. Their main component is the flavan structure, which consists of three linked rings of carbon atoms that may bind to several kinds of substituents. Flavonoids are categorized into six main subgroups based on their chemical structure and properties: isoflavones, flavonols, anthocyanins, flavones, and flavanones (Heim et al., 2002; Schnarr et al., 2022; Tao et al., 2024). **Table 5.2** shows the main flavonoid compounds present in the clove plant (Hassan et al., 2021; Ryu et al., 2016a, 2016b).

According to Table 5.2, rhamnocitrin ($C_{16}H_{12}O_6$) is the 7-methyl ether derivative of kaempferol. It belongs to the class of trihydroxyflavones, specifically the flavonols, and it can also be classified as a monomethoxyflavone. It shares functional similarities with kaempferol (Bhouri et al., 2011). Luteolin ($C_{15}H_{10}O_6$) is a tetrahydroxyflavone with hydroxy groups at positions 3', 4', 5, and 7. Luteolin is utilized in medicine for treating diverse diseases, including hypertension, inflammatory disorders, and cancer. This natural compound has gained increasing attention in modern research for its diverse biological effects, which include anti-inflammatory, antiallergic, and anticancer properties (Lin et al., 2008; Shukla et al., 2019). Quercetin ($C_{15}H_{12}O_7$), on the other hand, is a type of flavonol that, according to the International Union of Pure and Applied Chemistry

TABLE 5.2
Flavonoids present in clove.

Compound	Molecular formula	Molecular weight (g/mol)
Rhamnocitrin	$C_{16}H_{12}O_6$	300.26
Luteolin	$C_{15}H_{10}O_6$	286.24
Quercetin	$C_{15}H_{12}O_7$	302.23
Kumatakenin	$C_{17}H_{14}O_6$	314.29
Rutin	$C_{27}H_{30}O_{16}$	610.5
Myricetin	$C_{15}H_{10}O_8$	318.23
Naringenin	$C_{15}H_{12}O_6$	272.25
Kaempferol	$C_{15}H_{10}O_6$	286.24
Apigenin	$C_{15}H_{10}O_5$	270.24

(IUPAC), has nomenclature of 3, 3′, 4′, 5, 7-pentahydroxyflvanone, indicating the presence of OH groups attached at positions 3, 5, 7, 3′, and 4′. Quercetin has antioxidant and anti-infective properties that can help destroy cancer cells, control blood sugar, and prevent heart disease. Also, quercetin is a potent antioxidant flavonoid that has protective capability against tissue injury originating from diverse drug toxicities (David et al., 2016; Li et al., 2016). Rutin ($C_{27}H_{30}O_{16}$) or 3,3',4',5,7-Pentahydroxyflavone-3-rutinoside is a type of polyphenolic flavonoid found in nature with molecular weight 610.5 g/mol, and it can dissolve in organic solvents. Its stability and bioavailability are low, primarily because of its limited solubility in water. Rutin exhibits significant antioxidant effects due to its ability to reduce various oxidizing species (Negahdari et al., 2021). Myricetin (3,5,7,3′,4′,5′-hexahydroxyflavonol) is a polyhydroxy flavonol compound with the chemical formula $C_{15}H_{10}O_8$, which is easily soluble in polar solvents. Myricetin has been shown to have multiple biological functions. It exhibits a therapeutic effect on various diseases, including different types of tumors, diabetes, Alzheimer's disease, and pathogenic microbial infections (Imran et al., 2021; Song et al., 2021). Naringenin is an important flavonoid belonging to the flavanone category with the chemical formula $C_{15}H_{12}O_6$. Naringenin has various bioactive effects on human health, acting as an antioxidant, free radical scavenger, and modulator of the immune system. Furthermore, this substance has been shown to repair DNA (Salehi et al., 2019a, 2019b). Kaempferol ($C_{15}H_{10}O_6$) belongs to flavonols, which is known as 3,5,7-trihydroxy-2-(4-hydroxyphenyl)-4H-1-benzopyran-4-one (M Calderon-Montano et al., 2011). Apigenin ($C_{15}H_{10}O_5$) formally belongs to the flavone sub-class, which is known as 4′,5,7-trihydroxyflavone. Apigenin is a product of the phenylpropanoid pathway that can be derived from phenylalanine and tyrosine (Salehi et al., 2019a, 2019b). Some important flavonoids present in clove are represented in Figure 5.3.

5.3.3 Phenolic Acids

Phenolic acids are a type of organic compound present in various plant-based foods and drinks with strong antioxidant properties; they shield cells from harm induced by free radicals (Lafay & Gil-Izquierdo, 2008). Phenolic acids possess

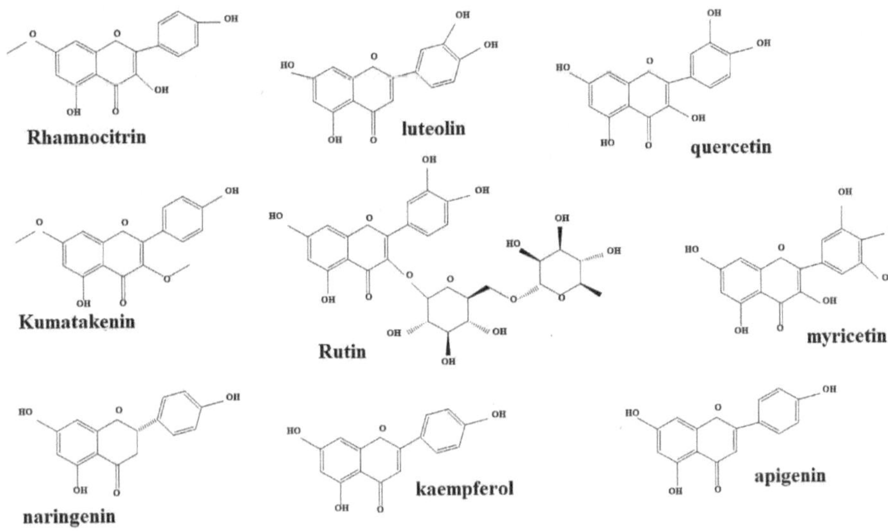

FIGURE 5.3 The main flavonoid compounds of clove (*S. aromaticum L.*).

biological activities such as anti-inflammatory, antimicrobial, and anticancer abilities (Abotaleb et al., 2020; Fu et al., 2021; Sieniawska et al., 2013). Phenolic acids are part of the phenolics category and can be identified by having an aromatic ring containing one or more hydroxyl groups. They can be further classified into hydroxybenzoic acids and hydroxycinnamic acids (Santana-Gálvez & Jacobo-Velázquez, 2018). These compounds are plentiful in different plant materials including fruits, vegetables, whole grains, and organic drinks. Derivatives of phenolic acids are modified forms of these compounds, chemically adjusted to boost bioavailability or biological effects. Esterified derivatives such as methyl gallate and ethyl ferulate have shown promise as pharmaceutical agents, courtesy of their enhanced stability and solubility (Ghasemzadeh & Ghasemzadeh, 2011; Kumar & Goel, 2019; Robbins, 2003). Table 5.3 shows the main phenolic acid compounds present in the clove plant (Charles et al., 1998; Hassan et al., 2021; Yoshimura et al., 2011).

Based on Table 5.3, gallic acid (3,4,5-trihydroxybenzoic acid) is a trihydroxybenzoic acid compound with chemical formula $C_7H_6O_5$. The scientist Carl Wilhelm Scheele discovered gallic acid in plants in the year 1786. One possible method of gallic acid production is a primary conversion of phenylalanine in caffeic acid, as a result in acid 3,4,5 trihydroxycinnamic and consequently in gallic acid. It is found in various natural sources, including fruits (such as grapes, apples, and strawberries), nuts (such as almonds and walnuts), herbs, and green tea (Fernandes & Salgado, 2016; Verma et al., 2013). Catechol (1,2-dihydroxybenzene) is a benzenediol composed of chemical formula $C_6H_6O_2$ (Fiege et al., 2000) while vanillic acid ($C_8H_8O_4$) is a monohydroxybenzoic acid. Vanillic acid is the intermediate compound produced during the bioconversion of ferulic acid to vanillin. Vanillin, when oxidized, forms chlorogenic acid. Vanillic acid also results from the metabolism of caffeic acid (Kaur et al., 2022; Sharma et al., 2020). Caffeic acid ($C_9H_8O_4$)—or 3,4-dihydroxy-cinnamic

TABLE 5.3
The main phenolic acids compounds present in the clove plant.

Compound	Molecular formula	Molecular weight (g/mol)
Gallic acid	$C_7H_6O_5$	170.12
Catechol	$C_6H_6O_2$	110.11
Vanillic acid	$C_8H_8O_4$	168.15
Caffeic acid	$C_9H_8O_4$	180.16
Syringic acid	$C_9H_{10}O_5$	198.17
Ferulic acid	$C_{10}H_{10}O_4$	194.18
Ellagic acid	$C_{14}H_6O_8$	302.19
Salicylic acid	$C_7H_6O_3$	138.12
Cinnamic acid	$C_9H_8O_2$	148.16
Benzoic acid	$C_7H_6O_2$	122.12

acid—is a hydroxycinnamic acid, and plants synthesize caffeic acid and its derivatives through the shikimic acid pathway, involving the deamination of phenylalanine to cinnamic acid and its subsequent transformation (Silva et al., 2014; Yousefian et al., 2020). Syringic acid ($C_9H_{10}O_5$)—or 3,5-dimethoxy-4-hydroxybenzoic acid—belongs to the group of benzoic acids and phenols and it is functionally related to gallic acid (Srinivasulu et al., 2018). Ferulic acid (4-Hydroxy-3-methoxycinnamic acid) is a compound consisting of trans-cinnamic acid with chemical formula $C_{10}H_{10}O_4$. Ferulic acid is known for its various health benefits such as antithrombotic and anticancer properties. Also, ferulic acid helps in preventing heart disease, reducing cholesterol levels, and improving sperm health. Additionally, it is used in the production of vanillin. Moreover, it is included in sports nutrition products and skincare products for its protective properties (Kumar & Pruthi, 2014; Ou & Kwok, 2004).

Ellagic acid ($C_{14}H_6O_8$) is considered a dimeric gallic acid derivative and it has a molecular weight of 302.19 g/mol., which is primarily produced in plants through the hydrolysis of ellagitannins. Ellagic acid is a bioactive polyphenolic compound that is gaining attention because of its antimutagenic and antiproliferative properties. Additionally, it is well-documented for its cardioprotective and hepatoprotective properties (Ríos et al., 2018; Sharifi-Rad et al., 2022). Salicylic acid (2-hydroxybenzoic acid) is a monohydroxybenzoic acid with formula $C_7H_6O_3$. Plants synthesize salicylic acid from cinnamate, which is produced through phenylalanine ammonia-lyase. The main function of salicylic acid is as a signaling molecule in plant defense responses. Additionally, it is commonly used as a food preservative, bactericide, and antiseptic (Chen et al., 2009; Hayat et al., 2013). Cinnamic acid (3-Phenylacrylic acid) is a carboxylic acid with chemical formula $C_9H_8O_2$ and a molecular weight of 148.16 g/mol. Cinnamic acid can be produced by enzymatically removing the amine group from phenylalanine. Several studies have shown that cinnamic acid has potential anticancer, antidiabetic, and neuroprotective properties (Ruwizhi & Aderibigbe, 2020). Similarly, Benzoic acid (benzene carboxylic acid) is a phenolic acid compound with formula C_7H_6O and molecular weight 122.12 g/mol that possesses diverse pharmacological functions (Maki & Takeda, 2000).

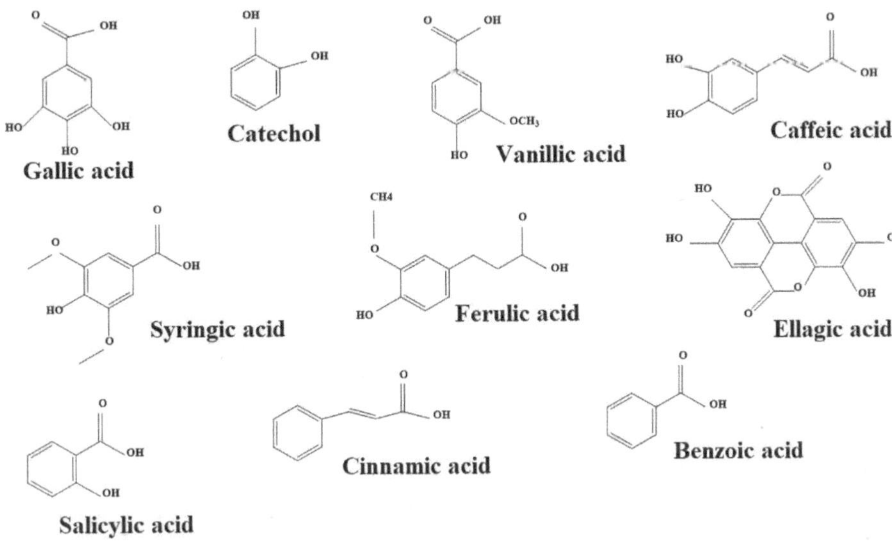

FIGURE 5.4 The main phenolic acid compound of clove (*S. aromaticum L.*).

5.3.4 TANNINS

Tannins are considered polyphenolic compounds with high molecular weight. In aquatic environments, tannins can bind to proteins. This property enables tannins to interact with various other substances such as polysaccharides, nucleic acids, steroids, alkaloids, and saponins (De Bruyne et al., 1999; Kardel et al., 2013). Tannins are essential to plants' defensive systems against insects, birds, and herbivorous mammals. Tannins cause a darker color/pigmentation in the plant due to their specific color imparting property (Adamczyk et al., 2017; Hassanpour et al., 2011). According to studies, tannins have antimicrobial and antioxidant properties and can be beneficial in preventing diseases (Chung et al., 1998; Hagerman et al., 1998; Sharma et al., 2021). Tannins are a broad family of polyphenolic substances categorized into two main groups: condensed tannins and hydrolyzable tannins. Hydrolyzable tannins are esters of gallic acid or ellagic acid with sugar, while they are known as polymers of flavanols, such as catechins and epicatechins (Field & Lettinga, 1992; Geissman, 1963; Okuda & Ito, 2011). Eugenin is the main ellagitannin found in clove flower buds (Nonaka et al., 1980). In a study, researchers identified some hydrolyzable tannins in the aqueous acetone extract of dried flower buds of cloves (Bao LiMing et al., 2012). The compounds isolated in the study were aromatinin A, alienanin B, rugosin C, syzyginin A, rugosin D, euprostin A, rugosein F, squarrosanin A, alunusnin A, and platycaryanin A.

5.4 CONCLUSIONS

Clove is a spice that has been widely used for its aromatic flavor and medicinal properties. Apart from its rich content of macronutrients and micronutrients like potassium, calcium, magnesium, manganese, and zinc, it also contains numerous

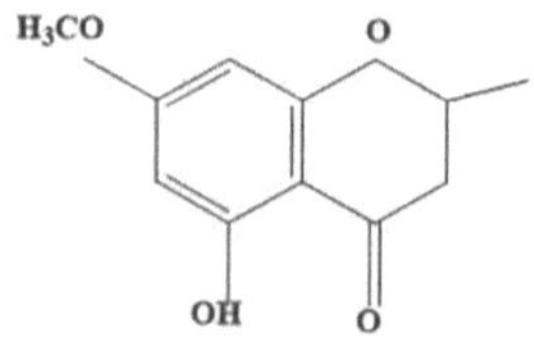

FIGURE 5.5 Eugenin present in clove (*S. aromaticum L.*).

bioactive compounds that contribute to its various biological activities. In this chapter, we have discussed the complete biochemistry and phytochemical composition of clove. One of the key bioactive compounds found in clove is its essential oil, which contains a complex mixture of aromatic compounds. These essential oils give cloves their distinct flavor and aroma and also contribute to their medicinal properties. Some of the major components of clove essential oil include eugenol, eugenyl acetate, and β-caryophyllene. In addition to essential oils, cloves also contain flavonoids, phenolic acids, and tannins, which are all known for their biological effects. Also, the chemical structure and biosynthesis pathway, particularly the shikkimic acid pathway and role of L phenylalanine to produce eugenol (the major pharmacological active constituent of Clove), are covered in this section. Overall, the diverse range of phytochemicals present in clove underscores their potential for various uses in the fields of medicine, nutrition, and cosmetics. Their antioxidant, antimicrobial, and anti-inflammatory properties make them valuable ingredients for natural health products and supplements.

REFERENCES

Abotaleb, M., Liskova, A., Kubatka, P., & Büsselberg, D. (2020). Therapeutic potential of plant phenolic acids in the treatment of cancer. *Biomolecules*, *10*(2), 221. https://doi.org/10.3390/biom10020221.

Adamczyk, B., Simon, J., Kitunen, V., Adamczyk, S., & Smolander, A. (2017). Tannins and their complex interaction with different organic nitrogen compounds and enzymes: Old paradigms versus recent advances. *ChemistryOpen*, *6*(5), 610–614. https://doi.org/10.1002/open.201700113.

Alashti, F. J., Sohbatzadeh, F., Ahmadian, S., Kenari, R. E., & Nazifi, E. (2024). Impact of atmospheric cold plasma pretreatment on morphology, structure, and chemical properties of clove (Syzygium aromaticum). *LWT*, *191*, 115639. https://doi.org/10.1016/j.lwt.2023.115639.

Armenta, S., Garrigues, S., Esteve-Turrillas, F. A., & de la Guardia, M. (2019). Green extraction techniques in green analytical chemistry. *TrAC Trends in Analytical Chemistry*, *116*, 248–253. https://doi.org/10.1016/j.trac.2019.03.016.

Bao LiMing, B. L., Eerdunbayaer, E., Nozaki, A., Takahashi, E., Okamoto, K., Ito, H., & Hatano, T. (2012). Hydrolysable tannins isolated from Syzygium aromaticum: Structure of a new C-glucosidic ellagitannin and spectral features of tannins with a tergalloyl group. https://doi.org/10.3987/COM-11-12392.

Batra, P., & Sharma, A. K. (2013). Anti-cancer potential of flavonoids: Recent trends and future perspectives. *3 Biotech*, *3*, 439–459. https://doi.org/10.1007/s13205-013-0117-5.

Bhouri, W., Sghaier, M. B., Kilani, S., Bouhlel, I., Dijoux-Franca, M.-G., Ghedira, K., & Ghedira, L. C. (2011). Evaluation of antioxidant and antigenotoxic activity of two flavonoids from Rhamnus alaternus L.(Rhamnaceae): Kaempferol 3-O-β-isorhamninoside and rhamnocitrin 3-O-β-isorhamninoside. *Food and Chemical Toxicology*, *49*(5), 1167–1173. https://doi.org/10.1016/j.fct.2011.02.011.

Bhuiyan, M. N. I., Begum, J., Nandi, N. C., & Akter, F. (2010). Constituents of the essential oil from leaves and buds of clove (Syzigium caryophyllatum (L.) Alston). *African Journal of Plant Science*, *4*(11), 451–454.

Carpentieri, S., Soltanipour, F., Ferrari, G., Pataro, G., & Donsì, F. (2021). Emerging green techniques for the extraction of antioxidants from agri-food by-products as promising ingredients for the food industry. *Antioxidants*, *10*(9), 1417. https://doi.org/10.3390/antiox10091417.

Chang, H.-J., Kim, J.-M., Lee, J.-C., Kim, W.-K., & Chun, H. S. (2013). Protective effect of β-caryophyllene, a natural bicyclic sesquiterpene, against cerebral ischemic injury. *Journal of Medicinal Food*, *16*(6), 471–480. https://doi.org/10.1089/jmf.2012.2283.

Charles, R., Garg, S. N., & Kumar, S. (1998). An orsellinic acid glucoside from Syzygium aromatica. *Phytochemistry*, *49*(5), 1375–1376. https://doi.org/10.1016/S0031-9422(98)00091-0.

Chen, Z., Zheng, Z., Huang, J., Lai, Z., & Fan, B. (2009). Biosynthesis of salicylic acid in plants. *Plant Signaling & Behavior*, *4*(6), 493–496. https://doi.org/10.4161/psb.4.6.8392.

Chung, K.-T., Wong, T. Y., Wei, C.-I., Huang, Y.-W., & Lin, Y. (1998). Tannins and human health: A review. *Critical Reviews in Food Science and Nutrition*, *38*(6), 421–464. https://doi.org/10.1080/10408699891274273.

Cotzias, G. C. (1958). Manganese in health and disease. *Physiological Reviews*, *38*(3), 503–532. https://doi.org/10.1152/physrev.1958.38.3.503.

David, A. V. A., Arulmoli, R., & Parasuraman, S. (2016). Overviews of biological importance of quercetin: A bioactive flavonoid. *Pharmacognosy Reviews*, *10*(20), 84. https://doi.org/10.4103%2F0973-7847.194044.

De Bruyne, T., Pieters, L., Deelstra, H., & Vlietinck, A. (1999). Condensed vegetable tannins: Biodiversity in structure and biological activities. *Biochemical Systematics and Ecology*, *27*(4), 445–459. https://doi.org/10.1016/S0305-1978(98)00101-X.

Di Sotto, A., Mancinelli, R., Gullì, M., Eufemi, M., Mammola, C. L., Mazzanti, G., & Di Giacomo, S. (2020). Chemopreventive potential of caryophyllane sesquiterpenes: An overview of preliminary evidence. *Cancers*, *12*(10), 3034. https://doi.org/10.3390/cancers12103034.

Ebrahimi, P., & Lante, A. (2022). Environmentally friendly techniques for the recovery of polyphenols from food by-products and their impact on polyphenol oxidase: A critical review. *Applied Sciences*, *12*(4), 1923. https://doi.org/10.3390/app12041923.

Fernandes, F. H. A., & Salgado, H. R. N. (2016). Gallic acid: Review of the methods of determination and quantification. *Critical Reviews in Analytical Chemistry*, *46*(3), 257–265. https://doi.org/10.1080/10408347.2015.1095064.

Fiege, H., Voges, H. W., Hamamoto, T., Umemura, S., Iwata, T., Miki, H.,. . . Paulus, W. (2000). Phenol derivatives. *Ullmann's Encyclopedia of Industrial Chemistry*. https://doi.org/10.1002/14356007.a19_313.

Field, J., & Lettinga, G. (1992). Toxicity of tannic compounds to microorganisms. In *Plant polyphenols: Synthesis, properties, significance* (pp. 673–692). Springer.

Fu, L., Pei, D., Yu, M., Shang, H., Si, J.-G., Zhang, H.-W.,. . . Zou, Z.-M. (2021). New phenolic acids from the whole herb of Elephantopus scaber Linn. and their anti-inflammatory activity. *Natural Product Research*, *35*(21), 3667–3674. https://doi.org/10.1080/14786419.2020.1723086.

Geissman, T. (1963). Flavonoid compounds, tannins, lignins and, related compounds. In *Comprehensive biochemistry* (pp. 213–250). Elsevier.

Gertsch, J., Leonti, M., Raduner, S., Racz, I., Chen, J.-Z., Xie, X.-Q.,. . . Zimmer, A. (2008). Beta-caryophyllene is a dietary cannabinoid. *Proceedings of the National Academy of Sciences*, *105*(26), 9099–9104. https://doi.org/10.1073/pnas.0803601105.

Ghasemzadeh, A., & Ghasemzadeh, N. (2011). Flavonoids and phenolic acids: Role and biochemical activity in plants and human. *Journal of Medicinal Plants Research*, *5*(31), 6697–6703. https://doi.org/http://doi.org.10.5897/JMPR11.1404.

Golmakani, M.-T., Zare, M., & Razzaghi, S. (2017). Eugenol enrichment of clove bud essential oil using different microwave-assisted distillation methods. *Food Science and Technology Research*, *23*(3), 385–394. https://doi.org/10.3136/fstr.23.385.

Gonzalez-Rivera, J., Duce, C., Campanella, B., Bernazzani, L., Ferrari, C., Tanzini, E.,. . . Tinè, M. R. (2021). In situ microwave assisted extraction of clove buds to isolate essential oil, polyphenols, and lignocellulosic compounds. *Industrial Crops and Products*, *161*, 113203. https://doi.org/10.1016/j.indcrop.2020.113203.

Guan, W., Li, S., Yan, R., Tang, S., & Quan, C. (2007). Comparison of essential oils of clove buds extracted with supercritical carbon dioxide and other three traditional extraction methods. *Food Chemistry*, *101*(4), 1558–1564. https://doi.org/10.1016/j.foodchem.2006.04.009.

Guo, Q., Peng, Q.-Q., Chen, Y.-Y., Song, P., Ji, X.-J., Huang, H., & Shi, T.-Q. (2022). High-yield α-humulene production in Yarrowia lipolytica from waste cooking oil based on transcriptome analysis and metabolic engineering. *Microbial Cell Factories*, *21*(1), 271. https://doi.org/10.1186/s12934-022-01986-z.

Hagerman, A. E., Riedl, K. M., Jones, G. A., Sovik, K. N., Ritchard, N. T., Hartzfeld, P. W., & Riechel, T. L. (1998). High molecular weight plant polyphenolics (tannins) as biological antioxidants. *Journal of Agricultural and Food Chemistry*, *46*(5), 1887–1892. https://doi.org/10.1021/jf970975b.

Haro-González, J. N., Castillo-Herrera, G. A., Martínez-Velázquez, M., & Espinosa-Andrews, H. (2021). Clove essential oil (Syzygium aromaticum L. Myrtaceae): Extraction, chemical composition, food applications, and essential bioactivity for human health. *Molecules*, *26*(21), 6387. https://doi.org/10.3390/molecules26216387.

Hassan, S. M., El-Bebany, A. F., Salem, M. Z., & Komeil, D. A. (2021). Productivity and post-harvest fungal resistance of hot pepper as affected by potassium silicate, clove extract foliar spray and nitrogen application. *Plants*, *10*(4), 662. https://doi.org/10.3390/plants10040662.

Hassanpour, S., MaheriSis, N., & Eshratkhah, B. (2011). Plants and secondary metabolites (Tannins): A review. *International Journal of Forest, Soil and Erosion*, *1*(1), 47–53.

Havsteen, B. H. (2002). The biochemistry and medical significance of the flavonoids. *Pharmacology & Therapeutics*, *96*(2–3), 67–202. https://doi.org/10.1016/S0163-7258(02)00298-X.

Hayat, S., Ahmad, A., & Alyemeni, M. N. (2013). *Salicylic acid*. Springer International Publishing. https://doi.org/10.1007/978-94-007-6428-6.

Hazafa, A., Rehman, K.-U., Jahan, N., & Jabeen, Z. (2020). The role of polyphenol (flavonoids) compounds in the treatment of cancer cells. *Nutrition and Cancer*, *72*(3), 386–397. https://doi.org/10.1080/01635581.2019.1637006.

Heim, K. E., Tagliaferro, A. R., & Bobilya, D. J. (2002). Flavonoid antioxidants: Chemistry, metabolism and structure-activity relationships. *The Journal of Nutritional Biochemistry*, *13*(10), 572–584. https://doi.org/10.1016/S0955-2863(02)00208-5.

Imran, M., Saeed, F., Hussain, G., Imran, A., Mehmood, Z., Gondal, T. A.,. . . Arshad, M. U. (2021). Myricetin: A comprehensive review on its biological potentials. *Food Science & Nutrition*, *9*(10), 5854–5868. https://doi.org/10.1002/fsn3.2513.

Jirovetz, L., Buchbauer, G., Stoilova, I., Stoyanova, A., Krastanov, A., & Schmidt, E. (2006). Chemical composition and antioxidant properties of clove leaf essential oil. *Journal of Agricultural and Food Chemistry, 54*(17), 6303–6307. https://doi.org/10.1021/jf060608c.

Jucá, M. M., Cysne Filho, F. M. S., de Almeida, J. C., Mesquita, D. D. S., Barriga, J. R. D. M., Dias, K. C. F.,. . . Ribeiro, J. E. (2020). Flavonoids: Biological activities and therapeutic potential. *Natural Product Research, 34*(5), 692–705. https://doi.org/10.1080/14786419.2018.1493588.

Kardel, M., Taube, F., Schulz, H., Schütze, W., & Gierus, M. (2013). Different approaches to evaluate tannin content and structure of selected plant extracts-review and new aspects. *Journal of Applied Botany and Food Quality, 86*(1). http://doi.org/10.5073/JABFQ.2013.086.021.

Kaur, J., Gulati, M., Singh, S. K., Kuppusamy, G., Kapoor, B., Mishra, V.,. . . Jha, N. K. (2022). Discovering multifaceted role of vanillic acid beyond flavours: Nutraceutical and therapeutic potential. *Trends in Food Science & Technology, 122*, 187–200. https://doi.org/10.1016/j.tifs.2022.02.023.

Kaur, K., Kaushal, S., & Rani, R. (2019). Chemical composition, antioxidant and antifungal potential of clove (Syzygium aromaticum) essential oil, its major compound and its derivatives. *Journal of Essential Oil Bearing Plants, 22*(5), 1195–1217. https://doi.org/10.1080/0972060X.2019.1688689.

Kumar, N., & Goel, N. (2019). Phenolic acids: Natural versatile molecules with promising therapeutic applications. *Biotechnology Reports, 24*, e00370. https://doi.org/10.1016/j.btre.2019.e00370.

Kumar, N., & Pruthi, V. (2014). Potential applications of ferulic acid from natural sources. *Biotechnology Reports, 4*, 86–93. https://doi.org/10.1016/j.btre.2014.09.002.

Lafay, S., & Gil-Izquierdo, A. (2008). Bioavailability of phenolic acids. *Phytochemistry Reviews, 7*, 301–311. https://doi.org/10.1007/s11101-007-9077-x.

Li, Y., Yao, J., Han, C., Yang, J., Chaudhry, M. T., Wang, S.,. . . Yin, Y. (2016). Quercetin, inflammation and immunity. *Nutrients, 8*(3), 167. https://doi.org/10.3390/nu8030167.

Lin, Y., Shi, R., Wang, X., & Shen, H.-M. (2008). Luteolin, a flavonoid with potential for cancer prevention and therapy. *Current Cancer Drug Targets, 8*(7), 634–646. https://doi.org/10.2174%2F156800908786241050.

Machado, K. D. C., Islam, M. T., Ali, E. S., Rouf, R., Uddin, S. J., Dev, S.,. . . Das, A. K. (2018). A systematic review on the neuroprotective perspectives of beta-caryophyllene. *Phytotherapy Research, 32*(12), 2376–2388. https://doi.org/10.1002/ptr.6199.

Maki, T., & Takeda, K. (2000). Benzoic acid and derivatives. *Ullmann's Encyclopedia of Industrial Chemistry, 3*. https://doi.org/10.1002/14356007.a03_555.

Marchese, A., Barbieri, R., Coppo, E., Orhan, I. E., Daglia, M., Nabavi, S. F.,. . . Ajami, M. (2017). Antimicrobial activity of eugenol and essential oils containing eugenol: A mechanistic viewpoint. *Critical Reviews in Microbiology, 43*(6), 668–689. https://doi.org/10.1080/1040841X.2017.1295225.

Masuoka, N., Matsuda, M., & Kubo, I. (2012). Characterisation of the antioxidant activity of flavonoids. *Food Chemistry, 131*(2), 541–545. https://doi.org/10.1016/j.foodchem.2011.09.020.

M Calderon-Montano, J., Burgos-Morón, E., Pérez-Guerrero, C., & López-Lázaro, M. (2011). A review on the dietary flavonoid kaempferol. *Mini Reviews in Medicinal Chemistry, 11*(4), 298–344. https://doi.org/10.2174/138955711795305335.

Negahdari, R., Bohlouli, S., Sharifi, S., Maleki Dizaj, S., Rahbar Saadat, Y., Khezri, K.,. . . Raeesi, S. (2021). Therapeutic benefits of rutin and its nanoformulations. *Phytotherapy Research, 35*(4), 1719–1738. https://doi.org/10.1002/ptr.6904.

Nonaka, G.-I., Harada, M., & Nishioka, I. (1980). Eugeniin, a new ellagitannin from cloves. *Chemical and Pharmaceutical Bulletin*, *28*(2), 685–687. https://doi.org/10.1248/cpb.28.685.

Okuda, T., & Ito, H. (2011). Tannins of constant structure in medicinal and food plants—hydrolyzable tannins and polyphenols related to tannins. *Molecules*, *16*(3), 2191–2217. https://doi.org/10.3390/molecules16032191.

O'Neal, S. L., & Zheng, W. (2015). Manganese toxicity upon overexposure: A decade in review. *Current Environmental Health Reports*, *2*, 315–328. https://doi.org/10.1007/s40572-015-0056-x.

Ou, S., & Kwok, K. C. (2004). Ferulic acid: Pharmaceutical functions, preparation and applications in foods. *Journal of the Science of Food and Agriculture*, *84*(11), 1261–1269. https://doi.org/10.1002/jsfa.1873.

Ouadi, S., Sierro, N., Goepfert, S., Bovet, L., Glauser, G., Vallat, A.,. . . Ivanov, N. V. (2022). The clove (Syzygium aromaticum) genome provides insights into the eugenol biosynthesis pathway. *Communications Biology*, *5*(1), 684. https://doi.org/10.1038/s42003-022-03618-z.

Panche, A. N., Diwan, A. D., & Chandra, S. R. (2016). Flavonoids: An overview. *Journal of Nutritional Science*, *5*, e47. https://doi.org/10.1017/jns.2016.41.

Picot-Allain, C., Mahomoodally, M. F., Ak, G., & Zengin, G. (2021). Conventional versus green extraction techniques—A comparative perspective. *Current Opinion in Food Science*, *40*, 144–156. https://doi.org/10.1016/j.cofs.2021.02.009.

Rice-Evans, C. (2001). Flavonoid antioxidants. *Current Medicinal Chemistry*, *8*(7), 797–807. https://doi.org/10.2174/0929867013373011.

Ríos, J.-L., Giner, R. M., Marín, M., & Recio, M. C. (2018). A pharmacological update of ellagic acid. *Planta Medica*, *84*(15), 1068–1093. https://doi.org/http://doi.org.10.1055/a-0633-9492.

Robbins, R. J. (2003). Phenolic acids in foods: An overview of analytical methodology. *Journal of Agricultural and Food Chemistry*, *51*(10), 2866–2887. https://doi.org/10.1021/jf026182t.

Rodriguez Garcia, S. L., & Raghavan, V. (2022). Green extraction techniques from fruit and vegetable waste to obtain bioactive compounds—A review. *Critical Reviews in Food Science and Nutrition*, *62*(23), 6446–6466. https://doi.org/10.1080/10408398.2021.1901651.

Ruwizhi, N., & Aderibigbe, B. A. (2020). Cinnamic acid derivatives and their biological efficacy. *International Journal of Molecular Sciences*, *21*(16), 5712. https://doi.org/10.3390%2Fijms21165712.

Ryu, B., Kim, H. M., Lee, J. S., Lee, C. K., Sezirahiga, J., Woo, J.-H.,. . . Jang, D. S. (2016a). New flavonol glucuronides from the flower buds of Syzygium aromaticum (Clove). *Journal of Agricultural and Food Chemistry*, *64*(15), 3048–3053. https://doi.org/10.1021/acs.jafc.6b00337.

Ryu, B., Kim, H. M., Woo, J.-H., Choi, J.-H., & Jang, D. S. (2016b). A new acetophenone glycoside from the flower buds of Syzygium aromaticum (cloves). *Fitoterapia*, *115*, 46–51. https://doi.org/10.1016/j.fitote.2016.09.021.

Salehi, B., Fokou, P. V. T., Sharifi-Rad, M., Zucca, P., Pezzani, R., Martins, N., & Sharifi-Rad, J. (2019a). The therapeutic potential of naringenin: A review of clinical trials. *Pharmaceuticals*, *12*(1), 11. https://doi.org/10.3390/ph12010011.

Salehi, B., Venditti, A., Sharifi-Rad, M., Kręgiel, D., Sharifi-Rad, J., Durazzo, A.,. . . Novellino, E. (2019b). The therapeutic potential of apigenin. *International Journal of Molecular Sciences*, *20*(6), 1305. https://doi.org/10.3390/ijms20061305.

Santana-Gálvez, J., & Jacobo-Velázquez, D. A. (2018). Classification of phenolic compounds. In *Phenolic compounds in food* (pp. 3–20). CRC press.

Santolin, L., Fiametti, K. G., da Silva Lobo, V., Wancura, J. H., & Oliveira, J. V. (2021). Enzymatic synthesis of eugenyl acetate from essential oil of clove using lipases in liquid formulation as biocatalyst. *Applied Biochemistry and Biotechnology*, *193*(11), 3512–3527. https://doi.org/10.1007/s12010-021-03610-z.

Schnarr, L., Segatto, M. L., Olsson, O., Zuin, V. G., & Kümmerer, K. (2022). Flavonoids as biopesticides–Systematic assessment of sources, structures, activities and environmental fate. *Science of the Total Environment*, *824*, 153781. https://doi.org/10.1016/j.scitotenv.2022.153781.

Sharifi-Rad, J., Quispe, C., Castillo, C. M. S., Caroca, R., Lazo-Vélez, M. A., Antonyak, H.,. . . De Masi, L. (2022). Ellagic acid: A review on its natural sources, chemical stability, and therapeutic potential. *Oxidative Medicine and Cellular Longevity*, *2022*. https://doi.org/10.1155%2F2022%2F3848084.

Sharma, C., M Al Kaabi, J., M Nurulain, S., N Goyal, S., Amjad Kamal, M., & Ojha, S. (2016). Polypharmacological properties and therapeutic potential of β-caryophyllene: A dietary phytocannabinoid of pharmaceutical promise. *Current Pharmaceutical Design*, *22*(21), 3237–3264. https://doi.org/10.2174/1381612822666160311115226.

Sharma, K., Kumar, V., Kaur, J., Tanwar, B., Goyal, A., Sharma, R.,. . . Kumar, A. (2021). Health effects, sources, utilization and safety of tannins: A critical review. *Toxin Reviews*, *40*(4), 432–444. https://doi.org/10.1080/15569543.2019.1662813.

Sharma, N., Tiwari, N., Vyas, M., Khurana, N., Muthuraman, A., & Utreja, P. (2020). An overview of therapeutic effects of vanillic acid. *Plant Arch*, *20*(2), 3053–3059.

Shukla, R., Pandey, V., Vadnere, G. P., & Lodhi, S. (2019). Role of flavonoids in management of inflammatory disorders. In *Bioactive food as dietary interventions for arthritis and related inflammatory diseases* (pp. 293–322). Elsevier.

Sieniawska, E., Baj, T., Los, R., Skalicka-Wozniak, K., Malm, A., & Glowniak, K. (2013). Phenolic acids content, antioxidant and antimicrobial activity of Ligusticum mutellina L. *Natural Product Research*, *27*(12), 1108–1110. https://doi.org/10.1080/14786419.2012.698413.

Silva, M. M., Santos, M. R., Caroço, G., Rocha, R., Justino, G., & Mira, L. (2002). Structure-antioxidant activity relationships of flavonoids: A re-examination. *Free Radical Research*, *36*(11), 1219–1227. https://doi.org/10.1080/198-1071576021000016472.

Silva, T., Oliveira, C., & Borges, F. (2014). Caffeic acid derivatives, analogs and applications: A patent review (2009–2013). *Expert Opinion on Therapeutic Patents*, *24*(11), 1257–1270. https://doi.org/10.1517/13543776.2014.959492.

Singh, P. A., Bajwa, N., & Baldi, A. (2021). Possible role of traditional systems of medicine to manage COVID-19: A review. *Israel Journal of Plant Sciences*, *68*(1–2), 3–28.

Singh, P. A., Tushir, S., Hazra, S., Purewal, S. S., & Agarwal, G. (2024). Novel approaches for the extraction and identification of phytoconstituents from herbs and spices. *The Natural Products Journal*. http://doi.org/10.2174/0122103155293641240417072907.

Song, X., Tan, L., Wang, M., Ren, C., Guo, C., Yang, B.,. . . Pei, J. (2021). Myricetin: A review of the most recent research. *Biomedicine & Pharmacotherapy*, *134*, 111017. https://doi.org/10.1016/j.biopha.2020.111017.

Soto-Cabrera, D., Salazar, J. R., Nogueda-Gutiérrez, I., Torres-Olvera, M., Cerón-Nava, A., Rosales-Guevara, J.,. . . Rosas-Acevedo, H. (2016). Quantification of polyphenols and flavonoid content and evaluation of anti-inflammatory and antimicrobial activities of Stenocereus stellatus extracts. *Natural Product Research*, *30*(16), 1885–1889. https://doi.org/10.1080/14786419.2015.1084302.

Srinivasulu, C., Ramgopal, M., Ramanjaneyulu, G., Anuradha, C., & Kumar, C. S. (2018). Syringic acid (SA)–a review of its occurrence, biosynthesis, pharmacological and industrial importance. *Biomedicine & Pharmacotherapy*, *108*, 547–557. https://doi.org/10.1016/j.biopha.2018.09.069.

Tammannavar, P., Pushpalatha, C., Jain, S., & Sowmya, S. (2013). An unexpected positive hypersensitive reaction to eugenol. *Case Reports*, *2013*, bcr2013009464. https://doi.org/10.1136/bcr-2013-009464.

Tao, H., Li, L., He, Y., Zhang, X., Zhao, Y., Wang, Q., & Hong, G. (2024). Flavonoids in vegetables: Improvement of dietary flavonoids by metabolic engineering to promote health. *Critical Reviews in Food Science and Nutrition*, *64*(11), 3220–3234. https://doi.org/10.1080/10408398.2022.2131726.

Tunç, M. T., & Koca, İ. (2019). Ohmic heating assisted hydrodistillation of clove essential oil. *Industrial Crops and Products*, *141*, 111763. https://doi.org/10.1016/j.indcrop.2019.111763.

Ulanowska, M., & Olas, B. (2021). Biological properties and prospects for the application of eugenol—a review. *International Journal of Molecular Sciences*, *22*(7), 3671. https://doi.org/10.3390/ijms22073671.

Vassão, D. G., Gang, D. R., Koeduka, T., Jackson, B., Pichersky, E., Davin, L. B., & Lewis, N. G. (2006). Chavicol formation in sweet basil (Ocimum basilicum): Cleavage of an esterified C9 hydroxyl group with NAD (P) H-dependent reduction. *Organic & Biomolecular Chemistry*, *4*(14), 2733–2744. https://doi.org/10.1039/B605407B.

Verma, S., Singh, A., & Mishra, A. (2013). Gallic acid: Molecular rival of cancer. *Environmental Toxicology and Pharmacology*, *35*(3), 473–485. https://doi.org/10.1016/j.etap.2013.02.011.

Weaver, C. M. (2013). Potassium and health. *Advances in Nutrition*, *4*(3), 368S–377S. https://doi.org/10.3945/an.112.003533.

Wei, M.-C., Xiao, J., & Yang, Y.-C. (2016). Extraction of α-humulene-enriched oil from clove using ultrasound-assisted supercritical carbon dioxide extraction and studies of its fictitious solubility. *Food Chemistry*, *210*, 172–181. https://doi.org/10.1016/j.foodchem.2016.04.076.

Xu, J.-G., Liu, T., Hu, Q.-P., & Cao, X.-M. (2016). Chemical composition, antibacterial properties and mechanism of action of essential oil from clove buds against Staphylococcus aureus. *Molecules*, *21*(9), 1194. https://doi.org/10.3390/molecules21091194.

Yang, D., Michel, L., Chaumont, J.-P., & Millet-Clerc, J. (2000). Use of caryophyllene oxide as an antifungal agent in an in vitro experimental model of onychomycosis. *Mycopathologia*, *148*(2), 79–82. https://doi.org/10.1023/a:1007178924408.

Yang, J., Li, Z., Guo, L., Du, J., & Bae, H.-J. (2016). Biosynthesis of *β*-caryophyllene, a novel terpene-based high-density biofuel precursor, using engineered Escherichia coli. *Renewable Energy*, *99*, 216–223. https://doi.org/10.1016/j.renene.2016.06.061.

Yoshimura, M., Amakura, Y., & Yoshida, T. (2011). Polyphenolic compounds in clove and pimento and their antioxidative activities. *Bioscience, Biotechnology, and Biochemistry*, *75*(11), 2207–2212. https://doi.org/10.1271/bbb.110491.

Yousefian, S., Lohrasebi, T., Farhadpour, M., & Haghbeen, K. (2020). Effect of methyl jasmonate on phenolic acids accumulation and the expression profile of their biosynthesis-related genes in Mentha spicata hairy root cultures. *Plant Cell, Tissue and Organ Culture (PCTOC)*, *142*(2), 285–297. https://doi.org/10.1007/s11240-020-01856-9.

Zari, A. T., Zari, T. A., & Hakeem, K. R. (2021). Anticancer properties of eugenol: A review. *Molecules*, *26*(23), 7407. https://doi.org/10.3390/molecules26237407.

6 Phytochemical Screening

Hernani, Christina Winarti, Nurliani Bermawi, Iceu Agustinisari, and Sri Wahyuni

6.1 INTRODUCTION

Clove (*Syzygium aromaticum* L.) is indigenous to southeast Asia and is extensively grown globally, focusing on cultivation in India, Sri Lanka, Indonesia, Madagascar, Zanzibar, Philippines, Tanzania, and Brazil. The plant is a perennial tropical species that can reach a height of 50 feet, featuring prominent glossy green leaves. The plant bears white flowers that have a pleasant scent. These flowers have a fleshy hypanthium, which is encircled by sepals. Additionally, the plant produces a purple drupe fruit. The primary products derived from clove include whole or crushed clove buds, clove bud essential oil, clove stem essential oil, clove leaf essential oil, and oleoresin (Hossain & Shahidi, 2018). Cloves contain an abundance of secondary metabolite compounds, commonly known as phytochemistry, that have pharmacological effects on health. Cloves contain an abundance of secondary metabolite compounds that have pharmacological effects on health. The parts of the clove plant commonly utilized are the buds, leaves and twigs (Figure 6.1). Flower buds contain about 10–20% essential oil, while clove leaves and twigs contain about 1–4% and 5–10% essential oil, respectively. Clove oil is a highly utilized essential oil in the culinary, pharmaceutical, and fragrance sectors. (Guntarti et al., 2023). These uses are also the case with eugenol, which has many uses, such as perfumes, flavorings, essential oils, and medicine as a local antiseptic and anesthetic. Clove flower buds contain 15–20% of essential oil, which consists

FIGURE 6.1 Part of clove: flower bud, stalk/twigs and leaves.

DOI: 10.1201/9781003521785-6

of eugenol (70–85%), eugenyl acetate (15%), and β-caryophyllene (5–12%) (Mittal et al., 2014). The quantity and makeup of clove essential oil mostly depend on factors such as pre-treatments, variety, agro-ecological conditions, and extraction processes (Batiha et al., 2020). The phytoconstituents of clove consist of a diverse range of chemical components, including monoterpenes, sesquiterpenes, phenolics, and hydrocarbon compounds. Other phytochemical compounds contained in clove oil are vanillin, crategolic acid, tannins, gallotannic acid, methyl salicylate, flavonoids eugenin, kaempferol, rhamnetin, eugenitin, and triterpenoids like oleanolic acid (Mittal et al., 2014; Usman & Bashir, 2022)

Eugenol is the primary constituent of the essential oil, comprising around 72–90% of its content. (Lumingkewas & Unitly, 2023). Twigs contain the highest eugenol at 90–98%, followed by flower buds (85–89%) and leaves (80–88%). Eugenol is a clove marker compound responsible for cloves' specific aroma. Every part of the clove plant has benefits and uses as a spice in the pharmaceutical, cosmetic, or insecticide industries (Figure 6.2). In Indonesia, clove flowers, generally used as spices and cigarettes, leaves, and stems contain an essential oil with a specific component, whereas methyl eugenol is known as mosquito repellent. Clove has pharmacological properties such as being antibacterial, anti-inflammatory, anesthetic, analgesic, antioxidant, and anticancer. Other activities of this substance include being an aphrodisiac, mosquito repellent, insecticidal, and antipyretic (Mittal et al., 2014).

In the past, every productive part of the plant was extracted, and the results were used directly as raw materials for drugs/pharmaceuticals. However, with the development of isolation and identification techniques, the chemical compounds are produced synthetically (Kocabas, 2017; Shaikh & Patil, 2020). The advancement of the pharmaceutical sector has stimulated various studies on the screening and identification of phytochemical compounds with pharmacological activity. Characterizing and evaluating plants and their phytoconstituents can provide data that support the medicinal qualities of these plants in the treatment of various ailments (Shaikh & Patil, 2020). The need for information on phytochemicals is also driven by the current trend of interest in developing discoveries derived from bioactive natural products found in traditional medicines (Atanasov et al., 2021).

Phytochemical screening is a process used to qualitatively analyze and identify several classes of active chemical ingredients, including phenols, alkaloids, flavonoids, flavones aglycones, coumarins, saponins, tannins, steroids, glycosides, and triterpenoids (El Faqer et al., 2022; Madubuike et al., 2018). The diverse secondary metabolites in these plant parts can be isolated and identified using various analytical techniques such as chromatography and colour reaction with specific reagents, as well as qualitative and quantitative use standard procedures.

Some of the methods commonly used for qualitative screening, including chemical and quantitative compound profiling, are analysis by high-performance liquid chromatography (HPLC), thin layer chromatography (TLC), high performance thin layer chromatography (HPTLC), gas chromatography (GC), paper chromatography, optimum performance laminar chromatography (OPLC), Fourier transform infrared (FTIR), and nuclear magnetic resonance (NMR) (Coskun, 2016; Ingle et al., 2017; Shaikh & Patil, 2020).

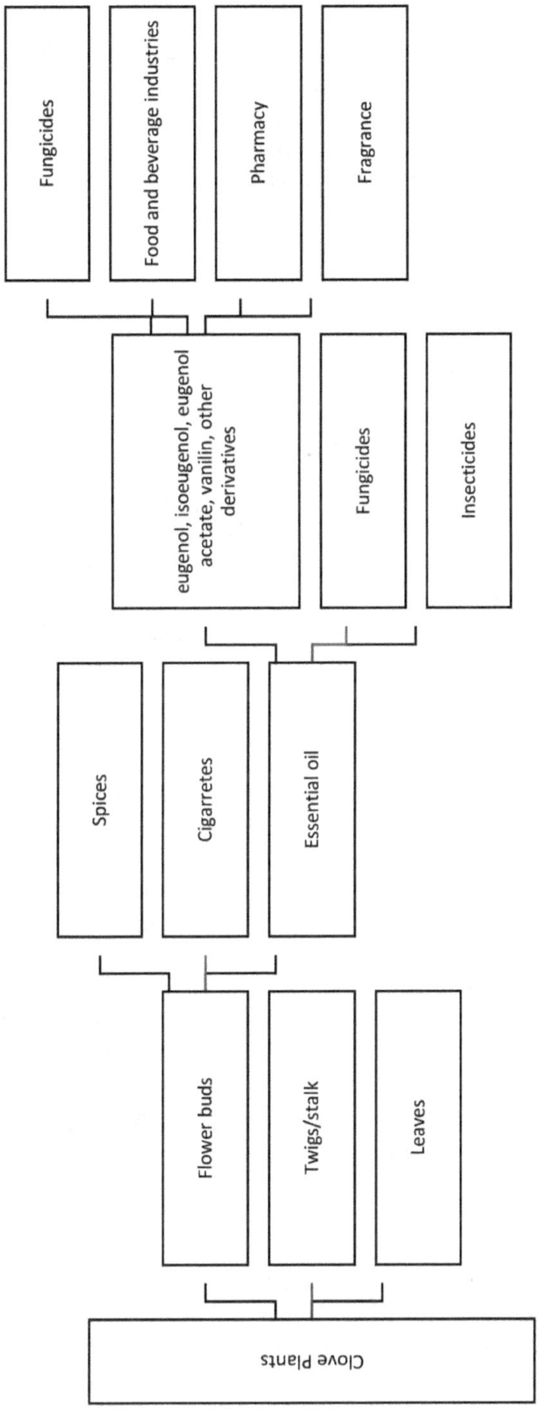

FIGURE 6.2 Uses of clove plant parts.

6.2 CHEMICAL COMPOUNDS

Phytochemicals are categorized as primary or secondary metabolites based on their function in plant metabolism. Primary metabolites encompass carbohydrates, amino acids, proteins, nucleic acids, chlorophyll, and other compounds. Plants can produce various secondary metabolites such as saponins, tannins, phenols, and alkaloids because these compounds provide certain plant activities (Patil et al., 2013; Singh et al., 2024; Choudhury et al., 2023). In addition, several compounds included in secondary metabolites are plant phenols, flavonoids, flavones aglycones, coumarins, steroids, glycosides and terpenoids (El Faqer et al., 2022).

6.2.1 Alkaloids

Alkaloids contain heterocyclic nitrogen atoms that have essential characteristics. This is also the basis for naming alkaloids. Alkaloids are essential for plant survival since they serve as a defense mechanism against microbes (via antibacterial and antifungal activities), insects, and herbivores (by acting as feeding deterrents). They also help plants defend themselves against other plants through allelopathy (Molyneux et al., 1996). Based on many studies, alkaloids are known to have some therapeutic effects, namely as anesthesia, in being cardioprotective, and as an anti-inflammatory (Heinrich et al., 2021), antihypertensive (indole alkaloids), antiarrhythmic, antimalaria, and anticancer (Wink et al., 1998).

6.2.2 Phenolic Compounds

Phenolic compounds are the most prevalent group of phytochemicals that are found extensively in plants such as cloves (Walton et al., 2003; Singh et al., 2016). Phenolic compounds consist of hydroxyl groups that bind directly to aromatic hydrocarbon groups. These chemical compounds have a crucial role as defense chemicals in plants. As a result, these chemicals also play many important roles that promote human well-being, such as acting as antioxidants. Antioxidants are protective agents that fight free radicals, which are the precursor to developing dangerous diseases in the human body. There are three crucial phenolic groups: flavonoids, phenolic acids and polyphenols (Koche et al., 2016). The type of high-concentration phenolic acid found in clove is gallic acid and its derivatives, such as hydrolyzable tannins. Other types of phenolic acids found in the clove are ellagic, ferulic, caffeic, and salicylic acids (Cheikhyoussef et al., 2022).

6.2.3 Flavonoids

Flavonoids represent the most extensive category of plant phenols and have received the most extensive research attention (Dai & Mumper, 2010). Anthocyanins, flavones, flavonols, flavanones, isoflavones, flavanonols, and other subclasses are the most commonly found types of flavonoids in nature (Zhang et al., 2022). Flavonoids such as kaempferol, quercetin, and their glycosylated derivatives are present in clove, albeit in lower amounts. The main phenolic compounds found include gallic acid, flavonol glycosides, phenolic volatile oils, such as eugenol and acetyl eugenol, and tannins. The remarkable capacity of clove as an effective radical scavenger and as a commercially feasible supplier of polyphenols was underscored (Cortés-Rojas et al., 2014).

6.2.4 Terpenoids

This class consists of natural compounds produced from isoprene units consisting of five carbon atoms. Natural lipids are present in all living organisms and are the most extensive secondary metabolites (Elbein & Molyneux, 1999). Terpenes and terpenoids are constituents of clove essential oil. These compounds are often used as flavours in food and fragrances in cosmetics (Harborne & Tomás-Barberán, 1992; Koche et al., 2016). Terpenes and terpenoids also have biological activities that benefit health: anticancer, antimicrobial, anti-inflammatory, antioxidant and antiallergic properties (Masyita et al., 2022).

6.2.5 Saponins

As a collective, clove saponins mostly consist of glycosylated steroids, triterpenoids, and steroid alkaloids at a chemical level (Bohlmann et al., 1998). They enhance the texture and stability of food products and possess possible health advantages, such as reducing cholesterol levels and exhibiting anticancer, anti-inflammatory, antibacterial, antiviral, and antiparasitic properties, among others. This compound also exhibits cytotoxicity against cancer cell lines and can function as an adjuvant, augmenting the immunological response to vaccines (Timilsena et al., 2023).

6.2.6 Tannins

Tannins, also known as tannic acid, belong to the group of water-soluble polyphenols that are also reported to be present in clove. This type of phytochemical compound has the potential to be anticarcinogenic and antimutagenic. This property is likely related to its antioxidant properties. Tannins also have antimicrobial activity, inhibiting the growth of several fungi, yeast, bacteria and viruses. Additionally, these compounds have been documented to have many physiological effects, including hastening blood coagulation, lowering blood pressure, reducing serum cholesterol levels, inducing liver necrosis, and influencing immunological responses (Chung et al., 1998).

6.3 PHYTOCHEMICAL SCREENING

Phytochemical screening is usually conducted using a standard method using specific reagents. Commonly, it is used to determine classed alkaloids, glycosides, terpenoids, flavonoids, phenols, tannins, and steroids. The abundance of phytochemical substances varies based on the specific solvent employed during the extraction procedure. The degree of solvent polarity affects the type and polarity of the extracted chemical compounds (Madubuike et al.,2018). Identification of phytochemical compounds qualitatively requires a standard procedure with specific reagents. The observed hue indicates the existence of the designated category of chemicals. The assessment is carried out if the colour during the reaction looks very strong and is given a positive value of 4 (+++), the colour is slightly less intense (++), or slightly weak (+) (Safeena & Kalinga, 2020).

Qualitative screening of phytochemical compounds can be done using specific chemicals. Several chemical compounds can be screened based on the methodologies used by researchers (Edeoga et al., 2005; Ogwuda et al., 2023), namely tannins, phlobatanins, saponins, steroids, terpenoids, flavonoids, alkaloids, and glycosides. This qualitative test is based on chemical reactions observed with specific colour changes. The following are chemicals or reagents and the colour changes observed. Some researchers also include carbohydrate, protein and lipid tests to complement phytochemical screening. Extract samples used in this test can be extracted during the extraction process. Preliminary qualitative screening is usually followed by further detection of phytochemical profiling using liquid chromatography mass spectroscopy (LCMS) or gas chromatography-mass spectroscopy (GCMS) for volatile compounds.

6.3.1 Preliminary Qualitative Screening

6.3.1.1 Test for Carbohydrates

Carbohydrates are present in clove, and, according to Lone and Jain, carbohydrates in a plant extract can be known through three types of tests with different reagents. The reaction of the extract with these reagents produces a particular colour change (Lone & Jain, 2022).

6.3.1.1.1 Molisch's Test

This test uses Molisch's reagent, which must be prepared fresh by dissolving 3.75 g of α-naphthol in 25 mL of 99% ethanol. A total of 1 mL of plant extract is added to 0.4 mL of Molisch's reagent. Subsequently, concentrated sulfuric acid was introduced. The principle of this test is based on the reaction between concentrated acid, which catalyzes the dehydration of sugar to form furfural (from pentose) or hydroxymethylfurfural (from hexose). The aldehydes condense with naphthol molecules to form a blue or purple ring on the interfacial layer of acid and test solution (Nagalingam et al., 2012).

6.3.1.1.2 Benedict's Test

This assay detects reducing sugars, which include monosaccharides and some disaccharides that have unbound ketone or aldehyde functional groups. Benedict's reagent is a deep-blue alkaline solution made by reacting anhydrous sodium carbonate (100 g), sodium citrate (173 g), and copper(II) sulfate pentahydrate (17.3 g). The test procedure is to react 1 mL of plant extract with 1 mL of Benedict's reagent, then heat it for 5 minutes. The formation of an orange precipitate indicates the presence of carbohydrates (disaccharides). Reducing sugars from carbohydrates reduces the copper (II) ions from Benedict's solution to copper (I) ions, causing a colour change. Copper (I) ions are insoluble in water, forming a precipitate. The higher the concentration of reducing sugar, the closer the colour formed is to the final colour of brick-red and the more precipitate is formed (Hernández-López et al., 2020).

6.3.1.1.3 *Fehling's Test*

This test is to distinguish between reducing sugars and non-reducing sugars). The working principle of this test is the reaction of reducing sugars in carbohydrates (carbonyl, aldehyde or ketone groups) with Fehling solution. The reduction of sugar will reduce copper (II) ions to cuprous (I) oxide (Cu2O), which is insoluble and forms yellow or red precipitation. Fehling solution consists of Fehling A and Fehling B. Fehling A solution is made by dissolving 7 g CuSO4.7H2O in 100 mL water. Fehling B is obtained by dissolving 24 g KOH and 34.6 g potassium sodium tartrate in 100 mL water. The test procedure is to boil and filter the extract with the addition of water. Then, the Fehling solution is added to the filtrate. The formation of a reddish-brown precipitate indicates the presence of carbohydrates (Scales, 1915).

6.3.1.2 Test for Proteins

Cloves are reported to possess proteins and, similar to the test for carbohydrates, a positive test result is indicated by a change in colour (Lone & Jain, 2022).

6.3.1.2.1 *Xanthoproteic Test*

This protein assay is based on detecting amino acids contained in phenolic or indolic groups or aromatic amino acids, such as tyrosine and tryptophan. A total of 0.25 mL of nitric acid was reacted with 1 mL of plant extract, then 1 mL of 40% NaOH solution was added to the mixture, and the colour change was observed. The aromatic groups in the amino acids contained in the plant extract will be nitrated by concentrated nitric acid, which produces a yellow colour. Adding an alkaline solution causes an orange colour change due to salt formation in the tautomeric form of the nitro compound (Sudha et al., 2020).

6.3.1.2.2 *Biuret Test*

This test aims to identify the existence of peptide bonds in the given material. Peptide bonds are covalent bonds between carbon atom number 1 (C1) of amino acids and nitrogen number 2 (N2) of adjacent amino acids. The test procedure is to react 1 mL of plant extract with 4% NaOH and 1% $CuSO_4$. The reaction causes hydrogen atoms to be released, and then dipeptide bonds are formed between each amino acid's carbon and nitrogen atoms. Cupric ions can use the free electrons in the peptide in the biuret reagent. In an alkaline environment, the existence of protein is showed by the production of violet-pink or purple colour (Martina & Vojtech, 2015).

6.3.1.3 Test for Lipids

The various techniques used to test lipids in clove include solubility testing, translucent spot testing, emulsification testing, saponification testing, Sudan IV testing, acrolein testing, dichromate testing, testing for free fatty acids, unsaturation testing, Burchard testing, and glycerol testing. This chapter only outlines three types of assays for lipid detection in cloves (Mumtaz et al., 2020).

6.3.1.3.1 Solubility Test

This assay is designed to identify the presence of all lipids. The underlying premise of this test is predicated on the solubility of lipids in different solvents. Lipids exhibit high solubility in chloroform, moderate solubility in polar solvents like ethanol, and negligible solubility in water. The extract is distributed among three test tubes, and subsequently chloroform, ethanol, and water solvents are introduced into each tube. The solution is agitated and left undisturbed for one minute, after which its solubility is examined.

6.3.1.3.2 Sudan IV Test

This test relies on the affinity and solubility of lipids in non-polar substances. Sudan IV is a dye that does not have a polarity. In the presence of lipids, the dye will form a chemical bond with the lipids, forming a red-orange colour (https://biologyreader.com/qualitative-analysis-of-lipids.html).

6.3.1.3.3 Dichromate Test

This test identifies glycerol by utilizing the notion of an oxidation reaction. Glycerol-containing samples will undergo a chemical reaction with dichromate ions, resulting in the formation of a solution with a brown colour. Furthermore, the chromic ions oxidize the glycerol until the colour of the solution turns blue. The sample does not contain glycerol if the solution does not turn blue. A few drops of 5% potassium dichromate solution were added into 2–3 mL of extract, then 5 mL concentrated nitric acid was added (https://biologyreader.com/qualitative-analysis-of-lipids.html). The glycerol test can also be done by reacting the sample using $CuSO_4.5H_20$ solution and adding a little NaOH solution. Test results are positive if the reaction produces a clear blue solution (Lone & Jain, 2022).

6.3.1.4 Test for Alkaloids

Alkaloids are nitrogen-containing chemical compounds commonly found in plants and animals. There are six types of alkaloid detection tests with different reagents used, namely Mayer's test, Wagner's test, Hager's test, Dragendroff's test, tannic acid test, and Murexide test (Mir et al., 2016). Three standard test methods are as follows:

6.3.1.4.1 Mayer's Test

The preparation of Mayer's reagent involved the dissolution of mercuric chloride and potassium iodide in water. The reagent (1 mL) is added to the plant extract (1 mL) and shaken. The formation of a white/cream-colored precipitate indicates a positive reaction results. This precipitate is a complex compound due to the reaction between electrons in alkaloid nitrogen atoms covalently bonded with potassium metal ions.

6.3.1.4.2 Wagner's Test

Wagner's reagent is made by dissolving iodine and potassium iodide in distilled water. The test procedure is to mix the plant extract with HCl solution and add a few drops of Wagner's reagent. Coordinate covalent bonds between potassium ions and nitrogen contained in alkaloids produce complex compounds in reddish-brown colored precipitates.

6.3.1.4.3 Dragendorff's Test

Dragendroff's reagent, also known as iodobismuthate, is a solution of bismuth nitrate in acid. The acids commonly used are acetic acid, tartaric acid, and hydrochloric or sulfuric acid). The solution is then mixed with potassium iodide solution. An affirmative outcome is denoted by the form of orange or reddish-brown precipitation. A complex is formed due to the bond between the electron pair of the N atom in the alkaloid and a positive ion of potassium.

6.3.1.5 Test for Flavonoids

The usual alkaloid test for plant extracts uses ammonia and sulfuric acid reagents (Lone & Jain, 2022). The examined plant extract is treated with ammonia solution, and concentrated sulfuric acid is added. The absence of the yellow hue in the solution suggests the existence of flavonoids in cloves.

6.3.1.6 Test for Tannins

Tannins are known as astringent polyphenolic biomolecules that are present in cloves. Tannins can bind to proteins and other molecules, such as amino acids or alkaloids, to form complexes, resulting in their precipitation. Some of the properties of tannins that can be the principle in tannin testing are soluble in water, alcohol, glycerol, acetone, and dilute alkali. Reaction with iron compounds produces purple, violet or black precipitation. Potassium dichromate and lead acetate also precipitate tannin. Some methods of tannin detection are the gelatin test, Goldbeater's skin test, the phenazone test, the catechin test, the chlorogenic acid test, and the vanillin-hydrochloric acid test (Salminen & Karonen, 2011). The three types of tests commonly used in plant extract testing are as follows.

6.3.1.6.1 Gelatine Test

Gelatine and sodium chloride solutions were added to the plant extract filtrate. The production of a white precipitate shows a positive reaction.

6.3.1.6.2 Ferric Chloride Test

The basis of this test relies on the inherent properties of tannin when it interacts with iron (Fe). Several small amounts of ferric chloride solution are introduced into the filtrate or sample obtained from the plant extract. If a blue-green colour is observed, the sample is confirmed to contain tannin.

6.3.1.6.3 Lead Acetate Test

This test is based on the reaction of tannin with lead acetate. Adding lead acetate solution to the filtrate of plant extracts containing tannin will produce a white residue.

6.3.1.7 Test for Terpenoids and Triterpenes

Triterpenes are a constituent of terpenoids that are present in clove oil. These chemical molecules have several methyl groups and can undergo oxidation to form alcohols, aldehydes, and carboxylic acids, contributing to their complexity and distinctiveness in biological systems. Triterpenes possess active sites that facilitate glycosylation, transforming them into a distinct category of chemicals known as saponins (triterpene glycosides) (Shagufta, 2018). The procedure for this assay involves combining the extract with chloroform and concentrated sulfuric acid to generate a distinct layer. The reddish-brown hue observed at the interface layer indicates the existence of terpenoids (Harborne, 1984). With a temperature of 40°C at the time of the addition of chloroform, the method is also applicable for testing triterpene compounds. Another method is to use acetic anhydride and sulfuric acid reagents. The presence of terpenoids in the extract is demonstrated by creating a blue or green ring due to chemical reactions (Ayoola et al., 2008).

6.3.1.8 Test for Saponin

Cloves are reported to have saponins, and the presence of saponins can be determined using foam tests, which can be performed with or without sodium bicarbonate. A foam test can be done by adding distilled water (5 mL) into a test tube containing 1 mL of extract, then vortexed for 10 minutes. The persistence of a foam column in the presence of HCl suggests the existence of saponins (Archana et al., 2012; Sati & Kumar, 2015). Another foam test method is adding the sodium bicarbonate to a plant extract and distilled water. The solution is then shaken vigorously for 5 minutes. The occurrence of foam is a sign of the existence of saponins.

6.3.1.9 Test for Anthraquinones

Anthraquinones are a plentiful group of chemicals found in many natural substances including some in cloves. They are categorized as quinones, and their derivatives constitute the most extensive collection of natural quinones. Benzoquinones and naphthoquinones are included in this category as well. Anthraquinones exhibit the feature of staining when subjected to alkaline environments (Diaz et al., 2018). This is done by mixing the plant extract with benzene and 10% ammonia solution. The solution is agitated to achieve the desired hue of anthraquinones. If the extract contains anthraquinones, it will result in the formation of a pink layer (Lone & Jain, 2022).

6.3.1.10 Test for Steroid and Sterol

Detection of steroids and sterols in clove is conducted using chloroform and sulfuric acid reagents in cloves. The steroid testing is performed by dissolving the sample in chloroform and supplementing it with concentrated sulfuric acid. A favorable response to the presence of steroids is characterized by a red colour shift in the upper layer and a yellow layer of sulfuric acid exhibiting green fluorescence. (Love & Gibbs, 1974; Niass et al., 2023). The sterol testing procedure remains consistent, with the sign being the appearance of a red hue in the chloroform layer (Lone & Jain, 2022).

6.3.1.11 Test for Cardiac Glycoside

Cardiac glycoside is a steroid that can adhere mainly to the heart muscle. This compound has a lactone ring and a special sugar called digitoxose. Several test methods for detecting this compound are Raymond's, legal, Keller Killiani, xanthidrol, and baljet tests. However, the generally utilized method for assessing plant extract is the Keller Killiani test. The plant extract was obtained through the process of evaporation at a temperature of 40°C. The remaining substance was mixed with water, and then a solution of glacial acetic acid containing ferric chloride was added. The introduction of sulfuric acid weakened the solution. The presence of cardiac glycoside is shown by creating a brown ring at the interface of the solution (Lone & Jain, 2022).

6.3.1.12 Test for Resins

Detection of the presence of resin in a particular sample of clove can be determined by several methods depending on the type of resin and its intended use. The test methods are solubility test, alcohol solubility test, smell test, heat test, and staining test (Stepp, 2003). The presence of resin can also be determined by adding concentrated acetic acid and sulfuric acid to the extract sample. A purple to violet colour change within 10 minutes indicates the presence of resin in the extract.

6.3.2 Advance Phytochemical Screening

6.3.2.1 Gas Chromatography (GC) and Mass Spectrometry (MS)

This approach is employed to isolate volatile chemicals present in clove. The kinetics of the chemical species are influenced by its dispersion in the gas phase. Gas chromatography converts a sample into vapor and injects it onto the top of the chromatographic column. The sample is conveyed through the column by the movement of the inactive, gaseous mobile phase. The column consists of a solid surface that is chemically inactive, on which a stationary phase in the form of a liquid is attached through adsorption (Ingle et al., 2017). The migration rate of the chemical species is dictated by its dispersion in the gas phase. Gas chromatography is equipped with mass spectrometry to identify unknown compounds as well as to quantify compounds that have been identified. The material's molecular weight can be determined using this MS spectrum (Sohilait, 2015). This method is mainly employed for elucidating the structure of organic molecules, sequencing peptides or oligonucleotides, and identifying the presence of previously identified chemicals.

6.3.2.2 Liquid Chromatography

High-performance liquid chromatography (HPLC) is an accurate analytical technique employed to separate and detect organic and inorganic compounds in various samples of clove, including those derived from biological, pharmaceutical, food, environmental, and industrial sources. HPLC is a valuable technique for analyzing substances that cannot be converted into vapor or undergo decomposition at high temperatures. Additionally, it serves as a beneficial supplement to gas chromatography. Research on the characteristics of phytochemical extracts from plants typically involves using liquid chromatography mass spectroscopy (LCMS) to analyze and identify their compounds. This analytical approach is employed for the purpose of segregating, characterizing, and determining the amount of both identified and unidentified substances. This technique is highly sensitive, allowing for the analysis of small and multicomponent molecules present in an extract (Mukherjee, 2019).

6.4 PHYTOCHEMICALS CONTAINED IN CLOVE EXTRACT AND THEIR ANALYSIS

Different types of cloves will give different screening results. Clove bud (Eugenia caryophyllus) extracted by distillation and isolation method with dichloromethane positively contained alkaloids, glycoside, steroids, carbohydrates, terpenoids, tannins, and phenolic compounds (Kumar et al., 2010). Clove bud (Synzigium aromaticum) extracted by distillation and isolation method with dichloromethane showed different results. The extract did not contain tannins and glycosides. The balsams detected in the extract were resinous compounds containing cinnamic and benzoic acids or their esters (Ogwuda et al., 2023) (Table 6.1). This type of phytochemical was not tested by Kumar et al. (Kumar et al., 2010). Clove oil is characterized by phenol saponins, alkaloids, flavonoids, cardiac glycosides, tannins, and steroids (Dahiya & Soni, 2014). In addition, tannins, alkaloids, terpenoids, carbohydrates, glycosides, ketones, aldehydes, and 46 phenolic components were found in the clove methanol extract (Jimoh et al., 2017).

The process of isolating phytochemical components from clove pods was conducted using hexane as the solvent. The test results show that clove pods contain saponins, tannins, and alkaloids, which are detected as slightly weak. Flavonoid compounds were detected as moderately strong, while glycosides and steroids were not detected (Elbashir et al., 2021)

Preliminary testing of phytochemical compounds on clove leaves has been carried out (Lumingkewas & Unitly, 2023; Suhendar & Sogandi, 2019). Both studies used the maceration method with methanol solvent with different concentrations. The preliminary test results indicated that clove leaves include alkaloids, flavonoids, saponins, tannins, steroids, triterpenoids, and phenolics (Suhendar & Sogandi, 2019). Further identification by GCMS showed four main compounds with the highest abundance (Table 6.4).

The methanol extracts of clove leaf and flower buds were analyzed for their phytochemical components, which included alkaloids, flavonoids, saponins, tannins, and triterpenoids. The concentrations of these components ranged from low (+) to incredibly high (+++) and were essentially the same in both the leaf and flower buds (Lumingkewas & Unitly, 2023). Clove leaves have higher alkaloid levels than clove flowers (data not shown). This study proves that clove leaves also have the potential to be developed as pharmaceutical raw materials as well as clove flowers. Preliminary testing of phytochemical compounds on clove leaf extract is tabulated in Table 6.2.

Further detection using GCMS can provide information on all chemical compounds in clove extract. Even if the initial screening of chemical compounds was not detected, the GCMS test results showed the derivatives of these chemical compounds. No sterol compounds were found during the initial screening of phytochemical components in the hexane extract of clove pods. Still, the GCMS results detected gamma-sitosterol compounds which belong to the sterol group (Elbashir et al., 2021).

Flower buds are the most processed and utilized part of the clove plant, as they have a different composition than other clove pats such as clove leaf and clove stalk. Clove bud oil has a more pungent and intense floral scent than clove leaf oil. Cloves possess a delightful aroma that originates from significant amounts of essential oils

TABLE 6.1
Phytochemical of preliminary screening results of the clove flower buds/pods.

Metabolites	Results							
	Extraction method							
	Distillation and dichloromethane (Ogwuda et al., 2023)	Hexane (Elbashir et al., 2021)	Aqueous (Sawant et al., 2022)	Ethanol (Sawant et al., 2022)	Chloroform (Sawant et al., 2022)	Infused/water (Niass et al., 2023)	Methanol (Niass et al., 2023)	Acetone (Niass et al., 2023)
	buds	pods	buds	buds	buds	buds	buds	buds
Tannins	–	+	+	+	–	+	+	+
Balsams	+	Not tested	Not tested	Not tested	Not tested	Not tested	Not tested	Not tested
Phlobatanins	–	Not tested	+	–	–			
Cardiac glycosides	–	Not tested	+	+	–	Not tested	Not tested	Not tested
Flavonoids	+	++	–	+	+			
Steroids/Sterols	+	–	+	+	+	+	+	+
Terpenoids	+	Not tested	–	+	+	–	+	+
Volatile oils	+	Not tested	Not tested	Not tested	Not tested	Not tested	Not tested	Not tested
Carbohydrate	+	Not tested	+	+	–			
Glycosides	–	–	Not tested	Not tested	Not tested	Not tested	Not tested	Not tested
Saponin	Not tested	+	+	–	–	+	+	–
Alkaloids	Not tested	+	+	+	–	Not tested	Not tested	Not tested
Phenolic compound	Not tested	Not tested	+	+	–	Not tested	Not tested	Not tested
Protein	Not tested	Not tested	–	–	–			
Anthocyanin	Not tested	Not tested	–	–	–	–	+	+
Laucoanthocyanin						+	–	–
Anthraquinones	Not tested	Not tested	+	–	–			
Coumarine	Not tested	Not tested	–	+	–	–	+	+
Resin	Not tested	Not tested	+	–	–			
Fatty Acid	Not tested	Not tested				+	+	+
Emodins	Not tested	Not tested				+	–	–

KEY: (+) = Present (–) = Absent
(–) Means absence of the main class; (+) means presence of the main class
(++) Means presence of the main class in relatively greater quantity.

TABLE 6.2
Phytochemical screening results of the clove (*Syzygium aromaticum*) leaf extract.

Phytochemicals	Result	
	Extraction method	
	Methanol 96% (Suhendar & Sogandi, 2019)	Methanol 70% (Lumingkewas & Unitly, 2023)
Alkaloid	+	+
Flavonoid	+	+
Saponin	+	+
Tannin	+	+
Steroid	+	Not tested
Triterpenoid	+	+
Fenolik	+	Not tested

present in the blooms (10–20%), stalks (5–10%), and leaves (1–4%) (Nurdjannah & Bermawie, 2012). Xu et al. (Xu et al., 2016) analyzed the chemical composition of clove bud essential oil using gas chromatography-mass spectrometry (GC-MS). The oil extracted from flower buds is primarily composed of eugenol (55.6%), followed by eugenol acetate (12.5%), chavicol (6.5%), methyl salicylate (5.9%), β-aryophyllene (5.2%), α-caryophyllene (4.9%), p-allyl phenol (2.8%), δ-cadinene (2.5%), caryophyllene oxide (1.4%), and linalool (1.1%). α-Pinene and p-cymene are present in very small amounts (Singh et al., 2018). Clove pod hexane extract has more chemical compounds than ethanol extract. GCMS results showed 20 chemical compounds detected. Some of the main chemical compounds based on the percentage area can be seen in Table 6.3. The percentage area of eugenol compounds in ethanol extracts is higher than in hexane extracts. The variety of clove plants or regions of origin influence the percentage of chemical compound content. There is a variation in the percentage area of clove bud phytochemicals detected in testing three varieties of cloves from Indonesia (Taher et al., 2019) (Table 6.3).

Clove leaves contain eucalyptol, caryophyllene, α-cardinol, and limonene compounds (Mohammed et al., 2015). The essential oil derived from clove leaves possesses a unique and pleasing scent and a light yellow hue. Another research detected eugenol (87.8%), β-caryophyllene (13.0%), and α-humulene (1.5%) in the essential oil extracted from clove leaves (Kapahi & Thappa, 1989). Another study found 23 compounds, which included eugenol (76.8%), β-caryophyllene (17.4%), eugenol acetate (1.2%), and α-humulene (2.1%) (Jirovetz et al., 2006). Meanwhile, Srivastava et al. (Srivastava et al., 2005) detected 22 chemical compounds in the oil, with eugenol, isoeugenol, β-caryophyllene, α-humulene, and γ-cadinene being the most significant constituents.

Clove stem oil is not commonly utilized in commercial applications like clove bud oil due to its lower concentration of the components responsible for the fruity odor of clove oil, resulting in a less pronounced aroma in clove stem oil. However,

TABLE 6.3
The main compounds identified by GC-MS from different parts of cloves extracted using various solvents.

Part of cloves	Solvent	Compounds	Reference
Buds	Water (steam distillation method)	Eugenol (81.13–84.44%), *β*-caryophyllene (0.38–0.50%), Humulene (3.45–4.60%), Eugenyl acetate (11.60–15.02%), Caryophyllene oxide (0.05–0.24%)	(Sohilait, 2015)
Pods	Hexane	Eugenol (26.18%), (5.45%), (27.63%), 2,3,4,-trimethoxyacetophenone (16.25%), Diisooctyl phthalate (4.23%)	(Elbashir et al., 2021)
	Ethanol	Eugenol (81.21%), Phenol,2-methoxy-4-(2-propenyl)-acetate (10.60%), caryophyllene (4.65), 1,2,3-Benzeneetriol (0.72%), Caryophyllene oxide (0.76%), 2,3,4-trimethoxyacetophenone (0.70%)	(Elbashir et al., 2021)
Flowers	Methanol	Eugenol (36.43–56.57%), Acetisoeugenol (4.62–22.61%), Aceteugenol (18.70%), Caryofilen (11.47%), trans Caryofilen (13.09–15.42%), (–) Caryofilen Oxide (0.81–1.82%)	(Taher et al., 2019)
Flower stalk	Methanol	Eugenol (85.50–88.93%), Caryofilen (7.50–9.68%), Caryofilen Oxide (2.01–2.68%)	(Taher et al., 2019)
Stem	Methanol	Eugenol (97.20–98.83%), *β*-caryophyllene (0.49–1.73%), Humulene (0.00–0.30%), Eugenyl acetate (0.68–0.73%), Caryophyllene oxide (0.00–0.89%)	(Sohilait, 2015)
Leaves	Methanol	Eugenol (91.18–92.51%), Caryofilen (2.60%), trans-Caryofilen (4.23%), (–) Caryofilen Oxide (1.78–4.01%)	(Taher et al., 2019)
Leaves	Methanol	3-allyl-6-methoxy phenol eugenol (57.04%), phenol, 2- methoxy-4-(2-propenyl) eugenol acetate (26.18%), caryophyllene (14.61%), dan 1,4,7-cycloundecatriene, 1,5,9,9-tetramethyl (2.17%)	(Suhendar & Sogandi, 2019)
Leaves	Methanol	Eugenol (81.06–86.04%), *β*-caryophyllene (11.95–16.16%), Humulene (1.22–1.80%), Eugenyl acetate (2.02–3.05%), Caryophyllene oxide (0.00–0.37%)	(Sohilait, 2015)

stem oil contained a significantly higher amount of free eugenol than bud oil (Patil & Dhale, 2013). The extraction of clove root oil by steam distillation produced about 6% of oil. The root oil has a bright yellow colour and contains 85–95% eugenol.

6.5 CONCLUSION

Cloves contain many secondary metabolites that have pharmacological effects on health. The various secondary metabolites present in these plant parts can be determined qualitatively and quantitatively. Qualitative screening is known as phytochemical screening of different types of active chemical constituents such as phenols, alkaloids, flavonoids, flavones, aglycones, coumarins, saponins, tannins, steroids, glycosides and triterpenoids, using standard procedure with specific reagents. The appearance of the resulting color indicates the presence of the identified group of compounds. The assessment is made if the color during the reaction appears very strong with a positive value of 4 (+++), if the color is slightly less intense (+), or if it is slightly weak (+). Clove buds typically have the highest essential oil content, followed by leaves and twigs. Essential oils range from 6–22% for buds, clove leaves about 2–5% and clove twigs 1–2%. The main compound in cloves is eugenol, which is present in the essential oil of all parts of the clove. To analyze the entire chemical content of essential oils, gas chromatography (GC) or high-performance liquid chromatography (HPLC) can be used.

REFERENCES

Archana, P., Samatha, T., Mahitha, B., Chamundeswari, C., & Ramaswamy, N. (2012). Preliminary phytochemical screening from leaf and seed extracts of Senna alata L. Roxb-an ethnomedicinal plant. *International Journal of Pharmaceutical and Biological Research*, 3(3), 82–89.

Atanasov, A. G., Zotchev, S. B., Dirsch, V. M., & Supuran, C. T. (2021). Natural products in drug discovery: Advances and opportunities. *Nature Reviews Drug Discovery*, 20(3), 200–216. https://doi.org/10.1038/s41573-020-00114-z.

Ayoola, G. A., Coker, H. B., Adesegun, S. A., Adepoju-Bello, A. A., Obaweya, K., Ezennia, E. C., & Atangbayila, T. O. (2008). Phytochemical screening and antioxidant activities of some selected medicinal plants used for malaria therapy in Southwestern Nigeria. *Tropical Journal of Pharmaceutical Research*, 7, 1019–1024.

Batiha, G. E., Alkazmi, L. M., Wasef, L. G., Beshbishy, A. M., Nadwa, E. H., & Rashwan, E. K. (2020). Syzygium aromaticum L. (Myrtaceae): Traditional uses, bioactive chemical constituents, pharmacological and toxicological activities. *Biomolecules*, 10(2). https://doi.org/10.3390/biom10020202.

Bohlmann, J., Meyer-Gauen, G., & Croteau, R. (1998). Plant terpenoid synthases: Molecular biology and phylogenetic analysis. *Proceedings of the National Academy of Sciences of the United States of America*, 95(8), 4126–4133. https://doi.org/10.1073/pnas.95.8.4126.

Cheikhyoussef, A., Cheikhyoussef, N., Rahman, A., & Hussein, A. A. (2022). Chapter 14—Clove (Syzygium aromaticum) phenolics: Extraction, compositions, and biological activities. In M. F. Ramadan (Ed.), *Clove (Syzygium Aromaticum)* (pp. 215–233). Academic Press.

Choudhury, A., Singh, P. A., Bajwa, N., Dash, S., & Bisht, P. (2023). Pharmacovigilance of herbal medicines: Concerns and future prospects. *Journal of Ethnopharmacology*, 309, 116383.

Chung, K. T., Wong, T. Y., Wei, C. I., Huang, Y. W., & Lin, Y. (1998). Tannins and human health: A review. *Critical Reviews in Food Science and Nutrition*, 38(6), 421–464. https://doi.org/10.1080/10408699891274273.

Cortés-Rojas, D. F., de Souza, C. R., & Oliveira, W. P. (2014). Clove (Syzygium aromaticum): A precious spice. *Asian Pacific Journal of Tropical Biomedicine*, 4(2), 90–96. https://doi.org/10.1016/s2221-1691(14)60215-x.

Coskun, O. (2016). Separation techniques: Chromatography. *Northern Clinics of Istanbul*, 3(2), 156–160. https://doi.org/10.14744/nci.2016.32757.

Dahiya, P., & Soni, A. (2014). Phytochemical analysis, antioxidant and antimicrobial activity of syzygium caryophyllatum essential oil. *Asian Journal of Pharmaceutical and Clinical Research*, 7, 202–205.

Dai, J., & Mumper, R. J. (2010). Plant phenolics: Extraction, analysis and their antioxidant and anticancer properties. *Molecules*, 15(10), 7313–7352. https://doi.org/10.3390/molecules15107313.

Diaz, G., Miranda, I. L., Sartori, S. K., De Rezende, D. C., & Diaz, M. A. (2018). Anthraquinones: An overview. *Studies in Natural Products Chemistry*, 58, 313–338.

Edeoga, H. O., Okwu, D. E., & Mbaebie, B. O. (2005). Phytochemical constituents of some Nigerian medicinal plants. *African Journal of Biotechnology*, 4, 685–688.

Elbashir, D., Kehail, M., Abdelrahim, Y., & Ali, A. (2021). Phytochemistry and GC-MS screening and biocidal potentiality of clove (Syzygium aromaticum L.) pods against mosquitos' larvae. *International Journal of Science and Research Archive*, 3, 083–089. https://doi.org/10.30574/ijsra.2021.3.2.0135.

Elbein, A. D., & Molyneux, R. J. (1999). Alkaloid glycosidase inhibitors. *Comprehensive Natural Products Chemistry*, 129–160.

El Faqer, O., Bendiar, S., Rais, S., Elkoraichi, I., Mohamed, D., Elouaddari, A.,. . . Mtairag, E. M. (2022). Phytochemical characterization and immunomodulatory effects of aqueous, ethanolic extracts and essential oil of Syzygium aromaticum L. on human neutrophils. *Scientific African*, 18. https://doi.org/10.1016/j.sciaf.2022.e01395.

Guntarti, A., Lestari, P., & Nurani, L. (2023). Profile analysis of clove leaf oil (syzygium aromaticum (l.) merr. & l. M. Perry) and products in the market. *Medical Sains: Jurnal Ilmiah Kefarmasian*, 8, 97–106. https://doi.org/10.37874/ms.v8i1.534.

Harborne, J. B. (1984). *Phytochemical Methods A Guide to Modern Techniques of Plant Analysis* (1st ed.). Springer Dordrecht.

Harborne, J. B., & Tomás-Barberán, F. A. (1992). Ecological chemistry and biochemistry of plant terpenoids. *Journal of Natural Products*, 55(6), 830–831.

Heinrich, M., Mah, J., & Amirkia, V. (2021). Alkaloids used as medicines: Structural phytochemistry meets biodiversity-an update and forward look. *Molecules*, 26(7). https://doi.org/10.3390/molecules26071836.

Hernández-López, A., Sanchez Felix, D. A., Zuñiga Sierra, Z., Garcia Bravo, I., Dinkova, T. D., & Avila-Alejandre, A. X. (2020). Quantification of reducing sugars based on the qualitative technique of Benedict. *ACS Omega*, 5(50), 32403–32410.

Hossain, A., & Shahidi, F. (2018). Bioactives in spices, and spice oleoresins: Phytochemicals and their beneficial effects in food preservation and health promotion, 3, 8–75. https://doi.org/10.31665/JFB.2018.3148.

Ingle, K. P., Deshmukh, A. G., Padole, D. A., Dudhare, M. S., Moharil, M. P., & Vc, K. (2017). Phytochemicals: Extraction methods, identification and detection of bioactive compounds from plant extracts. *Journal of Pharmacognosy and Phytochemistry*, 6, 32–36.

Jimoh, S., Arowolo, L., & Alabi, K. (2017). Phytochemical screening and antimicrobial evaluation of syzygium aromaticum extract and essential oil. *International Journal of Current Microbiology and Applied Sciences*, 6, 4557–4567. https://doi.org/10.20546/ijcmas.2017.607.476.

Jirovetz, L., Buchbauer, G., Stoilova, I., Stoyanova, A., Krastanov, A., & Schmidt, E. (2006). Chemical composition and antioxidant properties of clove leaf essential oil. *Journal of Agricultural and Food Chemistry*, 54(17), 6303–6307. https://doi.org/10.1021/jf060608c.

Kapahi, B. K., & Thappa, R. K. (1989). Some essential oil bearing plants under cultivation in Andaman Island. *Pafabi*, 11, 23–24.

Kocabas, A. (2017). Ease of phytochemical extraction and analysis from plants? *Anatolian Journal of Botany*, 1, 26–31. https://doi.org/10.30616/ajb.345430.

Koche, D., Shirsat, R., & Kawale, M. (2016). An overview of major classes of phytochemicals: Their types and role in disease prevention. *Hislopia Journal*, 9(1/2), 1–11.

Kumar, U., Kumar, B., Anil, B., & Kumar, Y. (2010). Phytochemical investigation and comparison of antimicrobial screening of clove and cardamom. *International Journal of Pharmaceutical Sciences and Research*, 1, 138–147.

Lone, Z., & Jain, N. (2022). Phytochemical analysis of clove (Syzygium aromaticum) dried flower buds extract and its therapeutic importance. *Journal of Drug Delivery and Therapeutics*, 12, 87–92.

Love, A., & Gibbs, R. D. (1974). Chemotaxonomy of flowering plants. *Taxon*, 23, 154.

Lumingkewas, A., & Unitly, A. (2023). Identification of phytochemical compounds antioxidant leaf extract and clove flower extract (Syzygium aromaticum). *Journal Penelitian Pendidikan IPA*, 9, 8533–8538. https://doi.org/10.29303/jppipa.v9i10.4927.

Madubuike, P., Ezemokwe, D., & Madubuike, N. (2018). Phytochemical screening of eugenia caryophyllata (clove leaves) and characterization of its essential oil. *World Wide Journal of Multidisciplinary Research and Development*, 4(2), 163–167.

Martina, V., & Vojtech, K. (2015). A comparison of Biuret, Lowry and Bradford methods for measuring the egg's proteins. *Mendel Net*, 2015, 394–398.

Masyita, A., Mustika Sari, R., Dwi Astuti, A., Yasir, B., Rahma Rumata, N., Emran, T. B.,. . . Simal-Gandara, J. (2022). Terpenes and terpenoids as main bioactive compounds of essential oils, their roles in human health and potential application as natural food preservatives. *Food Chem X*, 13, 100217. https://doi.org/10.1016/j.fochx.2022.100217.

Mir, M. A., Parihar, K., Tabasum, U., & Kumari, E. (2016). Estimation of alkaloid, saponin and flavonoid, content in various extracts of Crocus sativa. *Journal of Medicinal Plants Studies*, 4(5), 171–174.

Mittal, M., Gupta, N., Parashar, P., Mehra, V., & Khatri, M. (2014). Phytochemical evaluation and pharmacological activity of syzygium aromaticum: A comprehensive review. *International Journal of Pharmacy and Pharmaceutical Sciences*, 6, 67–72.

Mohammed, N. H., Ahmed, M. H., & Hussien, M. O. (2015). Qualitative analysis of the essential oil of syzygium aromaticum (l.) (clove) using gas chromatography-mass spectrometry (gc-ms). *International Journal of Research in Pharmacy and Chemistry*, 5(2), 350–354.

Molyneux, R. J., Nash, R. J., & Asano, N. (1996). Chapter four the chemistry and biological activity of calystegines and related nortropane alkaloids. In S. W. Pelletier (Ed.), *Alkaloids: Chemical and Biological Perspectives* (pp. 303–343). Pergamon.

Mukherjee, P. K. (2019). Chapter 11—LC–MS: A rapid technique for understanding the plant metabolite analysis. In P. K. Mukherjee (Ed.), *Quality Control and Evaluation of Herbal Drugs* (pp. 459–479). Elsevier.

Mumtaz, F., Zubair, M., Khan, F., & Niaz, K. (2020). Analysis of plants lipids. In *Recent Advances in Natural Products Analysis* (pp. 677–705). Elsevier.

Nagalingam, S., Sasikumar, C. S., & Cherian, K. M. (2012). Extraction and preliminary phytochemical screening of active compounds in Morinda citrifolia fruit. *Asian Journal of Pharmaceutical and Clinical Research*, 5(2), 179–181.

Niass, O., Samb, I., Diatta, B. D., Diop, A., Dieng, A., Diongue, K.,. . . Diop, Y. (2023). Phytochemical screening, secondary metabolites assay, antioxidant, antibacterial and antifungal activity of clove buds (Syzygiumaromaticum) extracts. https://doi.org/10.21275/20011903.

Nurdjannah, N., & Bermawie, N. (2012). 11—Cloves. In K. V. Peter (Ed.), *Handbook of Herbs and Spices* (2nd ed., pp. 197–215). Woodhead Publishing.

Ogwuda, U., Adewumi, C., Anyip, P., & Ochi, I. (2023). Phytochemical screening and chemical analysis of cloves (syzygium aromaticum) flower buds, 2582–5208. https://doi.org/10.56726/IRJMETS32302.

Patil, D. A., & Dhale, D. A. (2013). *Spices and Condiments Origin, History and Applications.* Daya Pub. House.

Patil, D. A., Gurav, K., Kadam, A., Thite, S., Thoke, R., & Kore, D. B. (2013). Qualitative analysis of secondary metabolites from some filicales members. *IJRPC*, 3, 300–302.

Safeena, M., & Kalinga, J. (2020). Qualitative and quantitative screening of secondary metabolites in selected medicinal plants of Sri Lanka. *Journal of Medicinal Plants Studies*, 8(5), 94–100. www.plantsjournal.com/.

Salminen, J. P., & Karonen, M. (2011). Chemical ecology of tannins and other phenolics: We need a change in approach. *Functional Ecology*, 25(2), 325–338.

Sati, S., & Kumar, P. (2015). Assessment of Himalayan juniper, Juniperus squamata buch–ham ex d. don for phytochemical screening and antimicrobial potential against some infection causing pathogens. *World Journal of Pharmaceutical and Medical Research*, 4, 998–1011.

Sawant, H. S., Khan, D., & Patel, D. R. (2022). Phytochemical screening of *Syzygium aromaticum* in different extracts and its antibacterial properties against disease causing pathogens. *International Journal of Creative Research Thoughts*, 10(4), c208–c218.

Scales, F. M. (1915). The determination of reducing sugars. *The Journal of Biological Chemistry*, 23(1), 81–87.

Shagufta, P. (2018). Introductory chapter: Terpenes and terpenoids. In P. Shagufta & A.-T. Areej (Eds.), *Terpenes and Terpenoids* (Ch. 1). IntechOpen.

Shaikh, J., & Patil, M. (2020). Qualitative tests for preliminary phytochemical screening: An overview. *International Journal of Chemical Studies*, 8, 603–608. https://doi.org/10.22271/chemi.2020.v8.i2i.8834.

Singh, P. A., Brindavanam, N. B., Kimothi, G. P., & Aeri, V. (2016). Evaluation of in vivo anti-inflammatory and analgesic activity of Dillenia indica f. elongata (Miq.) Miq. and Shorea robusta stem bark extracts. *Asian Pacific Journal of Tropical Disease*, 6(1), 75–81.

Singh, P. A., Tushir, S., Hazra, S., Purewal, S. S., & Agarwal, G. (2024). Novel approaches for the extraction and identification of phytoconstituents from herbs and spices. *The Natural Products Journal*. http://doi.org/10.2174/0122103155293641240417072907

Singh, V., Pahuja, C., Ali, M., & Sultana, S. (2018). Analysis and bioactivities of essential oil of the flower buds of Syzygium aromaticum (L.) Merr. et LM Perry. *Journal of Medicinal Plants Studies*, 6(6), 79–83.

Sohilait, H. (2015). Chemical composition of the essential oils in eugenia caryophylata, thunb from Amboina Island. *Science Journal of Chemistry*, 3. https://doi.org/10.11648/j.sjc.20150306.13.

Srivastava, A., Srivastava, S. K., & Syamsundar, K. V. (2005). Bud and leaf essential oil composition of Syzygium aromaticum from India and Madagascar. *Flavour and Fragrance Journal*, 20, 51–53.

Stepp, J. R. (2003). Plant resins. Chemistry, evolution, ecology, ethnobotany. *Economic Botany*, 57(3), 419–420.

Sudha, R., Philip, X. C., & Suriyakumari, K. V. P. (2020). Phytochemical constituents of leaves of *Moringa oleifera* grow in Cuddalore District, Tamil Nadu, India. *SBV J Basic, Clinical and Applied Health Science*, 3(4), 164–167.

Suhendar, U., & Sogandi. (2019). Identifikasi senyawa aktif ekstrak daun cengkeh (syzygium aromaticum) sebagai inhibitor streptococcus mutans, 12, 229–239. https://doi.org/10.15408/kauniyah.v12i2.12251.

Taher, D. M., Solihin, D. D., Cahyaningsih, U., & Sugita, P. (2019). Chemical components analysis of clove (syzygium aromaticum (l). Merr. & perry). *Global Scientific Journal*, 7(4), 379–391.

Timilsena, Y. P., Phosanam, A., & Stockmann, R. (2023). Perspectives on saponins: Food functionality and applications. *International Journal of Molecular Sciences*, 24(17). https://doi.org/10.3390/ijms241713538.

Usman, R. A., & Bashir, M. (2022). Phytochemical profile and antibacterial activity of clove (syzgium aromaticum) methanol extract against some food-borne pathogens. *Nigerian Journal of Microbiology*, 36(1), 6120–6130.

Walton, N. J., Mayer, M. J., & Narbad, A. (2003). Molecules of interest: Vanillin. *ChemInform*, 34.

Wink, M., Schmeller, T., & Latz-Brüning, B. (1998). Modes of action of allelochemical Alkaloids: Interaction with neuroreceptors, DNA, and other molecular targets. *Journal of Chemical Ecology*, 24(11), 1881–1937. https://doi.org/10.1023/A:1022315802264.

Xu, J.-G., Liu, T., Hu, Q.-P., & Cao, X.-M. (2016). Chemical composition, antibacterial properties and mechanism of action of essential oil from clove buds against staphylococcus aureus. *Molecules*, 21, 1194. https://doi.org/10.3390/molecules21091194.

Zhang, Y., Cai, P., Cheng, G., & Zhang, Y. (2022). A brief review of phenolic compounds identified from plants: Their extraction, analysis, and biological activity. *Natural Product Communications*, 17, 1934578X2110697. https://doi.org/10.1177/1934578X211069721.

7 Volatile Compounds and Affecting Factors

Samuel Adelani Babarinde, Olugbenga Solomon Bello, Olagoke Zachaeus Olatunde, Samuel Oluwatobi Agboola, and Toluwanimi Rachael Arowolo

7.1 INTRODUCTION

Generally, the quantitative and qualitative volatile compounds (VCs) in any botanical can be influenced by several factors including varietal difference, the plant parts, historical age, region of origin, extraction techniques, processing procedures, agronomic parameters/plant nutrients, period of harvest, postharvest handling, and the climatic factors (Singh et al., 2021b, 2022). Eugenol, the major chemical constituent in the clove EO, is often identified alongside other compounds like acetyleugenol, Beta-caryophyllene, alpha-humulene, and copaene. The production of volatile oils in cloves, such as the aromatic compound eugenol, is influenced by several key factors. First, the genetic variety of the clove plant plays a significant role, determining the inherent capability of oil production. Climatic conditions, including temperature, humidity, and rainfall, are crucial as they directly impact plant growth and metabolic processes essential for oil synthesis. Soil quality and nutrient availability also affect oil production, with well-drained, nutrient-rich soils supporting optimal growth (Singh et al., 2021a, 2024). Harvesting at the right maturity stage and careful post-harvest handling, including drying and storage conditions, are essential to preserve the volatile oil content and quality. These factors collectively contribute to the overall yield and composition of volatile oils in cloves, crucial for their aromatic and medicinal properties. Certain VC can degrade due to improper drying or storage, which can change the overall aroma profile (Singh et al., 2023).

7.2 FACTORS AFFECTING VOLATILE COMPOUNDS IN CLOVE

7.2.1 Effect of Varietal Difference in the Chemical Composition of Clove

In a study conducted in Indonesia with two varieties (namely, Tuni and Zanzibar) of *S. aromaticum*, there were difference in the EO yield (v/w) extracted from different plant parts (Hariyadi et al., 2020). The bud had 1.3% for each of the two varieties, while the stalk had 1.0% and 1.2%, and the leaf had 0.4% and 0.35% for Tuni and Zanzibar varieties, respectively. The chemical composition of the EOs extracted

DOI: 10.1201/9781003521785-7

were different from different plant parts. For instance, for Tuni variety, the eugenol. alpha-humulene, caryophyllene, and patchouli alcohol in the bud essential oil were 67.9%, 6.2%, 21.1%, and 1.2%, respectively; while the stalk yielded 80.6% eugenol, 13.0% caryophellen, 2.5% eugenyl acetate, and 1.2% caryophyllene oxide. The leaf yielded 60.5% eugeneol, 32.2% caryophyllene, 3.5% alpha-humulene, 2.1% eugenyl acetate, and 1.0% caryophyllene oxide. For the Zanbizar variety, 47.4% eugenol, 25.3% caryophyllene, 19.1% eugenyl acetatae, 3.3% alpha-humulene, 1.9% delta-cadinene, and 1.5% copaene were identified from the bud EO. The stalk yielded the 67.5% eugenol, 24.8% caryophyllene, 2.7% alpha-humulene, 1.5% patchouli alcohol and 1.0% caryophyllene oxide. The proportion of eugenol, caryophyllene, alpha-humulene, eugenyl acetate and caryophyllene oxide in the leaf was 63.5%, 29.1%, 3.2%, 2.8% and 1.0%, respectively (Hariyadi et al., 2020). The results of their work indicate that the highest yield of essential oil of clove was obtained from the buds while the highest proportion of eugenol was obtained from the stalk.

7.2.2 Effect of Different Plant Parts in the Chemical Composition and Yield of Clove

The most widely used portion, the buds, are picked just before they open, and the spice is made by drying them. The strong, distinctive scent of the oil derived from the buds is caused by its high eugenol concentration. These are the most valued part for essential oil extraction because they have the highest concentration of eugenol.

Most of the available literature alludes to the fact that the flower bud is the plant part with the highest level of eugenol in clove. Records from India show that the clove root yielded oil via steam distillation which was as high as about 6%. Eugenol was the major compound being 85–95% (Pruthi, 2001). The percentage yield of *S. aromaticum* oil extracted from the flower bud obtained from Osogbo, Nigeria by Dichloromethane-aided hydrodistillation technique was 21.20%. The dominant constituents present were eugenol (75.10%), eugenyl acetate (13.57%), and β-caryophyllene (5.27%), while limonene (1.45%) and α-terpinolene (1.12%) were the minor constituents (Jimoh et al., 2017). The study of Boughendjioua (2018) carried out in Algeria on the clove leaf and flower bud without the pedicel is silent on the EO yield but the chemical composition reveals that the EO had eugenol (80.00%) as the major constituent, with 5.01% and 2.9% eugenyl acetate and β-caryophyllene, respectively. There seems to be paucity of information on the essential oil obtained from the stem of *S. aromaticum,* possibly due to the reduced quantity and quality of eugenol that is present in the stem essential oil (Kaur & Kaushal, 2019; Patil & Dale, 2013). From a sample obtained from China, eugenol content was 57.95% in the *S. aromaticum* bud oil, while it was 39.26% in the fruit oil (Ma et al., 2019).

7.2.3 Effect of Geographical Origin on the Yield and Chemical Composition of Clove

In a study carried out in the Sultanate of Oman on clove obtained from the United Arab Emirates, Jordan, and Saudi Arabia, the yield was recorded as 0.7%–0.92%. Comparatively high numbers of constituents (22, 26, and 28 with high percentage) were identified using GC. (Hossain et al., 2014). The major constituents identified in the Saudi Arabia EO were eugenol > β-E-caryophyllene > α-humulene > acetyl eugenol being 57.17%, 29.94%, 3.71%, and 5.92%, respectively. The major compounds in the United Arab Emirates EO were in the order eugenol > β-E-caryophyllene> α-humulene > acetyl eugenol being 63.64%, 23.04%, 2.97%, and 4.60%, respectively, while the Jordan EO had eugenol (54.97%), β-Ecaryophyllene (31.45%), acetyl eugenol (2.44%), and α-humulene (4.25%), in its essential oil (Hossain et al., 2014). The results indicate that the three clove samples obtained from the Gulf Region had a commonality of eugenol, β-Ecaryophyllene, α-humulene and acetyl eugenol as the major compounds.

In another study by Razafimamonjison et al. (2014), 121 commercial EOs isolated from *S. aromaticum* leaf, stem, and bud were subjected to chromatographic analysis and principal component analysis (PCA). Ten constituents were identified from the EOl samples. Eugenol was the major constituent of Madagascar (72.08–80.71%) and Indonesia (77.32–82.36%) bud EOs. Significant difference was observed in eugenyl acetate (11.68–21.32% vs. 8.61–10.55%) and β-caryophyllene (2.76–6.38% vs. 5.34–8.64%) for Madagascar and Indonesia, respectively. The leaf EOs from Madagascar and Indonesia have the following variations: eugenol (80.87–83.58% vs. 75.04–77.54%), β-caryophyllene (11.65–15.02 vs. 17.04–19.53%), and eugenyl acetate (0.29–1.45% vs 0–0.06%), respectively. The dominant compounds of Zanzibar, Madagascar, and Indonesia stem EOs were eugenol (87.52–89.47%, 91.81–96.65%, and 88.76–89.28%, respectively) and β-caryophyllene 7.40–7.75%, 1.66–4.48%, and 7.19–9.70% for Zanzibar, Madagascar, and Indonesia stem EOs, respectively.

7.2.4 Effect of Phenological Stage in the Yield and Chemical Composition of Clove

The phenological stage of a plant can also have an effect on the chemical composition of the VCs obtained from it. In a study in Zanzibar, EOs were extracted from the clove bud at budding-3, full budding, and flowering of 3–4 years (young tree) and 45 years (matured tree). Sizeable compounds (11–12) were identified in the EO from young trees, while 9–12 compounds were identified in the EO from mature trees. The most complex chemical composition of the essential oil obtained from the bud was at the flowering stage. Eugenol was the dominant constituent with 82.3% and 81.3% in the bud at mature and young trees, respectively; while at budding-3 stage, mature and young trees had 15.5% and 10.3% eugenol acetate, respectively. β-cis-caryophyllene content (5.57%) was lowest in the bud at mature stage while the young trees had 7.70% β-cis-caryophyllene content (Alfikri et al., 2020).

The changes in the EO yield and composition from *S. aromaticum* leaf harvested at different phenological phases were studied by Razafimamonjison et al. (2016). The EO yield of the phenological phases was in order of: mature leaves (3.8%) < expanded leaves 2 (4.1%) < expanded leaves 1 (4.5%) < young leaves (5.1%). The proportion of eugenyl acetate and eugenol displayed a divergent evolution during the leaf developmental phases. Eugenol attained its peak at expanded leaves 2 (84.00–90.48%) and mature leaf stage (88.32–90.22%); eugenyl acetate reached the lowest ones with 0.96–7.16% and 0.36–1.64%, respectively. Eugenyl acetate's level in the young leaf stage was 61.44–65.52%, while eugenol reached a minimum of 25.43–30.38% at that stage. From their results, it is recommended that EO should be extracted from the expanded leaves 2 and mature leaves stage when optimal yield and maximum eugenol proportion are the primary focus.

7.2.5 Effect of Agronomic Parameters in the Yield and Chemical Composition of Clove

The quantity and quality of essential oil obtained from a botanical may also be influenced by the agronomic parameters offered to the plant during its growth. Setiawan et al. (2021) studied the effect of shading and watering levels on the productivity of eugenol in clove seedlings in Indonesia. The six-month field experiment revealed that the low watering rate drastically increased eugenol yield per seedling, while heavy shading consistently reduced eugenol levels. Therefore, it is recommended that clove be cultivated on the land with moderate water stress and not in a waterlodged area. Also, the plant density should not be so high as to cause mutual shading when eugenol production from the leaf is the focus.

7.2.6 Effect of Postharvest Handling of Vegetal in the Chemical Composition and the Yield of Clove

After harvesting of any botanical sample, the postharvest handling is usually influenced by the intended purpose, cost, and environmental conditions among other factors. In the extraction of EOs from any botanical material, drying is the common postharvest processing. This is due to the fact that drying usually reduces the bulkiness of the vegetal mass imposed by the inherent moisture content. Therefore, the method of drying should be carefully selected to ensure that the quality of the EO would not be compromised. Air drying, sun drying or oven drying should be carefully selected to achieve the desired goal. Hastuti et al. (2017) subjected fresh clove bud to sun drying till a moisture content of 13 ± 1% was obtained. The vegetal was subjected to steam distillation and EO obtained was stored in laboratory at 25°C for a period of 4 months. The EO obtained from the fresh clove, sun-dried clove, and the one obtained from a 4-month storage period were subjected to GC-MS. The EO sample had eugenol, caryophyllene, and eugenyl acetate as their major components. The proportion of caryophyllene in fresh clove was slightly lower than what was obtained in the sun-dried clove but decreased with storage. Also, the eugenyl acetate content diminished with storage period and drying but the content of eugenol increased. Other minor compounds of clove EO were affected by drying and storage.

Also, Tran et al. (2020), monitored the compositions of clove EO and their blends under different temperature and light conditions for a 4-month period. Although eugenol, β-caryophyllene, and acetyleugenol were retained in the EOs throughout the storage period, direct light condition and 45°C storage condition was contributory to rapid oxidation rate of the essential oil, contrasted by EOs stored in the dark and at 4°C. Storage period evoked increased eugenol content in the EOs. These results can serve as an index of the quality of clove EO.

In another study in Indonesia (Hastuti et al., 2017), samples of clove buds were collected and subjected to oven drying at 50°C till the moisture content reached 13 ± 1%. The EO extracted from the fresh and dried clove was analyzed by GC-MS. The major constituents were eugenol > caryophyllene > acetyleugenol. It was observed that the content of eugenol and chavicol and isoeugenol increased, while monoterpenes and esters decreased due to the drying process. There were slight changes on clove EO composition after 4-month storage.

7.2.7 Effect of Vegetal Grinding in the Yield and Chemical Composition of Clove

Grinding of vegetal material can also influence the EO yield and the chemical composition. A possible reason is that grinding increases the surface area of the particle due to smaller particle sizes. When *S. aromaticum* seeds were ground and compared with the unground seeds, the essential oil yield was higher in ground seeds (about 14.3%) than 7.14% obtained in unground seeds. However, Eugenol was lower in the ground seed (68.7%) compared with 87.39% obtained in the unground seed. The proportions of the other three compounds [Cyprene (20.52% vs 7.22%), cis-Thujopsene (1.94% vs 0.75%) and Phenethyl isovalerate (6.41% vs 3.63%)] identified in the two EO samples were also more abundant in the grounded than the unground seed (Ben Hassine et al., 2021). It also implies that grinding exposed the active components in the seed, allowing them to escape due to oxidation or other chemical reactions.

In another study, the EO yield of ground clove bud and unground clove bud were 14.45% and 14.50% w/w, respectively. After the 7-hour distillation period, eugenol obtained from the unground clove bud was 88.56%, while the ground clove bud yielded 87.85% eugenol. The composition of eugenol, eugenyl acetate, and β-caryophyllene for unground clove and ground were 77.53% vs. 72.40%, 12.07% vs. 12.61%, and 5.90% vs. 12.61%, respectively (Safrudin et al., 2015). This indicates that only eugenol decreased in proportion due to grinding of clove buds, while grinding increased the proportion of eugenyl acetate and β-caryophyllene.

7.3 ISOLATION OF CLOVE ESSENTIAL OIL AND OLEORESIN

Soxhlet extraction is the most common technique to obtain EOs, phenolic compounds, and other secondary metabolites from plant materials (Al Jitan et al., 2018; Weggler et al., 2020). In Soxhlet extraction, the EOs and other compounds extracted from plants dissolved in the selected solvent. It is a simple technique and consists of a heating mantle, Soxhlet apparatus, and a water bath connected to a Soxhlet unit for water circulation within the apparatus (Al Jitan et al., 2018; Weggler et al., 2020). Usually, the extraction times fall between 12–24 hours, which makes this a time-consuming method. Also, large volumes of solvents are required in this technique (Weggler et al., 2020).

Steam or hydro-distillation is an indigenous method, commonly used by industrial companies. In the hydro-distillation method, plant materials are dipped in water, while in steam distillation, the plant materials are not in contact with water, but the stream is allowed to pass through the plant materials before being cooled by a condenser (Fokou et al., 2020). However, a few challenges of distillation method include loss or decomposition of some EO constituents because of the thermal effects (Memarzadeh et al., 2015; Zeng et al., 2016) or hydrolysis during or after extraction and occurrence of photo-oxidation due to exposure of essential oil to the light when a glass Clevenger is being used. These shortcomings have brought about improvement and optimization of new or existing extraction methods such as supercritical CO_2 extraction, ultrasound-assisted extraction, and microwave-assisted extraction (Benmoussa et al., 2019; Damyeh et al., 2016).

The microwave-assisted extraction method (MAE) is a variant of the hydro-distillation or stream distillation technique where the heating source is electric, unlike MAE, which generates its heat from a microwave (Lucchesi et al., 2004). Microwave-assisted extraction has been known as a method with some advantages over other conventional extraction techniques such as reduction of the amount of solvent required and extraction time, minimizing energy consumption. By implication, MAE makes use of a fraction of energy consumed in the hydro-distillation or steam distillation technique. Consequently, the use of MAE implies a decrease in operating costs of extraction (Ferhat et al., 2006; Lucchesi et al., 2004) This extraction method is also eco-friendly through minimizing emissions of carbon(iv)oxide to the atmosphere (Cardoso-Ugarte et al., 2013; Farhat et al., 2009).

Supercritical carbon(iv)oxide extraction is the latest and most sophisticated method of isolation of EO from vegetal material (Fokou et al., 2020). It is eco-friendly, innovative, and a better alternative for obtaining therapeutic and aromatic EOs from plant materials (Fornari et al., 2012). Supercritical carbon(iv)oxide extraction as a green technology mainly uses pure or modified carbon(iv)oxide at its supercritical stage for the extraction of EOs with a higher percentage of valuable oxygenated compounds while saving cost in terms of plant materials, energy, and time (Pourmortazavi & Hajimirsadeghi, 2007; Yamini et al., 2008). Carbon(iv)oxide is commonly used for this purpose due to it being less reactive, non-inflammable, non-toxic, easy to liquify, and easy to remove from plant materials (Fokou et al., 2020).

The essential oil and oleoresins are different by-products of clove. Table 7.1 presents an abridged comparison of the two by-products.

7.4 CHARACTERIZATION OF CLOVE VOLATILE COMPOUNDS

Clove essential oil is high in volatile organic compounds, with eugenol making up between 70 and 90% of the overall oil content. A few more important constituents are eugenyl acetate, α-humulene, and β-caryophyllene. The actual constituent may differ according to several factors like the region of origin, harvest season, and extraction technique.

The most widely used technique for analyzing volatile organic compounds in cloves is the gas chromatography-mass spectrometry technique. The gas chromatograph is often fused with silica capillary column of 30 m × 0.25 mm and a film thickness of 0.25 μm. The temperature for the analysis ranges between 50 and 300°C while the injection port temperature is around 250°C. Helium is used as the carrier gas at a constant flow rate of 1 mL per minute. The constituents of the EO are identified based on their retention indices with a comparison relative to n-alkanes.

TABLE 7.1
Comparison of oleoresins and clove essential oil

Parameters	Essential oil	Oleoresin
Major constituents	Eugenol, Eugenyl acetate, β-Carophyllene, α-Humulene, α-pinene	eugenol, 4-allyl-2-methoxyphenyl acetate, O-acetyleugenol, trans-caryophyllene
Potential bioactivity	Food preservative, pharmacological, antioxidant, perfume, pesticide, and antimicrobial	Food preservative, pharmacological, antioxidant, perfume, pesticide, and antimicrobial
Stability	Depends on extraction method	Depends on extracting solvent
Acceptability/Popularity	More popular	Less popular
Plant parts mostly used	Bud, leaf, stem	Bud
Usual appearance	Pale to clear yellow	Green-brownish
Major extraction technique	Distillation, supercritical oil extraction	Hydrocarbon solvent extraction

The recorded mass spectra are compared with those reported in the NIST library for further identification of the compound. By using the GC peak area normalization method, the essential oil's percentage composition is determined (Babarinde et al., 2017; Ben Hassine et al., 2021; Johannah et al., 2015).

Clove EO is quantified using additional characterization methods, such as GC-FID (Alma et al., 2007). TLC and HPLC were used for the characterization of *S. aromaticum* flower bud EO (Rana et al., 2011). Rana et al. (2011) reported column chromatography as competent method for the purification of eugenol in *S. aromaticum*. When TLC and GC were used for the analysis of column purified fractions, the major quantity of eugenol was identified in fraction II.

Previously reported studies on the dominant constituents of clove EO that are extracted utilizing clove buds sourced from various geographical locations are listed in Table 7.2.

7.5 MAJOR CLOVE VOLATILE COMPOUNDS

Many VCs can be found in clove's well-known aromatic EO. As it contains flavonoids, hidroxicinamic acids, hidroxibenzoic acids, and hidroxiphenyl propionates, clove is a major botanical source of phenolic compounds. For fresh plant material, eugenol is the primary bioactive compound present in cloves (Neveu et al., 2010). The phenolic acids salicylic, ferulic, elagic, and caffeic acids are also present in clove. Clove contains smaller amounts of flavonoids like quercetin, kaempferol, and its glycosilated derivatives. Clove buds contain concentrations of essential oil that can reach up to 18%. Eugenol makes up about 89% of clove EO, with eugenol acetate and β-cariofileno making up the remaining 5% to 15% (Jirovetz et al., 2006). α-humulene is another significant compound that can be obtained from essential oil

TABLE 7.2
Main chemical constituents obtained from the GC-MS analysis of the essential oil of cloves.

S/N	Main compounds (%)	Geographical location	References
(1)	Eugenol (75.8) Eugenyl acetate (18.9) *β*-Carophyllene (4.2) *α*-Humulene (0.30)	Egypt	(El-Mesallamy et al., 2012)
(2)	Eugenol (87.0) Eugenyl acetate (8.01) *β*-Carophyllene (3.56) *α*-Humulene (0.40)	Turkey	(Alma et al., 2007)
(3)	Eugenol (82.6) Eugenyl acetate (8.03) *β*-Carophyllene (7.45) *α*-Humulene (–)	Italy	(Tomaino et al., 2005)
(4)	Eugenol (81.13) Eugenyl acetate (11.60) *β*-Carophyllene (3.45) *α*-Humulene (0.50)	Indonesia	(Sohilait, 2015)
(5)	Eugenol (55.28) Eugenyl acetate (–) *β*-Carophyllene (20.97) *α*-Humulene (7.08)	China	(Yin et al., 2023)
(6)	Eugenol (82.16) Eugenyl acetate (16.55) *β*-Carophyllene (0.79) *α*-Humulene (–)	Morocco	(Razzouk et al., 2022)
(7)	Eugenol (82.40) Eugenyl acetate (0.90) *β*-Carophyllene (14.00) *α*-Humulene (1.80)	Slovakia	(Kacániová et al., 2021)
(8)	Eugenol (97.98) Eugenyl acetate (1.01) *β*-Carophyllene (0.11) *α*-Humulene (–)	Brazil	(de Carvalho et al., 2021)
(9)	Eugenol (55.61) Eugenyl acetate (20.55) *β*-Carophyllene (14.85) *α*-Humulene (2.76)	Java	(Amelia et al., 2016)
(10)	Eugenol (74.65) Eugenyl acetate (8.71) *β*-Carophyllene (12.79) *α*-Humulene (1.54)	Manado	(Amelia et al., 2016)

TABLE 7.2 ***(Continued)***
Main chemical constituents obtained from the GC-MS analysis of the essential oil of cloves.

S/N	Main compounds (%)	Geographical location	References
(11)	Eugenol (81.84) Eugenyl acetate (0.93) β-Carophyllene (13.40) α-Humulene (1.51)	Madagascar	(Razafimamonjison et al., 2014)
(12)	Eugenol (80.00) β-caryophyllene (2.9) Eugenyl acetate (5.01)	Algeria	(Boughendjioua, 2018)
(13)	Eugenol (60.4) β-Carophyllene (24.0) Eugenyl acetate (10.0)	Egypt	(Alitonou et al., 2012)
(14)	Eugenol (68.7–87.4), β-caryophyllene (-) α-humulene (-)	Tunisa	Ben Hassine et al., 2021)
(15)	Eugenol (76.542) β-caryophyllene (4.3) Eugenyl acetate (11.2)	India	(Tran et al., 2020)
(16)	Eugenol (43.34) β-Carophyllene (40.90) α-Humulene (3.90) p-cineole (4.09)	Nigeria	(Babarinde et al., 2023)

in clove at concentrations of about 2.1%. Clove essential oil also contains smaller amounts of ethyl hexanoate, 2-heptanone, benzaldehyde, farnesol, limonene, and β-pinene among other volatile substances (Cortés-Rojas et al., 2014).

Three types of clove oil abound: leaf, bud, and stem oils. The flower buds of Eugenia caryophyllus are used to make bud oil. It mostly comprises 60–90% eugenol, along with caryophyllene, eugenyl acetate, and other trace amounts. The leaves of Eugenia caryophyllus are used to make leaf oil. It is primarily composed of eugenol (82–88%), with trace amounts of eugenyl acetate and other substances. Stem oils originate from Eugenia caryophyllus twigs. Eugenol makes up 90–95% of it, with traces of other substances (Uddin et al., 2017). One ingredient that greatly influences clove flavor is eugenol. Of the EO extracted from clove, 72–90% is eugenol, which is primarily responsible for the clove aroma (Kamatou & Viljoen, 2012). Additionally significant components of clove oil include crategolic acid; vanillin; acetyl eugenol; beta-caryophyllene; tannins like methyl salicylate and gallotannic acid; bicornin; flavonoids like rhamnetin, eugenitin, and kaempferol; triterpenoids like oleanolic acid campesterol, and stigmasterol; and a number of sesquiterpenes (Bao, 2012; Mittal et al., 2014; Razafimamonjison et al., 2014; Issac et al., 2015). Eugenol and β-caryophyllene are two of the main ingredients in clove oil. The aroma of cloves and their medicinal qualities are mainly attributed to these compounds. The clove plant parts that are used to extract the oil such as the flower buds, leaves, and stems have varying concentrations of these compounds (Amelia et al., 2016).

7.5.1 Eugenol

$C_{10}H_{12}O_2$ is the chemical formula for eugenol, a phenolic compound. With an oil content of 60–90%, it is the most prevalent volatile organic compound found in cloves. It has strong antiseptic, analgesic, and anti-inflammatory qualities. It smells spicy and clove-like. Around 70–90% of the compounds in clove EO are found in the bud, which are the richest source of eugenol. A significant component of clove leaves is eugenol, although it is present in smaller amounts than in flower buds. Eugenol is also present in the stems, although not as much as it is in the buds (Liñán-Atero et al., 2024).

7.5.2 β-Caryophyllene

β-caryophyllene, having the formula $C_{15}H_{24}$, is a bicyclic sesquiterpene. This hydrocarbon, which has a sesquiterpene content of 5–12%, is another significant volatile organic compound found in cloves. It possesses analgesic, anti-inflammatory, and anticancer qualities. It smells spicy and woodsy. Some structures of different clove volatile compounds in different parts of the plant are represented in Figure 7.1.

Typically found in flower buds, β-caryophyllene is present in smaller amounts (5–10%) than eugenol. The scent profile of leaves differs from that of buds due to their higher β-caryophyllene content. Like leaves, stems have a considerable concentration of β-caryophyllene, though the precise amount varies (Uddin et al., 2017).

7.6 PHARMACOLOGICAL ACTIVITY OF EUGENOL AND ALTERATION OF GROUPS IN BASIC MOIETY

Eugenol is the principal constituent of clove EO (Haro-González et al., 2021) and displays significant biological activities such as antimicrobial and antioxidant activities, which makes it a target compound for modification to improve its therapeutic properties (Nam & Kim, 2013). Eugenol derivatives could be prepared in two ways (i) by esterification reactions, which involve the reaction of the hydroxyl group of eugenol with carboxylic acids and (ii) by addition reactions, which take place at the allyl group of eugenol (da Costa et al., 2020). Eugenol derivatives obtained from esterification reactions [(1)–(11)] displayed no or random antibacterial activities, but the original eugenol still displayed better activities (da Costa et al., 2020). Compound (16) only displayed a promising antibiotic effect with MIC values of 500 µg/mL against *Klebsiella pneumoniae* and *Bacillus cereus* compared the original eugenol (MIC value = 1,000 µg/mL), among other derivatives [(12)–(15), (17)–(19)]

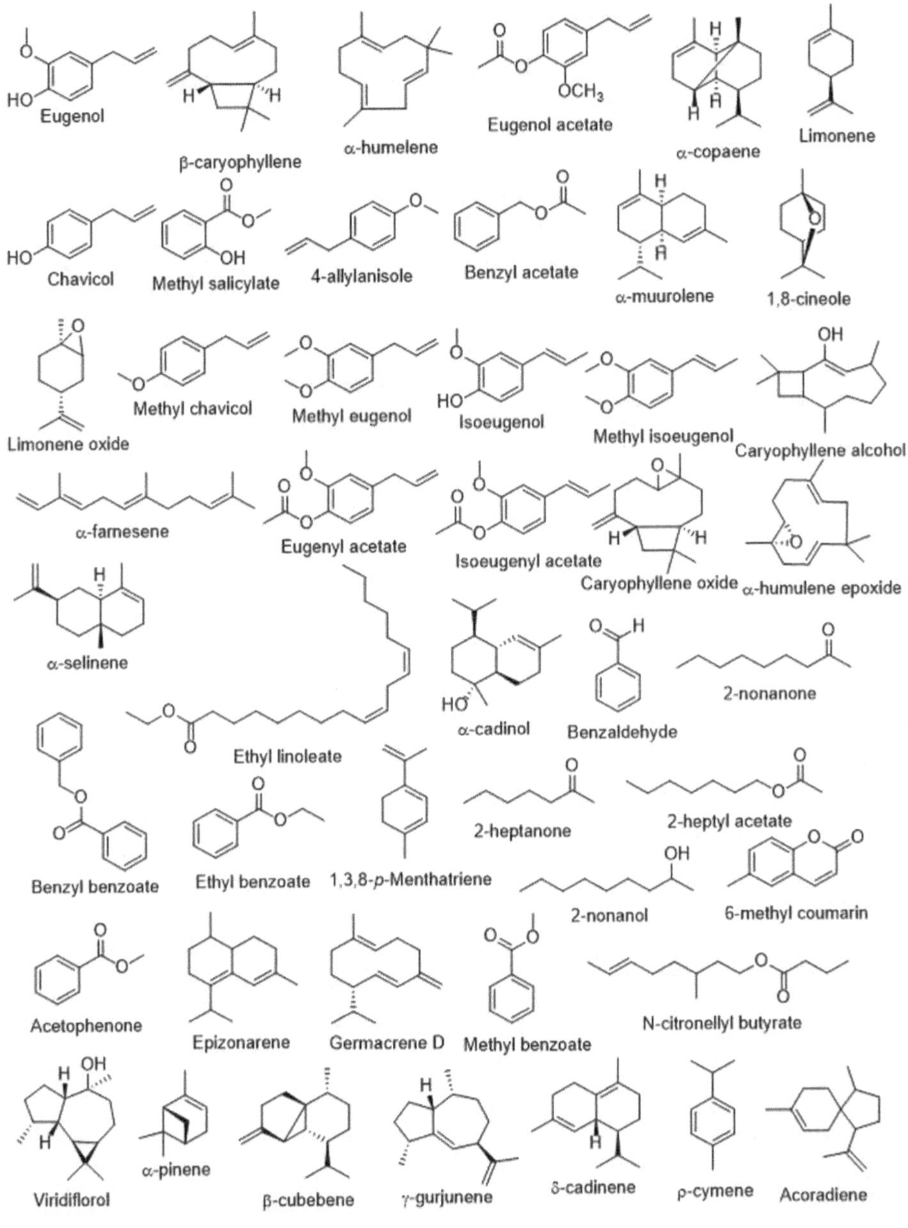

FIGURE 7.1 Structures of different clove volatile compounds in different parts of the plant.

prepared by addition reactions (da Costa et al., 2020). Chemical modification at the allyl group, regarding compounds (12), (14), (16), (18), and (19), exhibited higher antioxidant activity than other derivatives with IC_{50} values of 51.12, 20.00, 19.30, 32.00, and 30.37 μg/mL, respectively but lower than that of the starting compound (IC_{50} = 4.38 μg/mL) (da Costa et al., 2020; Singh et al., 2018). Figures 7.2 and 7.3 represent esterification reactions and addition reactions respectively.

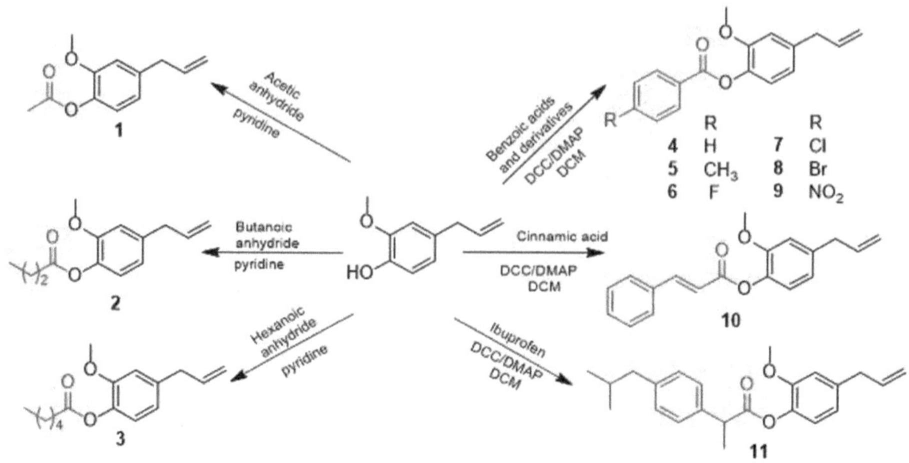

FIGURE 7.2 Eugenol derivatives [(1)–(11)] obtained from esterification reactions of the hydroxyl groups.

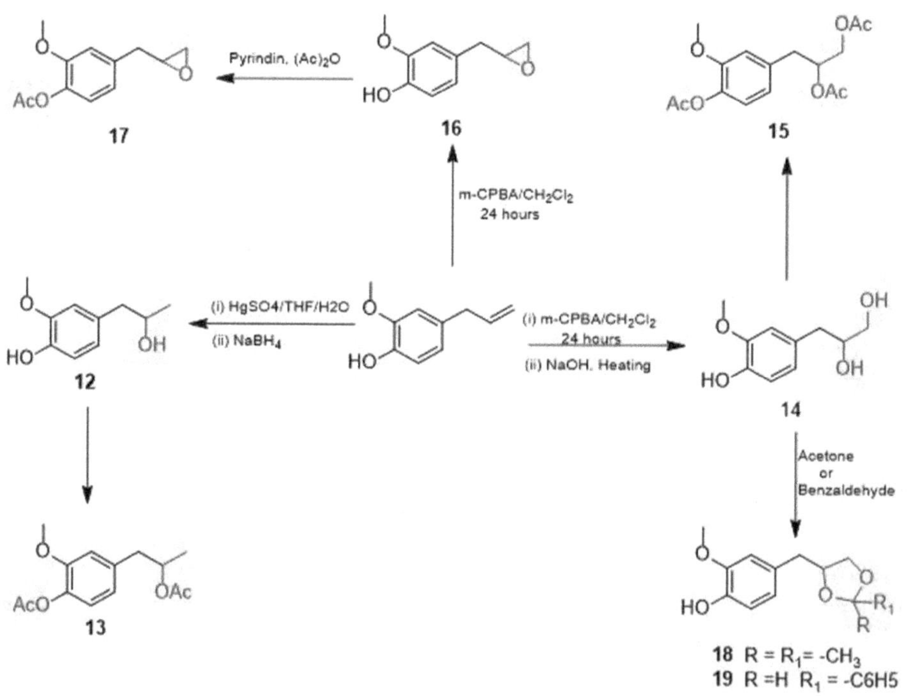

FIGURE 7.3 Eugenol derivatives [(12)–(19)] obtained from addition reactions of the allyl groups.

7.7 CHALLENGES OF CLOVE VOLATILE COMPOUNDS AND FUTURE TRENDS FOR MAXIMIZING POTENTIAL OF CLOVE VOLATILE COMPOUNDS

(1) The problem of authentication of plant species is a challenge. This is not peculiar to clove but to all botanical species. Often times, variations in the vernacular dialects makes accurate identification difficult in some species. According to the da Costa et al. (2020), about 1,800 species of *Syzygium* have been identified in tropical Asia. Accurate research on *Syzygium aromaticum* should start with correct identification and authentication of the species. This can be achieved with the help of experienced botanists and reference to a standard herbarium. While scientists would ordinarily embrace this position, potential end users of the research findings may find the issue of authentication of the botanical species a herculean task, especially where traditional herbal handlers are either misinformed or mischievous.

(2) The problems of volatile compounds of botanical origin include accurate sampling and low concentration. The problem of accurate sampling depends largely of the accuracy of the employed analytical method and the expertise of the operator of the detector. To tackle this challenge, analysis should be done in a reasonable number of replicates and the average values should be determined as the true representation of the available quantities and qualities of the compounds of interest. Also the expertise of the operator of the detecting device should be reckoned with to achieve accuracy in detecting the compound of interest.

(3) The volatility and reactivity of inherent constituents is another factor that poses challenges in the chemistry of the volatile compounds. Therefore, empirical studies should be geared toward ascertaining the impacts of storage and environmental conditions on the reactivity, stability, and bioactivity of the clove volatile compounds. Where instability of volatile compounds is identified as major challenges, nanoencapsulation of the oleoresins or EOs containing the volatile compounds should be explored. This will allow slow and target-oriented release of the bioactive compounds.

(4) In some literature materials, the authors emphasized the bioactive composition of clove EO and overlooked the quantity of the obtained oil. In majority of the articles where the yield was reported, the formula used to estimate the yield was not stated. According to Boutekedjiret et al. (2003), the yield was calculated using the dry botanical material as follows:

$REO\% = mEO/mS \times 100.$

Where mEO = essential oil mass (g), mS = mass of dry vegetal matter *s* (g), and REO = yield of essential oil (%).

Razzouk et al. (2022) recently stated that the yield should be calculated as:

$RHE\ (\%) = (VHE/DM) \times 100,$

where *RHE* is the EO yield in ml per 100 g dry vegetal matter; *VHE* is the EO volume recovered (mL), and *DM* is the quantity of dry vegetal material used for extraction (g).

It is necessary to include the qualitative and quantitative parameters (including the yield) while reporting the results of essential oil analysis.

(5) Where the sources of the vegetal matter were stated, it was not certain whether they were cultivated in the stated countries or imported to those countries from neighboring or far countries via trade. Therefore, it is expedient that the reports on geographical origin of the botanical materials used for EO extraction should be traced to the country of origin. Also, the coordinates of the place of collection should be provided where possible.

(6) Insufficient supply of the raw materials is another challenge in clove VCs research. The average yield of clove EO differs based on the factors listed in Section 7.1. This indicates that reasonable quantities of plant materials are needed to meet the demand for clove VCs. With the flower buds being the part with the highest yield of eugenol, the plant should be genetically manipulated to increase the EO yield obtainable from the buds. Domestication of genetically improved cultivars of *S. aromaticum* should be prioritized with supports from government and non-governmental organizations. Also, nanotechnology which involves the use of low levels of the VCs should be deployed in the pharmacological, agricultural and food industry.

(7) The conventional methods of obtaining the VCs of clove have been highlighted in Section 7.2 to include distillation, supercritical carbon (iv) oxide extraction, and Soxhlet- and microwave-assisted extraction. While the tendency to explore clove for pharmacological and pesticidal purposes would be progressive, research should focus on affordable methodologies that would not compromise the chemical composition of the VCs. To achieve this goal, collaborative works involving pharmacognosists, food scientists, pest control specialists and microbiologists, organic chemists, and engineers should be carefully designed and executed. The ultimate goal should be to make the volatile compounds affordable to end users for good health delivery and food security.

(8) A number of studies have established the bioactivity of the clove VCs in pharmacology, the food industry, and pest control. Eugenol, beta-caryophyllene, and alpha-humunene have insecticidal and food protection properties, while eugenol has antimicrobial compound against some oral pathogenic organisms. An aspect that seems to have received little or no attention is the negative impacts of the VCs on humans. This area should be investigated in future empirical studies.

(9) From a number of studies, post-harvest handling of botanicals influence the chemical characterization of the EOs or oleoresins. Therefore, the oleoresins or EOs obtained from freshly harvested clove and dried clove materials have different chemical compositions. Incidentally, it might not be understood whether such variations would amount to variant bioactivities when the products are subjected to the same bioassays. Therefore, it is imperative to make assay-guided empirical investigations. Where drying of the clove's part does not cause detrimental effect to the chemical composition and bioactivity of the EO or eleoresin, drying should be embraced to reduce the burden of handling bulky botanical matter.

(10) Although information abounds in literature on the variations of the clove VCs obtained from different countries, the deployment of geographic information systems to characterize the microclimate where the plant grows should be considered in the future research. The climatic conditions at the period of harvest should also be stated. Such data can be useful for country-dependent simulation studies on optimum clove growth conditions

(11) Despite the fact that the bioactivities of some specific clove VCs have been established, further works should be done to unearth new potential of the VCs. For instance, insects and microbial organisms that have not been assayed for their susceptibility to clove VCs should be assayed. Also, synergistic potential of such compounds with other bioactive compounds should be investigated. The factors that influence the bioactivity of the VC should also be investigated in synergistic/inhibitory studies.

(12) Since clove is a multipurpose plant, its emerging relevance in the SDGs is a certainty. Therefore, further works are needed on simulation, from its processing and storage to assay-guided bioactivity. Ultimately, the goal is mass production of its major bioactive constituents. Eugenol, caryophyllene, and alpha-humulene should be prioritized.

7.8 CONCLUSIONS

About 50 VOCs have been identified in clove globally, influenced by several factors. The major components are eugenol, caryophyllene, and α-humulene, being characterized by chromatographic methods. Supercritical CO_2 extraction and microwave-assisted extraction should be preferred to conventional distillation method where feasible. The choice between clove essential oil and oleoresin should depend on the extraction convenience, intended purpose, and stability of the inherent VOC. Future studies should focus on novel potential of clove VOC.

REFERENCES

Alfikri, F. N., Pujiarti, R., Wibisono, M. G., & Hardiyanto, E. B. (2020). Yield, quality, and antioxidant activity of clove (*Syzygium aromaticum* L.) bud oil at the different phenological stages in young and mature trees. *Scientifica*, *2020*(1), 9701701.

Alitonou, G. A., Noudogbessi, J. P., Sessou, P., Tonouhewa, A., Avlessi, F., Menut, C., & Sohounhloue, D. C. (2012). Chemical composition and biological activities of essential oils of *Pimenta racemosa* (Mill.) JW Moore from Benin. *International Journal of Biosciences*, *2*(9), 1–12.

Al Jitan, S., Alkhoori, S. A., & Yousef, L. F. (2018). Phenolic acids from plants: Extraction and application to human health. *Studies in Natural Products Chemistry*, *58*, 389–417.

Alma, M. H., Ertaş, M., Nitz, S., & Kollmannsberger, H. (2007). Chemical composition and content of essential oil from the bud of cultivated Turkish clove (*Syzygium aromaticum* L.). *BioResources*, *2*(2), 265–269.

Amelia, B., Saepudin, E., Cahyana, A. H., Rahayu, D. U., Sulistyoningrum, A. S., & Haib, J. (2016). GC-MS analysis of clove (*Syzygium aromaticum*) bud essential oil from Java and Manado. *International Symposium on Current Progress in Mathematics and Sciences 2016 (ISCPMS 2016) AIP Conference Proceedings*, *1862*, 1551–7616. http://doi.org/10.1063/1.4991186.

Babarinde, S. A., Dawodu, E. O., Ogundeji, D., Solomon, R., Audu, F. O., Adeyemi, A. A.,. . . Ibitoye, A. B. (2023). Chemical composition and toxicity of essential oils of clove flower buds and goat weed leaves against housefly larvae. *International Journal of Environmental Studies*, *80*(3), 635–648.

Babarinde, S. A., Pitan, O. O. R., Olatunde, G. O., & Ajala, M. O. (2017). Chemical composition of the essential oil of Nigeria grown *Hoslundia opposita* Vahl (Lamiaceae) dried leaves and its bioactivity against cowpea seed bruchid. *Chemistry & Biodiversity*, *14*(6), e1600418.

Bao, N. A. (2012). Hydrolysable tannins isolated from *syzygium aromaticum:* Structure of a new C-glucosidicellagitannin and spectral features of tannins with a tergalloyl group. *Heterocycles*, *85*(2), 365–381.

Ben Hassine, D., Kammoun El Euch, S., Rahmani, R., Ghazouani, N., Kane, R., Abderrabba, M., & Bouajila, J. (2021). Clove buds essential oil: The impact of grinding on the chemical composition and its biological activities involved in consumer's health security. *BioMed Research International*, *2021*(1), 9940591.

Benmoussa, H., Farhat, A., Romdhane, M., & Bouajil, J. (2019). Enhanced solvent-free microwave extraction of *Foeniculum vulgare* Mill. essential oil seeds using double walled reactor. *Arabian Journal of Chemistry*, *12*, 3863–3870.

Boughendjioua, H. (2018). Essential oil composition of *Syzygium aromaticum* (L.). *International Research Journal of Pharmacy and Medical Science*, *11*, 26–28.

Boutekedjiret, C., Bentahar, F., Belabbes, R., & Bessiere, J. M. (2003). Extraction of rosemary essential oil by steam distillation and hydrodistillation. *Flavour and Fragrance Journal*, *18*(6), 481–484.

Cardoso-Ugarte, G. A., Juárez-Becerra, G. P., SosaMorales, M. E., & López-Malo, A. (2013). Microwave-assisted extraction of essential oils from herbs. *Journal of Microwave Power and Electromagnetic Energy*, *47*(1), 63–72.

Cortés-Rojas, D. F., Fernandes de Souza, C. R., & Oliveira, W. P. (2014). Clove (*Syzygium aromaticum*): A precious spice. *Asian Pacific Journal of Tropical Biomedicine*, *4*(2), 90–96.

da Costa, J. S., da Cruz, E. D. N. S., Setzer, W. N., da Silva, J. K. D. R., Maia, J. G. S., & Figueiredo, P. L. B. (2020). Essentials oils from Brazilian Eugenia and *Syzygium* species and their biological activities. *Biomolecules*, *10*(8), 1155.

Damyeh, M. S., Niakousari, M., & Saharkhiz, M. J. (2016). Ultrasound pretreatment impact on Prangos ferulacea Lindl. and Satureja macrosiphonia Bornm. essential oil extraction and comparing their physicochemical and biological properties. *Industrial Crops and Products*, *87*, 105–115.

de Carvalho, S. Y. B., Almeida, R. R., Pinto, N. A. R., de Mayrinck, C., Vieira, S. S., Haddad, J. F.,. . .& Guimaraes, L. G. D. L. (2021). Encapsulation of essential oils using cinnamic acid grafted chitosan nanogel: Preparation, characterization and antifungal activity. *International Journal of Biological Macromolecules*, *166*, 902–912.

El-Mesallamy, A. M., El-Gerby, M., Azim, M. H. A. E., & Awad, A. (2012). Antioxidant, antimicrobial activities and volatile constituents of clove flower buds oil. *Journal of Essential Oil Bearing Plants*, *15*(6), 900–907.

Farhat, A., Ginies, C., Romdhane, M., & Chemat, F. (2009). Eco-friendly and cleaner process for isolation of essential oil using microwave energy: Experimental and theoretical study. *Journal of Chromatography A*, *1216*(26), 5077–5085.

Ferhat, M., Meklati, B., Smadja, J., & Chemat, F. (2006). An improved microwave clevenger apparatus for distillation of essential oils from orange peel. *Journal of Chromatography A*, *1112*(1–2), 121–126.

Fokou, J. B. H., Dongmo, P. M. J., & Boyom, F. F. (2020). Essential oil's chemical composition and pharmacological properties. In *Essential Oils-Oils of Nature*. IntechOpen. http://doi.org/10.5772/intechopen.86573. www.intechopen.com/chapters/68008

Fornari, T., Vicente, G., Vázquez, E., García-Risco, M. R., & Reglero, G. (2012). Isolation of essential oil from different plants and herbs by supercritical fluid extraction. *Journal of Chromatography A*, *1250*, 34–48.

Hariyadi, M. A., Yahya, S., & Wachjar, A. (2020). Morphological characters and essential oil constituents extracted of two clove varieties (*Syzygium aromaticum* (L.) Merr. & LM Perry.) from Ambon Island, Indonesia. *Plant Archives*, *20*(1), 2208–2214.

Haro-González, J. N., Castillo-Herrera, G. A., Martínez-Velázquez, M., & Espinosa-Andrews, H. (2021). Clove essential oil (*Syzygium aromaticum* L. Myrtaceae): Extraction, chemical composition, food applications, and essential bioactivity for human health. *Molecules*, *26*(21), 6387.

Hastuti, L. T., Saepudin, E., Cahyana, A. H., Rahayu, D. U. C., Murni, V. W., & Haib, J. (2017). The influence of sun drying process and prolonged storage on composition of essential oil from clove buds (*Syzygium aromaticum*). In *AIP Conference Proceedings* (Vol. 1862, No. 1). AIP Publishing.

Hossain, M. A., Harbi, S. R. A., Weli, A. M., Al-Riyami, Q., & Al-Sabahi, J. N. (2014). Comparison of chemical constituents and antimicrobial activities of three essential oils from three different brands' clove samples collected from Gulf region. *Asian Pacific Journal of Tropical Disease*, *4*(4), 262–268.

Issac, A., Gopakumar, G., Kuttan, R., Maliakel, B., & Krishnakumar, I. M. (2015). Safety and anti-ulcerogenic activity of a novel polyphenol-rich extract of clove buds (Syzygium aromaticum L). *Food & Function*, *6*(3), 842–852.

Jimoh, S. O., Arowolo, L. A., & Alabi, K. A. (2017). Phytochemical screening and antimicrobial evaluation of Syzygium aromaticum extract and essential oil. *International Journal of Current Microbiology and Applied. Science*, *6*, 4557–4567.

Jirovetz, L., Buchbauer, G., Stoilova, I., Stoyanova, A., Krastanov, A., & Schmidt, E. (2006). Chemical composition and antioxidant properties of clove leaf essential oil. *Journal Agricultural and Food Chemistry*, *54*(17), 6303–6307.

Johannah, N. M., Renny, R. M., Gopakumar, G., Maliakel, B., Sureshkumar, D., & Krishnakumar, I. M. (2015). Beyond the flavour: A de-flavoured polyphenol rich extract of clove buds (*Syzygium aromaticum* L.) as a novel dietary antioxidant ingredient. *Food & Function*, *6*(10), 3373–3382.

Kacániová, M., Galovicová, L., Borotová, P., Valková, V., Dúranová, H., Kowalczewski, P. Ł.,. . .& Savitskaya, T. (2021). Chemical composition. Vitro and in situ antimicrobial and antibiofilm activities of syzygium aromaticum (clove) essential oil. *Plants*, *10*, 2185.

Kamatou, V. I., & Viljoen, M. (2012). Eugenol-from the remote Maluku Islands to the international market place: A review of a remarkable and versatile molecule. *Molecules*, *17*(6), 6953–6981.

Kaur, K., & Kaushal, S. (2019). Phytochemistry and pharmacological aspects of Syzygium aromaticum: A review. *Journal of Pharmacognosy and Phytochemistry*, *8*(1), 398–406.

Liñán-Atero, R., Aghababaei, F., García, S. R., Hasiri, Z., Ziogkas, D., Moreno, A., & Hadidi, M. (2024). Clove essential oil: Chemical profile, biological activities, encapsulation strategies, and food applications. *Antioxidants*, *13*, 488.

Lucchesi, M., Chemat, F., & Smajda, J. (2004). Solvent-free microwave extraction of essential oil from aromatic herbs: Comparison with conventional hydrodistillation. *Journal of Chromatography A*, *1043*(2), 323–327.

Ma, Y. N., Xu, F. R., Chen, C. J., Li, Q. Q., Wang, M. Z., Cheng, Y. X., & Dong, X. (2019). The beneficial use of essential oils from buds and fruit of Syzygium aromaticum to combat pathogenic fungi of Panax notoginseng. *Industrial Crops and Products*, *133*, 185–192.

Memarzadeh, S. M., Pirbalouti, A. G., & AdibNejad, M. (2015). Chemical composition and yield of essential oils from Bakhtiari savory (Satureja bachtiarica Bunge.) under different extraction methods. *Industrial Crops and Products*, *76*, 809–816.

Mittal, M., Gupta, N., Parashar, P., Mehra, V., & Khatri, M. (2014). Phytochemical evaluation and pharmacological activity of *Syzygium aromaticum:* A comprehensive review. *International Journal of Pharmacy and Pharmaceutical Sciences*, *6*(8), 67–72.

Nam, H., & Kim, M. M. (2013). Eugenol with antioxidant activity inhibits MMP-9 related to metastasis in human fibrosarcoma cells. *Food and Chemical Toxicology*, *55*, 106–112.

Neveu, V., Perez-Jiménez, J., Vos, F., Crespy, V., du Chaffaut, L., Mennen, L.,. . . Scalbert, A. (2010). Phenol-explorer: An online comprehensive database on polyphenol contents in foods. *Database*, *2010*.

Patil, D. A., & Dhale, D. A. (2013). *Spices and Condiments: Origin, History and Applications*. Daya Publishing House, 76–80.

Pourmortazavi, S. M., & Hajimirsadeghi, S. S. (2007). Supercritical fluid extraction in plant essential and volatile oil analysis. *Journal of Chromatography A*, *1163*(1–2), 2–24.

Pruthi, J. S. (2001). *Minor Spices and Condiments Crop Management and Post-Harvest Technology*. Indian Council of Agricultural Research, 308–331.

Rana, I. S., Rana, A. S., & Rajak, R. C. (2011). Evaluation of antifungal activity in essential oil of the Syzygium aromaticum (L.) by extraction, purification and analysis of its main component eugenol. *Brazilian Journal of Microbiology*, *42*, 1269–1277.

Razafimamonjison, G., Boulanger, R., Jahiel, M., Ramanoelina, P., Fawbush, F., Lebrun, M., & Danthu, P. (2016). Variations in yield and composition of leaf essential oil from Syzygium aromaticum at various phases of development. *International Journal of Basic and Applied Sciences*, *5*(1), 90.

Razafimamonjison, G., Jahiel, M., Duclos, T., Ramanoelina, P., Fawbush, F., & Danthu, P. (2014). Bud, leaf and stem essential oil composition of Syzygium aromaticum from Madagascar, Indonesia and Zanzibar. *International Journal of Basic and Applied Sciences*, *3*(3), 224.

Razzouk, S., Mazri, M. A., Jeldi, L., Mnasri, B., Ouahmane, L., & Alfeddy, M. N. (2022). Chemical composition and antimicrobial activity of essential oils from three Mediterranean plants against eighteen pathogenic bacteria and fungi. *Pharmaceutics*, *14*(8), 1608.

Safrudin, I., Maimulyanti, A., & Prihadi, A. R. (2015). Effect of crushing of clove bud (Syzygium aromaticum) and distillation rate on main constituents of the essential oil. *American Journal of Essential Oils and Natural Products*, *2*(3), 12–15.

Setiawan, A., Ito, S., Mitsuda, Y., Hirata, R., Yamagishi, K., Umar, Y. P., & Kamei, I. (2021). Productivity of eugenol from clove (Syzygium aromaticum L.) under different light and soil water conditions. *AGRIVITA, Journal of Agricultural Science*, *44*(1), 96–104.

Singh, P. A., Bajwa, N., Chinnam, S., Chandan, A., & Baldi, A. (2022). An overview of some important deliberations to promote medicinal plants cultivation. *Journal of Applied Research on Medicinal and Aromatic Plants*, *31*, 100400.

Singh, P. A., Bajwa, N., Sharma, L., & Baldi, A. (2023). Farm-level medicinal plants production and economic analysis in the Indian state of Punjab. *Journal of Applied Research on Medicinal and Aromatic Plants*, *34*, 100486.

Singh, P. A., Dash, S., Choudhury, A., & Bajwa, N. (2024). Factors affecting long-term availability of medicinal plants in India. *Journal of Crop Science and Biotechnology*, *27*(2), 145–173.

Singh, P. A., Desai, S. D., & Singh, J. (2018). A review on plant antimicrobials of past decade. *Current Topics in Medicinal Chemistry*, *18*(10), 812–833.

Singh, P. A., Sood, A., & Baldi, A. (2021a). Determining constraints in medicinal plants adoption: A model geospatial study in the Indian state of Punjab. *Journal of Applied Research on Medicinal and Aromatic Plants*, *25*, 100342.

Singh, P. A., Sood, A., & Baldi, A. (2021b). An agro-ecological zoning model highlighting potential growing areas for medicinal plants in Punjab. *Indian Journal of Pharmaceutical Education and Research*, *55*(2), S492–S500.

Sohilait, H. J. (2015). Chemical composition of the essential oils in *Eugenia caryophylata*, Thunb from Amboina Island. *Science Journal of Chemistry, 3*, 95–99.

Tomaino, A., Cimino, F., Zimbalatti, V., Venuti, V., Sulfaro, V., De Pasquale, A., & Saija, A. (2005). Influence of heating on antioxidant activity and the chemical composition of some spice essential oils. *Food Chemistry, 89*(4), 549–554.

Tran, T. K. N., Tran, T. H., Le, X. T., Toan, T. Q., Le Ngoc, T. T., Anh, H. L. T., & Cang, M. H. (2020). Effect of storage conditions on the chemical composition of Vietnamese clove (Syzygium aromaticum) essential oil. *Journal of Global Pharma Technology, 12*, 447–454.

Uddin, M. A., Shahinuzzaman, M., Rana, M. S., & Yaakob, Z. (2017). Study of chemical composition and medicinal properties of volatile oil from clove buds (*Eugenia caryophyllus). International Journal of Pharmaceutical Sciences and Research, 8*(2), 895.

Weggler, B. A., Gruber, B., Teehan, P., Jaramillo, R., & Dorman, F. L. (2020). Inlets and sampling. In *Separation Science and Technology* (Vol. 12, pp. 141–203). Academic Press.

Yamini, Y., Bahramifar, N., Sefidkon, F., Saharkhiz, M. J., & Salamifar, E. (2008). Extraction of essential oil from Pimpinella anisum using supercritical carbon dioxide and comparison with hydrodistillation. *Natural Product Research, 22*(3), 212–218.

Yin, X., Hu, Q., Chen, X., Tan, S., Niu, A., Qiu, W., & Wang, G. (2023). Inclusion complexes of clove essential oil with sodium caseinate and gum Arabic prepared by high-pressure homogenization: Characterization and non-contact antimicrobial activity. *Food Control, 150*, 109765.

Zeng, Q. H., Zhao, J. B., Wang, J. J., Zhang, X. W., & Jiang, J. G. (2016). Comparative extraction processes, volatile compounds analysis and antioxidant activities of essential oils from *Cirsium japonicum* Fisch. ex DC and *Cirsium setosum* (Willd.) M. Bieb. *LWT-Food Science and Technology, 68*, 595–605.

8 Oleoresins

Tatang Hidayat, Christina Winarti,
Bagem Br Sembiring, and Hernani

8.1 INTRODUCTION

In the global market, the industry of spice nowadays occupies a massive segment with estimated at about 79 billion U.S. dollars and continues to expand. Moreover, the clove market was about at USD 4.1 Billion in 2023 globally and always expanding. The global market of oleoresin reached USD 1.44 billion in 2018. Imports of oleoresins in Europe increased at an average of4.1% from 2012–2016 (Procopio et al., 2022). Although the spice industry is continuing to grow, spices have several problems, especially shelf life and contamination. According to spice has been extremely influenced by coherent quality deterioration during storage and distribution. Therefore, among some efforts to protect quality degradation, processing dried spices into oleoresins is the best way to minimize such post-harvest losses (Madhusankha et al., 2023).

The market for spice products in the form of essential oils and oleoresins has grown significantly because customers are becoming more concerned about their health and seeking natural items that have some nutritional and health benefits. This trend is supported by the usage of oleoresins as synthetic chemical additive alternatives. Most oleoresin applications are attributed to its flavor, fragrance, and color characteristics, besides its therapeutics properties. Additional reasons for the increased interest include their numerous functional qualities, which have wide-ranging uses in the food, nutrition, and pharmaceutical industries. These qualities include antioxidant, anti-inflammatory, antibacterial, anticancer, and insect repellent (Habashy et al., 2018; Batiha et al., 2020).

Oleoresins are characterized as viscous plant extracts that consist of a volatile and a non-volatile component. The lipophilic chemicals, waxes, and pigments that make up the resinous fraction distinguish oleoresins from essential oils. According to Sowbhagya (2019), oleoresins are more complex extracts compared to essential oils comprising both volatile and non-volatile compounds. Additionally, this product's diverse chemical composition allows it to be used for a wide range of purposes. Oleoresin is made up of a collection of several secondary metabolites from plant species. Amongst the oleoresin compounds, phenylpropanoids, volatile lipophilic terpenoids, or low molecular weight aliphatic hydrocarbon derivatives make up the majority of the constituents of oleoresins (Shahidi & Hossain, 2018). Moreover, according to Procopio et al. (2022), oleoresins contain spicy compounds, essential oils, coloring matter, and resins, which can constitute the flavor, aroma, and color.

 DOI: 10.1201/9781003521785-8

Oleoresin offers a longer shelf life and excellent stability in storage without sacrificing quality. According to Soldan et al. (2021), oleoresins' moisture-free state gives them a competitive edge and a prolonged shelf life of 2–3 years. As a result, the concentrated oleoresins make handling and storage easier, and a small amount is all that is needed to give raw or dried spices the same sensory qualities. Furthermore, the ability to combine many oleoresins and their thermal stability make them more applicable in the food sector (Ferraz et al., 2021). Oleoresins have 5–20 times the pungency of similar dried spices, and their authenticity is higher than 90% (Galvin-King et al., 2020). As a result, utilizing oleoresins is financially advantageous for industry, and research on the spice extraction oleoresin and product development still much needed. This chapter will feature the various methods of clove oleoresin production derived from different kinds of clove parts, their phytochemical composition and benefits, as well as the utilization of clove oleoresin.

8.2 OLEORESIN EXTRACTION TECHNIQUES

Soxhlet and maceration, two conventional extraction techniques, as well as sophisticated extraction methods such as supercritical fluid extraction (SFE), microwave-assisted extraction (MAE), and ultrasound-assisted extraction (UAE), can all be used to extract oleoresin from cloves (Abdul Azis et al., 2023; Singh et al., 2024). The types of solvent, material particle size, ratio of sample to solvent, extraction temperature, and extraction duration have all been proven to affect extraction efficiency in addition to extraction techniques (Zhang et al., 2018). Furthermore, the process conditions and the extraction techniques are critical in influencing the composition of oleoresin (Procopio et al., 2022). Spice extracts can be of two types: volatile oil fractions, which are responsible for the flavor and aroma of the spice and non-volatile fractions, which are responsible for the taste or pungency of the spice (Shukla et al., 2019).

8.2.1 Conventional Extraction Techniques

8.2.1.1 Soxhlet Extraction

Due to its high extraction efficiency, Soxhlet extraction is one of the most commonly used techniques for obtaining oleoresin, especially from plant materials such as cloves (Abdul Azis et al., 2023). Compared to maceration or percolation, this automatic continuous extraction process is known for its excellent efficiency and requires less time and solvent consumption (Zhang et al., 2018). A thimble, usually made of cellulose or hard filter paper, is filled with the ground cloves. This thimble is then placed in the compartment of the Soxhlet apparatus. Meanwhile, the bottom flask is filled with the extraction solvent. The solvent vaporizes and rises to the sample thimble when heated, extracting oleoresin from the cloves. The extraction process is completed when the vapor condenses in the condenser at the top of the apparatus and drips back down. Compared to maceration-based extraction procedures, this process typically produces larger quantities (Bitwell et al., 2023). The resulting mixture is then concentrated to produce a more potent crude extract suitable for use in food, medicine, and cosmetics (Rizkiyah et al., 2023). Despite being time-consuming,

Soxhlet extraction remains highly effective for obtaining significant amounts of oleoresin and other bioactive compounds from cloves (Chemat et al., 2020).

8.2.1.2 Maceration Extraction

Plant materials like cloves, usually in a coarse or powdered state, are soaked in a selected solvent as part of the straightforward extraction process known as maceration (Bitwell et al., 2023). Depending on the solvent and intended result, the extraction process might take place at room temperature or higher and last for several hours to several days (Abdul Azis et al., 2023). According to Abdul Azis et al. (2023), the solvent dissolves bioactive compounds such as eugenol, acetyl eugenol, and caryophyllene when it permeates the clove material during maceration. Following extraction, the product and solvent are separated by centrifugation or filtration, and the resultant oleoresin is concentrated by vacuum evaporation to produce a potent solution. The choice of solvent used plays a critical function in establishing the bioactive compounds that are extracted. While maceration is an easy, affordable, and equipment-free method of obtaining oleoresin and bioactive compounds from cloves (Boyko et al., 2020), its disadvantage is that it is low efficiency and takes a long time to extract (Azwanida, 2015).

8.2.2 Advanced Extraction Techniques

8.2.2.1 Supercritical Fluid Extraction

When a substance reaches its critical point, it manifests properties of both a gas and a liquid, known as a supercritical fluid (SF). A substance enters its supercritical state due to two factors: temperature and pressure. SF has the solvating properties of a liquid but behaves more like a gas. One such supercritical fluid is CO_2, which reaches supercritical conditions at temperatures above 31.1°C and pressures above 7,380 kPa (Azwanida, 2015). Another technique for extracting clove oleoresin is supercritical fluid extraction using CO_2 (SC-CO_2). During SC-CO_2 extraction, carbon dioxide is compressed to a supercritical state, displaying characteristics of both a liquid and a gas. This makes supercritical CO_2 an excellent solvent to extract both hydrophobic and hydrophilic compounds (Idham et al., 2021). Clove oleoresin extraction works best at extraction pressures of at least 10 MPa and temperatures of about 40°C. Conversely, clove oil can be selectively recovered at lower temperatures and pressures (Reverchon & Marrone, 1997). Compared to conventional extraction techniques, this approach has several benefits, including high extraction efficiency, quicker extraction duration, and enhanced selectivity for target compounds (Rizkiyah et al., 2023).

8.2.2.2 Ultrasound-assisted Extraction

Electromagnetic waves of higher frequency than sound waves are known as ultrasonics, and they are generally used in the ultrasonic frequency range of 20 kHz to 2,000 kHz (Azwanida, 2015; Bitwell et al., 2023). Zhang et al. (2018) state that ultrasonic-assisted extraction (UAE), sometimes referred to as sonication or ultrasonic extraction, uses the energy of ultrasonic waves to aid in extraction procedures. UAE is particularly effective for extracting high yields of bioactive compounds from cloves, such as eugenol, caryophyllene, and acetyl eugenol (Idowu et al., 2021). In the UAE process, cloves are immersed in a solvent and exposed to high-frequency ultrasonic waves. The mechanical action of acoustic cavitation produced by ultrasonics increases temperatures and pressures, improving the permeability of cell walls and the surface area of interaction between solvents and plant samples (Azwanida, 2015; Bitwell et al., 2023). According to Abdul Azis et al. (2023), this enhances mass transfer, speeding up the extraction process. Lower solvent and energy consumption, reduced temperatures, shorter extraction durations, and higher yields are some of the benefits of employing UAE over conventional extraction techniques (Abdul Azis et al., 2023; Zhang et al., 2018).

8.2.2.3 Microwave-assisted Extraction

Microwave-assisted extraction (MAE), sometimes referred to as microwave extraction, is a relatively new technique that uses solvents and microwaves to extract bioactive compounds. The microwave frequency range is 300 MHz to 300 GHz (Bitwell et al., 2023). MAE employs microwave radiation to heat the solvent and extract the desired compounds from plant materials, such as cloves. Microwaves interact with polar substances like water and specific organic components in the plant matrix through processes like ionic conduction and dipole rotation, generating heat. In MAE, heat and mass transfer occur simultaneously, creating a synergistic effect that increases extraction yield and speeds up the extraction process (Zhang et al., 2018). Bioactive compounds in cloves, such as eugenol, acetyl eugenol, and caryophyllene, can be effectively extracted using MAE (Kennouche et al., 2015). Several factors, including the type of solvent used, the power and frequency of microwave radiation, the duration of the extraction process, and the size of the plant particles, can influence the effectiveness of MAE for extracting compounds from cloves (Abdul Azis et al., 2023). The benefits of MAE include higher extract yields, reduced thermal degradation, and selective heating of plant materials (Zhang et al., 2018). Compared to conventional maceration and Soxhlet extraction techniques, MAE is a more efficient and rapid extraction process.

8.3 OLEORESIN COMPOSITION

Plant extracts known as oleoresins consist of both volatile and non-volatile substances. Pigments, volatile/essential oils, fixed oils, pungent compounds, and antioxidants are the main components of oleoresin. Low molecular weight aliphatic hydrocarbon derivatives, phenylpropanoids, and volatile lipophilic terpenoids make up the majority of an oleoresin's composition (Shahidi & Hossain, 2018). Terpene compounds predominate in oleoresins, along with alcohols, esters, aldehydes, and certain phenylpropanoids as other volatile compounds. GC methods can be used to identify and quantify the volatile compounds.

8.3.1 Clove Parts Oleoresin Composition

Clove buds – or flowers – are the most processed and utilized part of the clove plant due to their significant commercial value attributed to their composition. However, cloves can also yield oleoresins – or essential oils – extracted from other plant parts such as leaves and flower stalks. The reason cloves are highly beneficial to the business is their pleasant perfume derived from essential oils present in sufficient amounts in the flowers (10–20%), stalks (5–10%), and leaves (1–4%) (Nurdjannah & Bermawie, 2012). In comparison to other parts like buds and stalks, which are extensively used as seasonings and raw materials in the cigarette industry, the clove tree's leaves are underutilized.

The physical characteristics of clove oleoresin are delineated by the U.S. Food and Drug Administration (U.S. FDA) in CFR Title 21, Section 184.1257 (Clove and its derivatives). According to FDA regulations (2021), clove oleoresin exhibits a specific gravity ranging from 1.036 to 1.060 g/mL, an optical rotation between −2° and 0°, and a refractive index of 1.527 to 1.538 at 20°C. It is devoid of residual solvents, except those falling within the tolerance levels outlined in Part 173, Subpart C or deemed generally recognized as safe (GRAS).

The chemical composition of clove oleoresin/extract is quite complex, containing both volatile and non-volatile substances. According to Xue et al. (2022), the volatile components found in clove extract mostly consist of aromatics, including eugenol, isoeugenol, and eugenyl acetate; terpenoids, including sesquiterpenoids, monoterpenoids, and diterpenes; and aliphatic compounds, including alkanes, ketones, alcohols, esters, and ethers. The main non-volatile compounds are flavonoids (such as luteolin, quercetin, rhamnocitrin, kumatakenin, and pachypodol), chromones (such as eugenin, 8-C-glucosylnoreugenin, biflorin, and isobiflorin), tannins (such as eugeniin, casuarictin, 1,3-di-O-galloyl-4,6-(S)-hexahydroxydiphenoyl-β-D-glucopyranose), and triterpenoids (such as Asiatic acid, arjunolic acid, and oleanolic acid) (Xue et al., 2022). Depending on the extraction solvent, clove oleoresin has a different chemical composition. Using liquid chromatography-mass spectrometry (LC-MS/MS) to determine the ethanolic extract of clove buds, the main compounds found were eugenol, fatty acids, and flavonoids. These included quinic acid, 2,4,6-trihydroxybenzoic acid, chlorogenic acid, quercetin 3-O-β-D-glucuronide, ellagic acid, quercetin, eugenol, acetylsalicylic acid, skullcap flavone I, dehydrodiisoeugenol, palmitic acid, and oleic acid. According to Xu et al. (2023), the clove extract contained 32.62% eugenol. Gallic acid, isobiflorin, biflorin, eugenol, trihydroxy triterpene acid, acetyleugenol, maslinic acid, coumaroylmaslinic acid, oleanolic acid, linoleic acid, palmitic acid, oleic acid, and stearic acid were identified in the clove buds extract obtained by ultrasonication using a dichloromethane-methanol mixture (7:13) (Ladurner et al., 2017). Additionally, LC-MS analysis of clove buds extracted with ethanol using the maceration method revealed the presence of rhamnetin, biflorin, gallic acid derivative, gallic acid, quinic acid, ellagic acid, syringic acid, caffeic acid, kaempferol, monogalloyl glucose, phenylacetic acid, p-hydroxybenzoic acid, eugenin, eugenitin, piperic acid, ferulic acid, myricetin, kaempferol-3-glucoside, delphinidin, phloroglucinol, digalloyl glucose, eugenol, quercetin, and kaempferol-3,5-dimethyl ether (Moradi et al., 2023). Furthermore, Fathoni et al. (2017) reported the presence of quinic acid, monogalloyl glucose, gallic acid, digalloyl glucose, isobiflorin, biflorin, ellagic acid, hydroxygallic acid, a derivative of gallic acid, luteolin, quercetin, naringenin, kaempferol, isorhamnetin, dimethoxyluteolin, rhamnetin, and three unidentified compounds in clove bud residues extracted with ethanol using the maceration method and examined by LC-MS/MS. Volatile compounds in clove buds were extracted using subcritical CO_2 and assessed by GC-MS, revealing the presence of α-pinene, eucalyptol, chavicol, α-cubebene, *a*-copaene, eugenol, trans-caryophyllene, *a*-humulene, germacrene D, eugenol acetate, and caryophyllene oxide (x0160;ernaitė et al., 2020).

Much research has demonstrated the presence of chemical compositions with antibacterial properties in clove leaves, such as flavonoids, triterpenoids, phenolics, and tannins. Analysis by LC-MS/MS of the ethanolic clove leaf extract revealed that it has more than ten components, namely 4-O-caffeoylquinic acid, ambronal, daturametelin H, quercetin-3-O-β-D-glucuronide, trichosanic acid, 3,3'-Di-O-methylellagic acid-4'-O-α-D-glucopyranoside, 5,2',6'-trihydroxy-7,8-dimethoxyflavone-2'-O-β-D-glucopyranoside, lucialdehyde B, luteolin-7-O-glucuronide, and candidate mass $C_{35}H_{36}N_4O_5$ that was believed to be involved in the pheophorbide component. Meanwhile, volatile compounds like 3-allyl-6-methoxyphenol-eugenol, caryophyllene 1,4,7,-cycloundecatriene, 1,5,9,9-tetramethyl, phenol, 2-methoxy-4-(2-propenyl), and eugenol acetate are found in clove leaf oleoresin that was extracted using methanol and subjected to GC-MS analysis (Suhendar & Sogandi, 2019).

Compared to clove buds and leaf extracts, there has been relatively less research on the chemical compositions of clove stem extract. According to one of the reviewed publications, GC-MS analysis of methanolic clove stem extract produced 13 compounds: beta myrcene, delta-3-carene, benzene, 1-methyl-2-(1-methylethyl)-; bicyclo[2.2.1]hept-2-ene, 1,7,7-trimethyl-; eucalyptol; gamma-terpinene; naphthalene; decahydro-2-methyl-; menthol; α-terpineol, and dodecane. With an area percentage of 84%, alpha-pinene was the predominant constituent (Safitri et al., 2022).

8.3.2 Oleoresin Composition Based on the Extraction Process

Oleoresin resulted from the extraction process containing pungent compounds, essential oils, colorants, and resins. The extraction method used will affect the chemical compound extracted in the oleoresin. Atawodi et al. (2011) measured many phenolic components in a clove methanol extract using a combination of LC/MSD and UV dual-array detectors after HPLC separation with a C-18 reverse phase column. These phenolic components include ellagic acid (9.1 mg/kg), its derivative (3.5 mg/kg), gallic acid (28.6 mg/kg), quercetin glucoside (3.5 mg/kg), and additional undiscovered phenolic compounds (Atawodi et al., 2011). Eugenol (4-allyl-2-methoxyphenol), a phenolic molecule from the phenylpropanoids class, was shown to be the major bioactive component of clove buds. Furthermore, Gonzalez-Rivera et al. (2021) found 48.9% eugenol and 46.5% caryophyllene in Indonesian clove buds extracted with water using a microwave process. Other studies employing microwave-assisted extraction of Indonesian clove buds extracted with water using the microwave technique obtained 48.9% of eugenol and 46.5% of caryophyllene (Gonzalez-Rivera et al., 2021). Extraction with microwave-assisted extraction of Indian clove buds using water consist of 90.1% of eugenol (Kapadiya et al., 2018). Meanwhile, ultrasound-assisted extraction of cloves using ethanol contained 64% of eugenol (Tekin et al., 2015). The variation of oleoresin compound indicated that the composition of the clove extracts is highly influenced by their variety, processing, and extraction methods (Nurdjannah & Bermawie, 2012). Other research by Li and Xu (2017) identified about 20 components of the oleoresin and revealed that eugenol (72.85%) is the main component, followed by eugenyl acetate (10.54%), β caryophyllene (4.54%), and d-cadinene. Furthermore, ethyl acetate extract of clove buds analyzed using GC-MS revealed eugenol (58.88%) as the most plentiful active component, followed by eugenyl acetate (23.86%) and trans-caryophyllene (14.44%)

(Yassin et al., 2020). Using different kinds of solvent to extract, El-Ghalab et al. (2020) concluded that n-hexane was efficient for extraction of phenols; the extracted fat appeared as brown viscous oil and smelled like clove volatile oil and consisted of a high amount of phenolic compound, almost similar to that of ethanol extract. The main factors in extracting organic extract with SFE include temperature, pressure, and extraction time. Extraction temperature and pressure alter the density of CO_2, thereby increasing the extraction yield due to the enhanced solubility of the clove components (Yang et al., 2014). The extraction methods, chemical composition and health effects of different clove parts are summarized at Table 8.1.

TABLE 8.1
Clove parts, extraction methods, chemical composition and health effects of clove oleoresin.

Material/extraction methods	Chemical composition	Percentage	Health effect	References
Clove buds	Eugenol Eugenyl acetate β caryophyllene d-cadinene	72.85% 10.54% 4.54%	Antioxidant Lipid peroxidation	Li and Xu (2017)
Methanol extract of defatted clove buds	Ellagic acid Ellagic acid derivative, gallic acid, Quercetin glucoside, other unidentified phenolic compounds	9.1 mg/kg 3.5 mg/kg 28.6 mg/kg 3.5 mg/kg	Antioxidant Chemo preventive for oxidation-induced diseases like diabetes, cancer, cardiovascular diseases	Atawodi et al. (2011)
Clove buds from Indonesia; in-situ microwave-assisted extraction using water	Eugenol β-caryophyllene α-caryophyllene α-cubebene α-copaene	48.9 ± 2.5% 46.5 ± 2.1% 3.7 ± 0.2% 2.3 ± 0.1% 2.2 ± 0.1%	–	Gonzalez-Rivera (2021)
Clove buds from India microwave-assisted extraction using water	Eugenol	90.1%	Antibacterial	Kapadiya et al. (2018)
Clove buds ethanol extract ultrasound-assisted extraction	Eugenol	64%	Antibacterial against *E coli* and *S. aureus*	Tekin et al. (2015)
Extract methanol clove leaves	3-allyl-6-methoxyphenol-Eugenol Caryophyllene Cycloundecatriene, 1,5,9,9-tetramethyl 2-methoxy-4-(2-propenyl) Eugenol acetate	57.04% 14.61% 2.17% 20.18%	Antibacterial for dental cavities	Suhendar and Sohandi (2019)

TABLE 8.1 *(Continued)*
Clove parts, extraction methods, chemical composition and health effects of clove oleoresin.

Material/extraction methods	Chemical composition	Percentage	Health effect	References
Methanol extract	Flavonoid Total phenolic content (TPC)	54.47 mg TE/g 160.55 mg GAE/g DW	Most potent antioxidant among 30 spices tested	Assefa et al. (2018)
Clove buds, Ethyl acetate extract	Eugenol Eugenyl acetate trans-caryophyllene	58.88% 23.86% 4.44%	Anti-candidiasis	Yassin et al. (2020)
Hexane extract	Eugenol Eugenol acetate	71.56% 8.99%	Antioxidant Hepatoprotective activity	Nassar et al. (2007)
Ethanolic extract	Total phenolic compound (TPC) Total flavonoid Compound (TFC) Eugenol Eugenol acetate and caryophyllene	10.53 ± 0.32 mg of GAE/g extract 1.13 ± 0.11 mg of QE/g extract 78.67% 11.77% 6.85%	Immunomodulatory effect of extracts and EO Bactericidal function of human PMNs	El Faqer et al. (2022)
Clove buds, SFE (SC-CO2), 120 min SC–CO2, 30°C 20 MPa	Eugenol *β*-caryophyllene *a*-humulene Eugenol acetate	54.58% 17.32% 2.26% 20.55%	–	Guan et al. (2007)
Clove buds, SFE (SC-CO2), 120 min SC–CO2 40°C 30 MPa	Eugenol *β*-caryophyllene α-humulene Eugenol acetate	55.14% 15.52% 2.02% 20.32%	–	
Clove buds, ultrasound-assisted SFE (SC–CO2), 115 min, 40°C 15 MPa	Eugenol *β*-caryophyllene *a*-humulene Eugenol acetate	59.18% 15.35% 1.93% 18.60%	–	Yang et al. (2014)
Clove buds, ultrasound-assisted SFE (SC–CO2), 170 min, 40°C 20 MPa	Eugenol *β*-caryophyllene *a*-humulene Eugenol acetate	55.63% 14.48% 1.81% 17.15%	–	

8.4 BIOACTIVITY OF CLOVE EXTRACT/OLEORESIN

Clove extract possesses numerous biological activities as follows: antioxidant, anti-inflammatory, antibacterial, anticancer, as well as neuroprotective properties (Huang et al., 2009). The presence of phenolic chemicals in clove extract is thought to be responsible for its antioxidant effect, as they are free radical scavengers and reduce oxidative stress. The phenolic compound is also responsible for anti-inflammatory properties, which inhibit the generation of inflammatory cytokines and enzymes.

Moreover, clove extract possesses an antibacterial effect against several kinds of microorganisms, which is mostly due to the existence of eugenol. Besides, oleoresin also has been found in studies to have anticancer attributes, causing cell death and reducing cancer cell proliferation (Bhagat & Chaturvedi, 2016). Furthermore, other research also suggests the neuroprotective effect of clove extract, which is also attributed to eugenol, proven to lower oxidative stress and inflammation (Abdul Aziz, 2023).

8.4.1 Antioxidant Activity

Clove has long been known to have strong antioxidant activity. The antioxidant activity of clove is mainly attributed to its phenolic compound such as eugenol as the main bioactive component. These components transfer a hydrogen atom from their hydroxyl group and inhibit free radicals. Oleoresin extracted from oven-dried clove buds using several solvents including ethyl acetates, 80% ethanol, and distilled water showed scavenging activities when tested by DPPH assay ranging from 25.3 to 91.4% and using ABTS assays between 49.45% to 99.49%. Among three solvents used, the high antioxidant capacity was found in ethanol and water extract, and its activity was comparable to TBHQ (El-Maaty et al., 2016). By using principal component analysis (PCA), Assefa et al. (2018) reported that methanolic extract of cloves buds had the most distinctive and potent antioxidant capacity, among 30 spices tested, with the highest content of total polyphenols. The high radical scavenging mechanism of clove extracts may be due to the ability of the eugenol to form dimers with two phenolic hydroxyl groups originating from eugenyl intermediate radicals (Cortés-Rojas et al., 2014).

Other research conducted by El-Ghalab et al. (2020) found a slightly different result. Evaluation of the antioxidant activity of different kinds of clove bud extracts revealed that the scavenging activity decreased in the following order using the DPPH assay; the highest activity was ascorbic acid, followed by eugenol-essential oil; then clove extracts, namely ethanol extract, hexane extract, ethyl acetate extract, and aqueous extract, respectively. Meanwhile testing with the ABTS assay in decreasing order are as follows: eugenol has the highest activity, followed respectively by trolox, hexane extract, essential oil, ethanol extract, and ethyl acetate extract, and the least was aqueous extract. Extracts with significant scavenging activities include high quantities of phenolic groups—particularly eugenol—and a low concentration of flavonoids. All extracts contain eugenol, except water extract; this phenol compound contributes mostly to scavenging activity and appears to be more active when combined with terpenes, such as essential oil.

8.4.2 Antimicrobial Activity

Clove extract was the most effective against multidrug resistance (Singh et al., 2018). Clove has been tested pharmacologically against a variety of harmful pathogens and microbes. Jirovetz et al. (2006) exhibit that clove bud extract has antibacterial activity against bacterial isolates of *Bacillus, Serratia marcescens*, and *Salmonella typhi*. Clove's inhibitory activity is attributable to the presence of several components, mostly eugenol. Other compounds are eugenyl acetate, β-caryophyllene, 2-heptanone, acetyl-eugenol, α-humulene, methyl salicylate, iso-eugenol, methyl-eugenol, phenyl propanoides, dehydrodieugenol, trans-confireryl aldehyde, biflorin, kaempferol, rhamnetin, myricetin, and gallic acid (Cai & Wu, 1996). Research by Suhendar and Sogandi (2019) found that clove leaf methanol extract inhibits the development of *Streptococcus mutans* bacteria, with an inhibitory zone of ±32 mm and a minimum inhibitory concentration (MIC) of 20% extract. The mechanism of action of inhibition of clove leaf extract against bacterial growth is thought to be by forming holes in the bacterial cell membrane.

8.4.3 Anticancer

Clove ethyl acetate extract suppresses tumor development while also promoting cell cycle arrest and apoptosis. One of the components of clove ethyl acetate extract, namely, oleanolic acid, was discovered to be responsible for its antitumor properties. Mechanism action was described as promoting Go/G1 cell cycle arrest and causing apoptosis in a dose-dependent manner (Liu et al., 2014).

Clove extract also has potential as an anticancer, as investigated by Kumar et al. (2014). Ethanol and water extract as well as essential oil have potential as in vitro anticancers at different doses using MTT and brine shrimp lethality test (BSLT) assays on MCF-7 human breast cancer cells. Results showed that essential oil had a strong cytotoxic effect on both MTT and BSLT, with an LD50 value of 37 µg/ml after 24 h in BSLT; followed by ethanol and aqueous extract. The IC50 values of the MTT assay after 24 and 48 hours were 36.43 and 17.6 µg/ml, respectively. In addition, research by Dwivedi et al. (2011) investigated the anticancer potential of clove oil, its ethanol and aqueous extract against DU-145 prostate cancer, HeLa cervical cancer, TE-13 esophageal cancer, and MDA-MB-231 (ER-ve) and MCF-7 (ER+ve) breast cancer with normal human peripheral blood lymphocytes to inhibit proliferation by MTT assay.

8.5 CLOVE OLEORESIN APPLICATION

Clove oleoresin has broad applications across several industries, including food and beverages to enhance flavor and aroma and as a natural preservative, as well as in perfumes and pharmaceuticals. It also has many therapeutic purposes due to its antioxidant and antibacterial properties. Utilization of clove oleoresin in the food industry is only in small quantities due to its very strong flavor. Clove is very popular in the cigarette industry ("cigarettes kretek"), especially in Indonesia as the biggest consumer for clove in this industry. In kretek cigarettes, the clove ingredient gives a distinctive aroma of cloves and a hot and biting taste, languid and bitter. Oleoresin is preferred over other clove products, because besides containing essential oils as a vaporized component, it also contains other non-evaporate materials such as resin, so it has a similar aroma and taste to clove flowers. Besides, oleoresin is barely contaminated by bacteria (Nurdjannah & Bermawie, 2012).

Clove oleoresin also has broad applications across several industries, including food and beverages to enhance flavor and aroma and as natural preservatives;

meanwhile in the pharmaceuticals it has many therapeutic effects. Utilization of oleoresin is almost similar to the original form, such as clove simplicial, powder, or clove oil. The study on clove bud extracts revealed active components with medicinal potential. Plants' therapeutic properties are often attributed to their secondary products, rather than a single component. In the food sector, clove bud extract and eugenol can be applied as a flavoring ingredient, condiment, and natural antioxidant and can be applied in several kinds of food products. Incorporation clove oleoresin in food products will improve food properties due to increasing its bioactivity, especially antioxidant activity, without reducing the quality and appearance of the products.

Research by Chatterjee and Bhattacharjee (2015) showed that incorporating lean fractions of clove eugenol in mayonnaise appears to have similar organoleptic features to commercial mayonnaise in terms of thickness, spoonability, and appearance. Even the mayonnaise-formulated clove extracts have better antioxidant activity, phenolic content, and reducing power compared to the commercial sample. Moreover, incorporation of clove oleoresin of 0.5% (w/v) into salted duck eggs revealed antioxidant activity, tested using free radical scavenging activity (DPPH technique), showing considerably greater scavenging value than the control (Harlina et al., 2018; Ribes et al., 2017; Jayanudin et al., 2019).

8.6 CONCLUSIONS

Oleoresin is a precious natural extract possessing many bioactive compounds. Due to broad application and benefits, research on clove oleoresin is very promising. Because, compared to clove essential oils, clove oleoresin research and development is quite limited. In facts oleoresins have more stability in terms of heat stability and flavor uniformity compared to essential oils. However, oleoresins are prone to disintegration and have a limited shelf life if not handled appropriately. Microencapsulation has been utilized to address this issue and ease the usage of oleoresins and essential oils. Microencapsulation techniques are primarily used to protect bioactive chemicals present in oleoresins from deterioration, enhance better release, improve shelf life and stability, as well as increase bioavailability of such bioactive substances. These extracts are increasingly being used directly as antibacterial and antioxidant agents in films and foods. Furthermore, clove oleoresins are a promising source of bioactive chemicals. Incorporating oleoresin into food can improve sensory properties while also providing health advantages. More research into the acceptable amount of oleoresin incorporated into meals or other products without posing health problems and disturbing the food taste still need to be explored. Another matter is the incorporation of spice extracts/oleoresins and essential oils as natural ingredients in food and packaging has its own problems regarding efficacy and consumer acceptance due to their strong aroma and the possibility of allergic reactions and hypersensitivity in some individuals. Moreover, although the toxicity of natural compound is low, their production costs are quite high, which hinders their commercialization/widespread application. Besides, the extraction process usually employs strong chemicals solvent which creates environmental problems regarding handling. Therefore, developing research using green extraction processes and apparatus is urgently needed.

REFERENCES

Abdul Aziz, A. H., D. N. Rizkiyah, L. Qomariyah, I. Irianto, M. A. Che Yunus, and N. R. Putra. 2023. Unlocking the full potential of clove (Syzygium aromaticum) spice: An overview of extraction techniques, bioactivity, and future opportunities in the food and beverage industry. *Processes* 11 (August): 1–17. https://doi.org/10.3390/pr11082453.

Assefa, A. D., Y.-S. Keum, and R. K. Saini. 2018. A comprehensive study of polyphenols contents and antioxidant potential of 39 widely used spices and food condiments. *Journal of Food Measurement and Characterization* 12: 1548–1555. https://doi.org/10.1007/s11694-018-9770-z.

Atawodi, S. E., J. C. Atawodi, B. Pfundstein, B. Spiegelhalder, H. Bartsch, and R. Owen. 2011. Assessment of the polyphenol components and in vitro antioxidant properties of *Syzygium aromaticum* (L.) Merr. & Perry. *Electronic Journal of Environmental, Agricultural & Food Chemistry* 10, no. 3: 1970–1978.

Azwanida, N. N. 2015. A review on the extraction methods use in medicinal plants, principle, strength and limitation. *Medicinal and Aromatic Plants* 4, no. 3 (July): 1–6. https://doi.org/10.4172/2167-0412.1000196.

Batiha, G. E.-S., L. M. Alkazmi, I. G. Wasef, A. M. Beshbishy, E. H. Nadwa, and E. K. Rashwan. 2020. *Syzygium aromaticum* L. (Myrtaceae): Traditional uses, bioactive chemical constituents, pharmacological and toxicological activities. *Biomolecules* 10 (January): 1–17. https://doi.org/10.3390/biom10020202.

Bhagat, N., and A. Chaturvedi. 2016. Spices as an alternative therapy for cancer treatment. *Systematic Reviews in Pharmacy* 7, no. 1: 46–56.

Bitwell, C., S. S. Indra, C. Luke, and M. K. Kakoma. 2023. A review of modern and conventional extraction techniques and their applications for extracting phytochemicals from plants. *Scientific African* 19 (February): 1–19. https://doi.org/10.1016/j.sciaf.2023.e01585.

Boyko, N. N., E. T. Zhilyakova, D. I. Pisarev, et al. 2020. A novel method for the extraction of the main compounds from the essential oil of clove buds. *Farmacia* 68, no. 1: 170–175. https://doi.org/10.31925/farmacia.2020.1.24.

Cai, L., and C. D. Wu. 1996. Compounds from *Syzygium aromaticum* possessing growth inhibitory activity against oral pathogens. *Journal of Natural Products* 59, no. 10: 987–990.

Chatterjee, D., and P. Bhattacharjee. 2015. Use of eugenol-lean clove extract as a flavoring agent and natural antioxidant in mayonnaise: Product characterization and storage study. *Journal of Food Science and Technology* 52, no. 8: 4945–4954. http://doi.org/10.1007/s13197-014-1573-6.

Chemat, F., M. A. Vian, A.-S. Fabiano-Tixier, et al. 2020. A review of sustainable and intensified techniques for extraction of food and natural products. *Green Chemistry* 22: 2325–2353.

Cortés-Rojas, D. F., de Souza, C. R. F., & Oliveira, W. P. (2014). Clove (*Syzygium aromaticum*): A precious spice. *Asian Pacific Journal of Tropical Biomedicine* 4, no. 2: 90–96. http://doi.org/10.1016/S2221-1691(14)60215-X.

Dwivedi, V., R. Shrivastava, S. Husain, C. Ganguly, and M. Bharadwaj. 2011. Comparative anticancer potential of clove (*Syzygium aromaticum*)—an Indian species—against cancer cell lines of various anatomical origin. *Asian Pacific Journal of Cancer Prevention* 12, no. 8: 1989–1993.

El Faqer, O., S. Bendiar, S. Rais, et al. 2022. Phytochemical characterization and immunomodulatory effects of aqueous, ethanolic extracts and essential oil of *Syzygium aromaticum* L. on human neutrophils. *Scientific African* 18 (October): 1–9. https://doi.org/10.1016/j.sciaf.2022.e01395.

El-Ghalab, Y., A. Al Jahid, J. J. Eddine, et al. 2020. *Syzygium aromaticum* L.: Phytochemical investigation and comparison of the scavenging activity of essential oil, extracts and eugenol. *Oriental Pharmacy and Experimental Medicine* 20: 153–158. https://doi.org/10.1007/s13596-019-00416-7.

El-Maati, M. F. A., S. A. Mahgoub, S. M. Labib, et al. 2016. Phenolic extracts of clove (*Syzygium aromaticum*) with novel antioxidant and antibacterial activities. *European Journal of Integrative Medicine* 8, no. 4: 494–504. https://doi.org/10.1016/j.eujim.2016.02.006.

Fathoni, A., E. Saepudin, A. H. Cahyana, D. U. C. Rahayu, and J. Haib. 2017. Identification of non-volatile compounds in clove (Syzygium aromaticum) from Manado. *AIP Conference Proceedings* 1862: 030079. https://doi.org/10.1063/1.4991183.

FDA. 2021. *Clove and Its Derivatives*, 21 C.F.R. 184. 1257(a).

Ferraz, M. C., F. R. Procópio, G. de Figueiredo Furtado, A. M. T. M. Moya, C. B. B. Cazarin, and M. D. Hubinger. 2021. Cinnamon and paprika oleoresin emulsions: A study of physicochemical stability and antioxidant synergism. *Food Research International* 150: 1–9. https://doi.org/10.1016/j.foodres.2021.110777.

Galvin-King, P., S. A. Haughey, and C. T. Elliott. 2020. The detection of substitution adulteration of paprika with spent paprika by the application of molecular spectroscopy tools. *Foods* 9, no. 7: 944. http://doi.org/10.3390/foods9070944.

Gonzalez-Rivera, J., C. Duce, B. Campanella, et al. 2021. In situ microwave assisted extraction of clove buds to isolate essential oil, polyphenols, and lignocellulosic compounds. *Industrial Crops and Products* 161: 1–14. https://doi.org/10.1016/j.indcrop.2020.113203.

Guan, W., S. Li, C. Hou, R. Yan, and J. Ma. 2007. Determination and correlation of solubilities of clove oil components in supercritical carbon dioxide. *CIESC Journal* 58(5): 1077.

Habashy, N. H., M. M. Abu Serie, W. E. Attia, and S. A. M. Abdelgaleil. 2018. Chemical characterization, antioxidant and anti-inflammatory properties of Greek Thymus vulgaris extracts and their possible synergism with Egyptian *Chlorella vulgaris*. *Journal of Functional Foods* 40: 317–328. https://doi.org/10.1016/j.jff.2017.11.022.

Harlina, P. W., M. Ma, R. Shahzad, M. M. Gouda, and N. Qiu. 2018. Effect of clove extract on lipid oxidation, antioxidant activity, volatile compounds and fatty acid composition of salted duck eggs. *Journal of Food Science and Technology* 55: 4719–4734.

Huang, W. Y., Y. Z. Cai, and Y. Zhang. 2009. Natural phenolic compounds from medicinal herbs and dietary plants: Potential use for cancer prevention. *Nutrition and Cancer* 62 (December): 1–20. https://doi.org/10.1080/01635580903191585.

Idham, Z., N. R Putra, A. H. A. Aziz, et al. 2021. Improvement of extraction and stability of anthocyanins, the natural red pigment from roselle calyces using supercritical carbon dioxide extraction. *Journal of CO2 Utilization* 56 (December): 101839. https://doi.org/10.1016/j.jcou.2021.101839.

Idowu, S., A. E. Adekoya, O. O. Igiehon, and A. T. Idowu. 2021. Clove (*Syzygium aromaticum*) spices: A review on their bioactivities, current use, and potential application in dairy products. *Journal of Food Measurement and Characterization* 15: 3419–3435. https://doi.org/10.1007/s11694-021-00915-9.

Jayanudin, M. Fahrurrozi, S. K. Wirawan, and Rochmadi. 2019. Antioxidant activity and controlled release analysis of red ginger oleoresin (*Zingiber officinale* var rubrum) encapsulated in chitosan cross-linked by glutaraldehyde saturated toluene. *Sustainable Chemistry and Pharmacy* 12 (January): 1–7. https://doi.org/10.1016/j.scp.2019.100132.

Jirovetz, L., G. Buchbauer, I. Stoilova, A. Stoyanova, A. Krastanov, and E. Schmidt. 2006. Chemical composition and antioxidant properties of clove leaf essential oil. *Journal of Agricultural and Food Chemistry* 54: 6303–6307. https://doi.org/10.1021/jf060608c.

Kapadiya, S. M., J. Parikh, and M. A. Desai. 2018. A greener approach towards isolating clove oil from buds of Syzygium aromaticum using microwave radiation. *Industrial Crops and Products* 112: 626–632. https://doi.org/10.1016/j.indcrop.2017.12.060.

Kennouche, A., F. Benkaci-Ali, G. Scholl, and G. Eppe. 2015. Chemical composition and antimicrobial activity of the essential oil of *Eugenia caryophyllata* cloves extracted by conventional and microwave Techniques. *Journal of Biologically Active Products from Nature* 5 (April): 1–11. https://doi.org/10.1080/22311866.2014.961100.

Kumar, P. S., R. M. Febriyanti, F. F. Sofyan, D. E. Luftimas, and R. Abdulah. 2014. Anticancer potential of *Syzygium aromaticum* L. in MCF-7 human breast cancer cell lines. *Pharmacognosy Research* 6, no. 4: 350–354. https://doi.org/10.4103/0974-8490.138291.

Ladurner, A., M. Zehl, U. Grienke, et al. 2017. Allspice and clove as source of triterpene acids activating the G protein-coupled bile acid receptor TGR5. front. *Pharmacology* 8 (July): 1–13. https://doi.org/10.3389/fphar.2017.00468.

Li, X., and J. G. Xu. 2017. Chemical composition and antioxidant activities of the oleoresin from the clove buds (*Syzygium aromaticum*). *European Journal of Food Science and Technology* 4, no. 5: 12–20.

Liu, H., J. C. Schmitz, J. Wei, et al. 2014. Clove extract inhibits tumor growth and promotes cell cycle arrest and apoptosis. *Journal of Oncology Research* 21, no. 5: 247–259. https://doi.org/10.3727/096504014X13946388748910.

Madhusankha, G. D. M. P., L. F. Siow, and Y. Y. Thoo. 2023. Efficacy of green solvents in pungent, aroma, and color extractions of spice oleoresins and impact on phytochemical and antioxidant capacities. *Food Bioscience* 56 (September): 1–12. https://doi.org/10.1016/j.fbio.2023.103171.

Moradi, E., H. Rakhshandeh, V. R. Baradaran, et al. 2023. HPLC/MS characterization of *Syzygium aromaticum* L. and evaluation of its effects on peritoneal adhesion: Investigating the role of inflammatory cytokines, oxidative factors, and fibrosis and angiogenesis biomarkers. *Physiological Reports* 11 (January): 1–13. https://doi.org/10.14814/phy2.15584.

Nassar, M., A. Gaara, A. El-Ghorab, et al. 2007. Chemical constituents of clove (*Syzygium aromaticum,* Fam. Myrtaceae) and their antioxidant activity. *Revista Latinoamericana de Química* 35, no. 3: 47–57.

Nurdjannah, N., and N. Bermawie. 2012. Cloves. In *Handbook of herbs and spices*, ed. K. V. Peter, 197–215. CRC Press.

Procopio, F. R., M. C. Ferraz, B. N. Paulino, P. J. do Amaral Sobral, and M. D. Hubinger. 2022. Spice oleoresins as value-added ingredient for food industry: Recent advances and perspectives. *Trends in Food Science & Technology* 122: 123–139. https://doi.org/10.1016/j.tifs.2022.02.010.

Reverchon, E., and C. Marrone. 1997. Supercritical extraction of clove bud essential oil: Isolation and mathematical modelling. *Chemical Engineering Science* 52, no. 20: 3421–3428. https://doi.org/10.1016/S0009-2509(97)00172-3.

Ribes, S., A. Fuentes, P. Talen, and J. M. Barat. 2017. Application of cinnamon bark emulsions to protect strawberry jam from fungi. *LWT- Food Science and Technology* 78: 265–272. http://doi.org/10.1016/j.lwt.2016.12.047.

Rizkiyah, D. N., N. R. Putra, M. A. C. Yunus, et al. 2023. Insight into green extraction for roselle as a source of natural red pigments: A review. *Molecules* 28 (January): 1–15. https://doi.org/10.3390/molecules28031336.

Safitri, Y. D., A. Amalia, Fatimah, and A. Muadifah. 2022. The identification of antibacterial compounds in clove stem extract (*Syzygium aromaticum*) and its effectiveness in inhibiting the growth of *Escherichia coli*. *The Journal of Biosciences* 8, no. 1: 24–29. https://doi.org/10.24114/jbio.v8i1.28089.

Shahidi, F., and A. Hossain. 2018. Bioactives in spices, and spice oleoresins: Phytochemicals and their beneficial effects in food preservation and health promotion. *Journal of Food Bioactives* 3: 8–75. https://doi.org/10.31665/JFB.2018.3148

Šernaitė, L., N. Rasiukevičiūtė, and A. Valiuškaitė. 2020. The extracts of cinnamon and clove as potential biofungicides against strawberry grey mould. *Plants* 9 (May): 1–11. https://doi.org/10.3390/plants9050613.

Shukla, A., S. N. Naikb, V. V. Gouda, and C. Das. 2019. Supercritical CO2 extraction and online fractionation of dry ginger for production of high-quality volatile oil and gingerols enriched oleoresin. *Industrial Crops and Products* 130: 352–362. https://doi.org/10.1016/j.indcrop.2019.01.005.

Singh, P. A., S. D. Desai, and J. Singh. 2018. A review on plant antimicrobials of past decade. *Current Topics in Medicinal Chemistry* 18, no. 10: 812–833.

Singh, P. A., S. Tushir, S. Hazra, S. S. Purewal, and G. Agarwal. 2024. Novel approaches for the extraction and identification of phytoconstituents from herbs and spices. *The Natural Products Journal*. http://doi.org/10.2174/0122103155293641240417072907

Soldan, A. C. F., S. Arvelos, E. O. Watanabe, and C. E. Hori. 2021. Supercritical fluid extraction of oleoresin from Capsicum annuum industrial waste. *Journal of Cleaner Production* 297 (March): 1–11. https://doi.org/10.1016/j.jclepro.2021.126593.

Sowbhagya, H. 2019. Value-added processing of by-products from spice industry. *Food Quality and Safety* 3, no. 2: 73–80. https://doi.org/10.1093/fqsafe/fyy029.

Suhendar, U., and S. Sogandi. 2019. Identification of bioactive compounds in clove leaves (*Syzygium aromaticum*) extract as inhibitor Streptococcus mutans. *Al-Kauniyah: Jurnal Biologi* 12, no. 2: 229–239. http://doi.org/10.15408/kauniyah.v12i2.12251.

Tekin, K., M. K. Akalin, and M. G. Şeker. 2015. Ultrasound bath-assisted extraction of essential oils from clove using central composite design. *Industrial Crops and Products* 77: 954–960. http://doi.org/10.1016/j.indcrop.2015.09.071.

Xu, L., X. Li, H. Chen, et al. 2023. Antibacterial and antioxidant properties of clove extract applied in the production of dry-cured duck. *LWT-Food Science and Technology* 185 (August): 1–8. https://doi.org/10.1016/j.lwt.2023.115153.

Xue, Q., Z. Xiang, S. Wang, Z. Cong, P. Gao, and X. Liu. 2022. Recent advances in nutritional composition, phytochemistry, bioactive, and potential applications of *Syzygium aromaticum* L. (Myrtaceae). *Frontiers in Nutrition* 9 (October): 1–25. https://doi.org/10.3389/fnut.2022.1002147

Yang, Y. C., M. C. Wei, and S. J. Hong. 2014. Ultrasound-assisted extraction and quantitation of oils from *Syzygium aromaticum* flower bud (clove) with supercritical carbon dioxide. *Journal of Chromatography A* 1323: 18–27. https://doi.org/10.1016/j.chroma.2013.10.098.

Yassin, M. T., A. A. F. Mostafa, and A. A. Al-Askar 2020. In vitro anticandidal potency of *Syzygium aromaticum* (clove) extracts against vaginal candidiasis. *BMC Complementary Medicine and Therapies* 20 (January): 1–9. https://doi.org/10.1186/s12906-020-2818-8.

Zhang, Q. W., L. G. Lin, W. C. Ye. 2018. Techniques for extraction and isolation of natural products: A comprehensive review. *Chinese Medicine* 13, no. 20: 1–26. https://doi.org/10.1186/s13020-018-0177-x.

9 Clove Oil
Processing and Preservation

Heri Septya Kusuma, Ganing Irbah Al Lantip, and Xenna Mutiara

9.1 INTRODUCTION

The rapid development of technology does not rule out the possibility of innovations in food preservation. Food preservation is the process of stopping or slowing down spoilage by microorganisms (Blackburn, 2006). Several natural methods can be used to preserve some types of foodstuffs. Recently, natural antimicrobial ingredients have been considered to have potential solutions as preservatives. Essential oils are one of the ingredients that can be used as preservatives because they have active components that are very promising due to their low toxicity and antioxidant and effective antibacterial properties (Fan et al., 2024). Essential oils consist of a complex mixture of several hydrocarbon components and oxygenated compounds (Temelli et al., 2012). Essential oils are obtained from aromatic plants by isolating flowers, fruits, leaves, roots, buds, twigs, and seeds (Das & Mukherjee, 2023).

Cloves (*Syzigium aromaticum* L) are native to Indonesia, especially the Maluku Islands (Baliga et al., 2015). Clove oil is one of the green gold commodities in Indonesia. Almost all parts of the clove tree can be isolated into essential oils. The total content of clove oil is 15–20% (w/w), with most of the compound eugenol being 70–95% (Guldiken et al., 2021). This eugenol content has many benefits, especially in the field of food preservation. Eugenol is famous for its pharmacological properties, namely antimicrobial, anticancer, antioxidant, anti-inflammatory, and analgesic (Sharma et al., 2024).

Various conventional and modern extraction methods can be used to isolate clove oil. However, differences in the use of extraction methods cause differences in the quality of clove oil obtained (Singh et al., 2024; Singh & Baldi, 2023). This is because each extraction process has its treatment of the isolated compounds and certain operating conditions. In addition to reviewing extraction methods, clove oil storage procedures need to be considered because of the content of volatile eugenol compounds (Sharma et al., 2022).

Therefore, this chapter will further explain the extraction methods that can be applied in the isolation of clove oil and their implementation. It will also explain the clove oil storage techniques that follow the nature of clove oil. The application of clove oil as a preservative agent can be a preservation solution using natural materials.

DOI: 10.1201/9781003521785-9

9.2 STATE OF CLOVE FARMING AND REFINING IN INDONESIA

The clove plant is known as a native plant of Indonesia and is one of the traded spice groups. Besides Indonesia, the two largest clove-producing countries are Zanzibar and Madagascar. In Indonesia, there are several points of distribution of clove plant growth, namely East Nusa Tenggara, Central Java, East Java, North Sulawesi, and Maluku, as shown in Figure 9.1. The average clove yield was 140,797 tons in 2021, with a total land area of 573,873 ha (Kusuma et al., 2023).

Currently, the clove plant has a high commercial value as a rapidly growing international trade commodity. The parts of the plant that are often traded are its dried flower buds (cloves), clove oil, and oleoresin. The world market is estimated to require 63,700 tons of clove commodities, and Indonesia can contribute 66% of the world's needs. The top five clove-importing countries are India, the United States, Germany, France, and Singapore (Sarma et al., 2014).

Several varieties of cloves can be grown in Indonesia. However, some superior varieties are often planted in Indonesia, such as (Kementerian Pertanian, 2018):

(1) Siputih cloves have large leaf blades that are yellow or light green in color, have fewer leafy branches, a yellow flower color, a large flower size, and dozens per clump.
(2) Cengkih Sikotok has small leaflets, a green to dark green-blackish and glossy color, lush and low branches, and reddish-yellow flowers, with each clump having 20 to 50 flowers.
(c) Zanzibar cloves have long, slender leaves, are dark green in color, have redder flowers with high flower production, and are the best variety.

Different types of cloves cause different levels of eugenol in clove oil. Pollinated clove flowers will give a higher eugenol yield than whole cloves. Bad clove leaves or deciduous clove leaves that experience natural drying through sunlight and are then wetted again by rainwater, rotten, dried again by the sun, and so on, make the oil

FIGURE 9.1 Distribution of clove plants in Indonesia.

not have a typical clove fragrance. Good-quality fallen clove leaves will have a clove fragrance and a better yield (Kementerian Pertanian, 2018).

The development of clove oil production in Indonesia, especially in the Maluku region, mostly uses simple technology. Clove oil is produced through a distillation process (steam distillation) of leaves, stalks, and flowers. The distilling equipment is an iron kettle with a capacity of 500 kg. The heating process uses fuel in the form of wood, with distillation times ranging from 10 to 24 hours. The vapor coming out of the tank will be flown into a pipe that passes through the cooling basin, and then the oil and water will be separated using a separating funnel. The designed plant, with a refining capacity of up to 18 tons/day, will produce 504 kg/day of oil (Bustaman, 2011).

9.3 METHODS FOR CLOVE OIL EXTRACTION

9.3.1 Hydrodistillation

Hydrodistillation is a popular and massive, simple, classical extraction method used to extract essential oils because it is the easiest to operate. Hydrodistillation is a type of distillation used to separate volatile compounds by boiling samples and then condensing the vapors. This method consists of a spherical container containing a sample with water as solvent. The sample mixture and water are then heated using an electric coil and steam is formed that moves through the condenser for condensation. The condensed liquid contains oil and water to be further separated to make pure essential oils (Katekar et al., 2023). Hydrodistillation also produces by-products in the form of hydrolats (suspended mixtures of water and oil) and wastewater (Madhura et al., 2024).

Kapadiya and Desai (2017) optimize hydrodistillation extraction conditions to obtain optimal conditions for clove bud extraction. The Taguchi method is used to optimize with four optimization parameters, namely solid loading, extraction time, sample size, and solvent volume (water). The results showed that the optimal parameters for the extraction of clove buds by the hydrodistillation method were within 45 min, 20 g of solid loading size (565 microns), and using 300 mL of water, which resulted in a clove oil yield of 11.35% w/w (Kapadiya & Desai, 2017).

9.3.2 Soxhlet Extraction

Soxhlet extraction is a classic extraction method with a condenser, extraction chamber, cellulose thimble, and round bottom flask. Sometimes, a heating coat is added to facilitate the extraction process due to its role in heating the solvent flask. When the solvent is heated and then reaches its boiling point, the solvent will evaporate and then pass through the condenser, and condensation occurs. The condensed vapor will pass through the sample in the cellulose thimble and separate the essential oils contained in the sample. The extracted mixture of solvent and oil will then come together inside a round-bottomed flask (Rashd et al., 2024).

Gowri and Manimegalai (2019) extracted 10 g of clove powder by Soxhlet extraction method for 8 hours with the addition of 500 mL of solvent, then compared chemical constituents and phytochemicals using GC-MS analysis. Three different solvents were used in this study, namely petroleum ether, chloroform, and ethanol. Different types of solvents result in differences in polarity and molecular structure and thus will affect the extraction yield (Bontzolis et al., 2024). The results of the study showed that cloves extracted with ethanol as a solvent contain more eugenol as the main component, whereas eugenol has various benefits such as antioxidants, antiseptics, and antifungals (Gowri & Manimegalai, 2019).

Yunusa S et al. (2019) compared the chemical composition and yield of extracted clove oil with Soxhlet extraction and ultrasonic-assisted extraction. 60 g of clove buds were extracted by the Soxhlet method with 300 mL of n-hexane solvent for 6 hours. Whereas in the ultrasonic-assisted extraction method, 20 g of clove powder mixed with 150 mL of n-hexane was then extracted for 20 min. The results showed that cloves extracted by the Soxhlet extraction method produced a yield of 54.70% with a eugenol acetate content of 5.72%, while cloves extracted by ultrasonic-assisted extraction method produced a yield of 71.55% with a eugenol acetate content of 9.24%. This study concludes that Soxhlet extraction is a traditional extraction method that takes longer with a larger amount of solvent but produces less oil yield compared to the newer extraction method, ultrasonic-assisted extraction. So today, there are more and more innovative extraction methods to maximize the yield of essential oils (Yunusa S et al., 2019).

9.3.3 Microwave-Assisted Extraction

Microwave-assisted extraction (MAE) is increasingly being applied massively due to its environmentally friendly nature and ability to save time and energy. Heat energy produced by microwaves consists of two mechanisms, namely dipole rotation and ion conduction. This heat will be absorbed by water in the system resulting in a rapid increase in pressure and temperature. Great pressure will affect the swelling of plant cells. Over time, plant cells such as cellulose, hemicellulose, pectin, and other substances that experience swelling will rupture and release active constituents from the cell, thereby increasing the yield of phytoconstituents (Patil & Rathod, 2023). This cell split will reduce the solvent extraction resistance between the cell wall and cytoplasm, thereby shortening the dissolution rate of the active ingredient (Chen et al., 2022).

Microwave-assisted hydrodistillation (MAHD) is an MAE derivative used for the extraction of heat-sensitive compounds. MAHD combines distillation with water as a solvent (hydrodistillation) with heating using a microwave. MAHD combines mass transfer and internal heat dependent on microwave power which can significantly increase extraction rates (Nour et al., 2024. This extraction rate is due to the production of radiation by electromagnetic waves, which continues to increase as thermal stress increases and internal pressure in the cell, which can result in rupture of the cell wall. The rupture of the cell wall will make it easier for volatile compounds in the material to move outward (Benmoussa et al., 2023). MAHD is proven to shorten the extraction time and effective heating rate, use less solvent, and produce quality oil at low cost so that it can be applied to laboratory and even industrial scales. Moreover, MAHD is an excellent alternative extraction method compared to other solvent extraction techniques such as the Soxhlet method due to its environmentally friendly nature (Boudraa et al., 2021; Suttiarporn et al., 2024).

Leon-Méndez et al. (2018) compare conventional hydrodistillation methods with MAHD to extract clove plants for the development of cosmetic formulas in the form of essential oil microcapsules. Based on the extraction kinetics obtained, cloves weighing 500 g extracted by the MAHD method for 40 min have the same recovery rate as cloves extracted by conventional hydrodistillation methods for 3 hours. This shows that the MAHD technique requires less time compared to the hydrodistillation technique to extract cloves. This phenomenon is influenced by heat transfer that occurs in materials that experience three types of heat transfer, namely, radiation, conduction, and convection, simultaneously. This accumulated heat can heat the entire sample simultaneously with high speed and less energy requirement but is capable of much better performance, whereas in the conventional hydrodistillation method, the heat that occurs in the system is only conduction and convection heat, so the heating is less than optimal (Leon-Méndez et al., 2018).

9.3.4 Ultrasound-Assisted Extraction

Most of the active compounds in plant cells are unstable. High heating during the extraction process can result in damage to plant tissue. Ultrasound-assisted extraction (UAE) offers several options that can be considered to overcome this problem. The UAE method can improve extraction efficiency. This is because the UAE has the capability of solvent penetration into plant cells through cavitation and mechanical vibration using ultrasonic waves. The amplitude of ultrasonic waves plays an important role in the energy supplied to the material and the level of cavitation. This wave has a range of 20 kHz–40 kHz so that it can extract components from plants without degrading existing cells (Tekin et al., 2015). In addition, the UAE has several advantages such as being environmentally friendly, easy to operate, safe because it is operated at low atmospheric pressure and temperature, shorter extraction time, and good extract quality. The application of the UAE for clove oil extraction is reported only for ultrasound-assisted hydrodistillation and UAE combined with supercritical carbon dioxide extraction.

(1) Ultrasound-assisted hydrodistillation
Ultrasonic-assisted hydrodistillation (UAHD), also called sonohydrodistillation, is an extraction method that combines hydrodistillation with ultrasound. Ultrasound mechanisms are fragmentation, erosion, capillarity, detexture, and sonoporation. Sonic waves will propagate through a liquid medium (in this case water) and form cavitation bubbles. The phenomenon of cavitation bubbles will cause high shear forces on the medium during the extraction process. This mechanism will result in fragmented particles and then induce erosion of plant structures. Erosion that occurs in the plant structure will increase the depth and speed of fluid penetration into the plant pores. The cell pores will then release cellular content and extractive media so that the plant experiences detexture or damage to the plant structure (Chemat et al., 2017).

Jadhav et al. (2023) conducted UAHD trials from clove buds and optimized with Box-Behnken design as a response surface methodology with four parameters, namely extraction time, capacity, heating coat power, and pulse ratio. The results showed that the optimal conditions needed to maximize the yield of clove oil with the UAHD method were an extraction time of 49.154 min, a capacity of 14.512 g, a power of 489.372 W, and a pulse ratio of 20:40. In this condition, the yield of clove oil obtained was as much as 12.5% with a eugenol content of 64.37%. Compared to conventional hydrodistillation methods, UAHD is proven to accelerate extraction time by up to 42%, save power by up to 75%, and save cooling water requirements by up to 58%. This shows that UAHD is a viable method for extracting cloves as an alternative to sustainable and environmentally friendly methods (Jadhav et al., 2023).

(2) Ultrasound-assisted supercritical carbon dioxide

Ultrasound-assisted supercritical carbon dioxide (USC-CO_2) is a combination of ultrasonic wave extraction with the addition of supercritical CO_2 solvent. This combination was developed to obtain faster extraction time and have high time efficiency. Ultrasonic waves result in swelling of the cell wall so that the cell pores undergo enlargement. Enlargement of these pores will make it easier for supercritical CO_2 to penetrate cells. Supercritical CO_2 penetrating the cell will result in rapid mass transfer because supercritical CO_2 has low viscosity surface tension properties and high diffusivity. USC-CO_2 undergoes three periods that describe the condition of the material when extracted, namely a rapid initial extraction period, a slower extraction period, and an unchanged extraction period. The rapid initial extraction period is caused by the oil being concentrated on the outside of the particle where the supercritical fluid meets the resistance of the matrix (Wei, Lin et al., 2016).

USC-CO_2 does not degrade the final product and offers several advantages such as shorter time, environmental friendliness, and higher yield. Wei et al. (2016) compared three methods for clove bud extraction, namely, heat reflux extraction, steam distillation, and USC-CO_2. Clove extraction using the heat reflux extraction method resulted in an oil yield of 16.51% with an extraction time of 240 min. Clove extraction by this method produces an oil that is yellow-brown in color and a specific odor that is different from the smell of clove oil which indicates that this oil contains more residues or unwanted constituents such as organic solvents. Extraction by steam distillation method yielded 20.04% yield within 300 hours, while extraction using the USC-CO_2 method resulted in a yield of 23.19% for 114 min. Clove extract obtained by the USC-CO_2 method is pale yellow and has a characteristic smell of cloves. The study proved that extraction with the USC-CO_2 method is more efficient because the extraction rate is high and requires a shorter time and higher yield compared to the steam distillation and heat reflux extraction methods (Wei, Lin et al., 2016).

9.3.5 Supercritical Fluid Extraction

Essential oils are usually produced using traditional methods such as hydrodistillation, steam distillation, and solvent distillation. The use of these methods is due to their low cost. However, these methods have some drawbacks, such as the thermal degradation of some components contained in essential oils (Guan et al., 2007). In addition, some solvent residues can pollute the environment and contaminate the essential oil. This also applies to the extraction process of clove oil. So some of the disadvantages caused by traditional extraction methods need to be overcome. The disadvantages that occur can be overcome by using the supercritical fluid extraction method (Khajeh et al., 2004). Supercritical fluid extraction has been around since the 1980s and is a new method with good compound isolation quality (Bogusz, 2008). The method has the advantage that a small change in temperature and pressure at critical conditions will cause a very large change in the density and solubility of the solvent (Lo & Baird, 2003). However, there is a disadvantage: the cost incurred to be able to use supercritical fluid extraction methods is more expensive than traditional methods (Lehotay & Schenck, 2000).

The solvent used in the supercritical fluid extraction process is usually CO_2. The use of CO_2 in terms of safety is more secure; its availability in nature is quite plentiful; it has good extraction properties; and it has a low critical point at 73 atm and 31°C. Supercritical fluid extraction with CO_2 solvent is suitable for hydrophobic compounds because these compounds have low polarity while CO_2 has high diffusivity (Yamada et al., 2017). The extraction of clove oil by the supercritical fluid method using a solvent in the form of CO_2 has been carried out by several authors: Ivanovic et al. (2013), Yazdani et al. (2005), and Della Porta et al. (1998). Conditions explored ranged from 10 to 50 MPa and temperatures from 30 to 40°C. It was found that as the pressure conditions increased, more clove oil was obtained (Gopalakrishnan et al., 1990). The best results were obtained at 220 bar pressure and 40°C with a 29.84% eugenol compound content. The critical fluid method with CO_2 solvent is more efficient than Soxhlet extraction. This is because, at low temperatures, it takes a shorter time and does not require additional steps to remove solvents to increase oil yield (Frohlich et al., 2019).

Many methods can be used to isolate clove oil. The method used can adjust to the condition or quality of the clove oil needed. The latest methods used to isolate clove oil can be seen in Table 9.1.

9.4 CLOVE OIL STORAGE

Clove oil is usually colorless or pale yellow with a characteristic clove aroma. Clove oil contains three main compounds: eugenol, eugenol acetate, and caryophyllene (Nurdjannah & Bermawie, 2012). Clove oil is included as an essential oil export commodity. Indonesia can export 2,500 MT to meet the needs of the global market. This is because the demand for clove oil is increasing because of the use of oil as a cosmetic, food, and body care product. So the storage process of clove oil needs to be considered during the shipping process. The content of volatile compounds in clove oil makes the storage or packaging process an urgency in itself (Pandey et al., 2024).

Hastuti et al. (2017) made observations regarding the storage of clove oil for 4 months at a temperature of 25°C, placed in an aluminum polyethylene bag. It was found that the stored oil increased in the content of eugenol compounds but decreased in the content of eugenol acetate and caryophyllene compounds. The reduction of the eugenol acetate

TABLE 9.1
Clove oil extraction method.

No	Tittle	Method	Procedure	Work overview	Research results	Reference
(1)	Enhanced extraction of clove essential oil by ultrasound and microwave-assisted hydrodistillation and their comparison in antioxidant activity	Ultrasound- and microwave-assisted hydrodistillation	The material is a dried clove bud that has been pulverized. Furthermore, the material will be added to distilled water in a ratio of 1:20 (g/mL). The pretreatment process uses ultrasonic waves at 40°C for 45 min. Extraction was done using a microwave power of 400 W for 45 min.	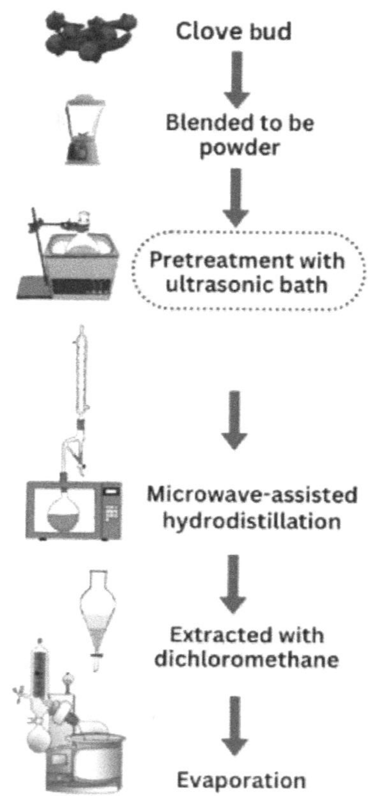	The oil obtained contained 78.49% eugenol and an antioxidant content of 49.77 ± 1.88 mM Trolox/mg.	(Suttiarporn et al., 2024)

TABLE 9.1 *(Continued)*
Clove oil extraction method.

No	Tittle	Method	Procedure	Work overview	Research results	Reference
(2)	Ultrasound-assisted hydrodistillation for extraction of essential oil from clove buds—A step toward process improvement and sustainable outcome	Ultrasound-assisted hydrodistillation	The material is dried and mature clove buds with an average moisture content of 8% and pulverized to 1,000 microns. The material will be rehydrated for 15 min. Next, the extraction process is ultrasonic (750 W, frequency 20 kHz) for 50 min with a ratio of 15 g of material to 500 mL of distilled water.	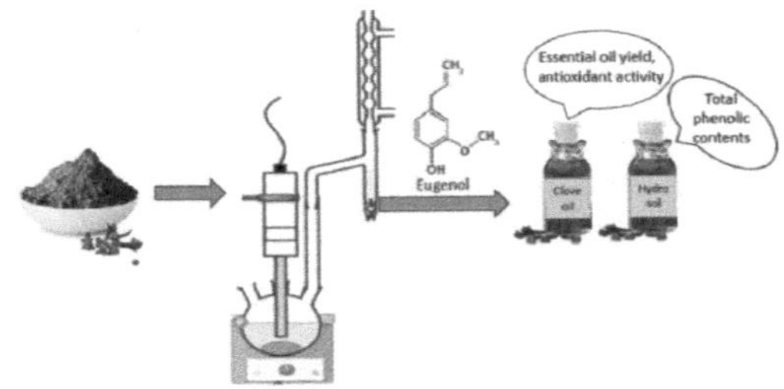	The oil yield was 15.23% (w/w), with a eugenol content of 70.29%, a phenolic content of 101.21mg GAE/g, and an oil antioxidant activity of 91.25%.	(Jadhav et al., 2023)
(3)	In situ microwave-assisted extraction of clove buds to isolate essential oil, polyphenols, and lignocellulosic compounds	In-situ microwave-assisted extraction	The material is dry clove buds. The extraction process took 90 min at 150 W with an ingredient ratio of 1:40 w/w. The solvent used was distilled water.	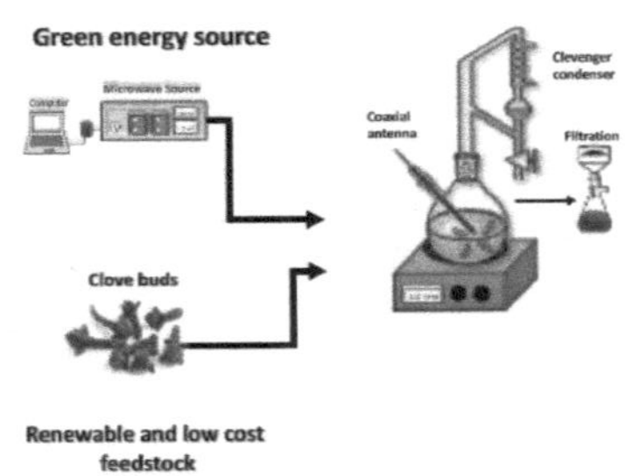	The oil yield obtained was 16%, with a eugenol content of 48.9%, β-caryophyllene content of 42.8%, and α-caryophyllene content of 3.7%.	(Gonzalez-Rivera et al., 2021)

(Continued)

Table 9.1 *(Continued)*
Clove oil extraction method.

No	Tittle	Method	Procedure	Work overview	Research results	Reference
(4)	Supercritical fluid extraction assisted by cold pressing from clove buds: Extraction performance, volatile oil composition, and economic evaluation	Supercritical fluid extraction assisted by cold pressing	The material used is clove buds that have been stored in a freezer at −18°C. Next, the material is pulverized to 16–80 mesh. 12 g of material will be extracted at 40°C with 150 bar pressure for 20 min.	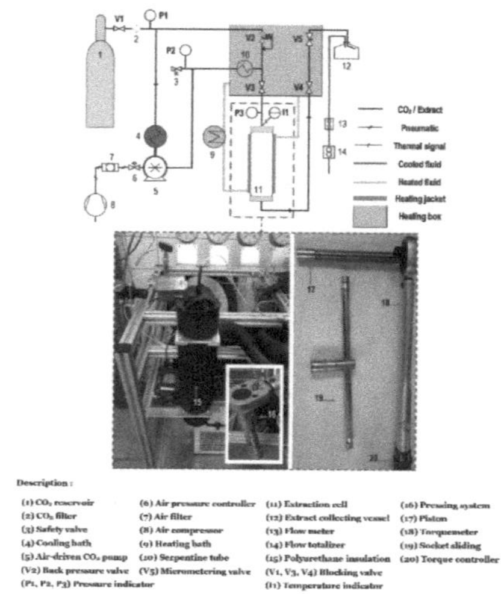	The oil yield obtained was 18.2 ± 0.1% (w/w), with 55.4% eugenol content, 7.77% β-caryophyllene content, and 0.86% α-Humulene content.	(Hatami et al., 2019)
(5)	Extraction of α-humulene-enriched oil from clove using ultrasound-assisted supercritical carbon dioxide extraction and studies of its fictitious solubility	Ultrasound-assisted supercritical carbon dioxide extraction	The dried clove buds amounted to 20 g and had been pulverized to a size of 0.355 mm. The extraction process was carried out at 44°C, 18.5 MPa pressure, 21.4 g/min CO_2 flow rate, and 115 mL extraction time.	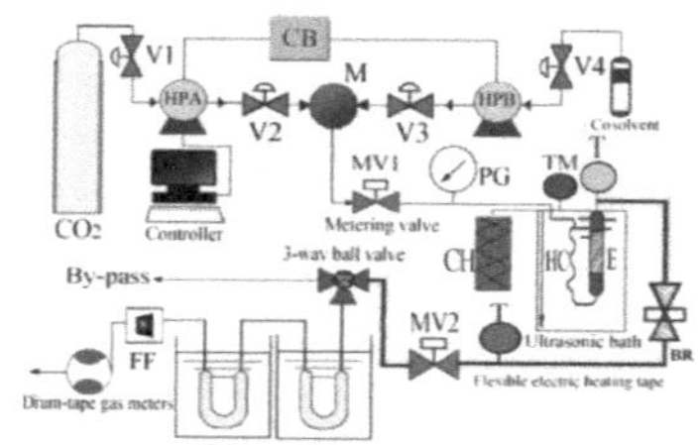	The oil yield was 23.31 ± 0.77 (g/g cloves, w/w%), and the α-humulene content was 2.75 ± 0.09 (g/g oil, w/w%).	(Wei, Xiao et al., 2016)

compound in clove oil changes the quality of clove oil, such as the aroma and flavour of clove oil becoming less sharp, reducing the effectiveness of clove oil in medical or aromatherapy applications, and reducing the market value of clove oil (Hastuti et al., 2017).

Mai (2020), after conducting research, found the best storage results: clove oil can be stored in dark bottles in the refrigerator at a temperature of 4°C. Under these storage conditions, the content of eugenol increases, although the compound eugenol acetate decreases but not significantly. If the oil is stored at high temperatures, it causes an increase in unidentified oxidation products, leading to deterioration in the quality of the oil (Mai, 2020).

9.5 CLOVE OIL APPLICATION FOR PRESERVATION

Clove oil is used as a preservative because it is antimicrobial, antibacterial, antifungal, antioxidant, and antiviral. In addition, clove oil can inhibit the production of enzymes such as amylase and proteases by binding to proteins to extend the shelf life of a product (Lekjing, 2016). Lekjing (2016) experimented by comparing chitosan without clove oil with chitosan modified with clove oil for the preservation of pork sausages. Chitosan 2.0% v/v was combined with three volume variations of clove oil, namely, 0.5, 1.0, and 1.5% v/v. The results showed that chitosan combined with 1.5% v/v clove oil was most effective for extending the shelf life of pork sausages. Clove oil has been shown to inhibit the growth of *L. monocytogenes, Aeromonas hydrophila, E. coli, Salmonella typhimurium, S. aureus*, and native putrefactive flora in various meat products (Lekjing, 2016).

Liu et al. (2020) analyzed the combination of clove oil with kojic acid (KA) in a fully deacetylated chitosan (FDCH) edible coating to coat white shrimp in cold storage for 15 days. The results showed that the combination of clove oil and kojic acid in FDCH at the ratio of FDCH:KA:clove oil 1:0.25:0.25 effectively suppressed the growth of microorganisms by killing some bacteria in the head and shell of white shrimp, increasing shelf life, delaying color change, and maintaining the texture and structure of white shrimp (Liu et al., 2020). Pandey et al. (2023) studied clove oil combined with edible coating for preservation purposes of fresh apple pieces with the help of Box-Behnken design optimization modeling. Seventeen experimental combinations with 3 variables, namely, clove oil nanoemulsion concentration, time, and temperature, were carried out. Significantly, a nanoemulsion coating modified with clove oil improved the physicochemical characteristics of apples by a value of $p < 0.05$. Clove oil nanoemulsion was proven to extend the shelf life of fresh apple pieces at 4°C for 14 days with a clove oil content of 1% and showed good antioxidant activity of 50–80% (Pandey et al., 2023).

Clove oil, with its phenolic compounds, is basically able to inhibit microbial growth because it can affect cell metabolism. Due to its properties, fortifying with clove oil encapsulation is possible to preserve the nutritional value of a food product (Özdikicierler & Günç Ergönül, 2022). The concept of encapsulation involves coating liquid droplets with polymers as environmental protection. There are three methods of clove oil encapsulation:

(1) Spray or lyophilization (freeze-drying) method
This is a frequently applied method for oil encapsulation. The principle is that the spray dryer is based on the emulsification of the core material with

a solution of wall material. This method is known for its low cost, being suitable for a wide selection of walls, providing good efficiency, being suitable for large scales, being suitable for thermosensitive substances that are unstable to water, and having stability in the encapsulated product. The most important indicator during the encapsulation process is efficiency, which shows how much mass of clove oil can be entrapped inside the capsule (Talón et al., 2019).

(2) Emulsification method
The oil elution process involves the dispersed phase of an aqueous solution, which is water, and the dominant phase, which is oil or lipid. Emulsion can be performed continuously to reduce the interfacial tension of the dispersed phases and produce a stable emulsion. Emulsification can be performed in many ways, such as the dripping method (Hasheminejad et al., 2019), the nanoemulsion method (Ruíz-González et al., 2019), and the ultrasonic emulsification method (Shahavi et al., 2015). Color assessment, electron microscopy observation, surface charge measurement, particle size, and encapsulation efficiency can be used to determine the encapsulation performance.

(3) Entrapment method in liposomes
Liposomes are spherical vesicles with one double lipid layer that can form in an aqueous solution. The advantages of using the liposome method are that the particles are biodegradable, non-immunogenic, biocompatible, and non-toxic. However, there are drawbacks, such as expensive process costs and low solubility. Clove oil liposomes can be obtained by the ethanol injection method. Clove oil, phospholipids, and cholesterol will be dissolved in ethanol while stirring. Furthermore, it will be injected into ultrapure water, and liposomes will be formed after the ethanol solution is in contact with the aqueous solution. In determining the performance of the method, morphological characterization by electron microscopy, particle distribution, and zeta potential testing can be done (Cui et al., 2015).

9.6 FUTURE PROSPECTS OF CLOVE OIL

The content of clove oil makes the oil still soar in the world market. Table 9.2 shows some of the components contained in clove oil and their benefits (Jacob Pulikkotil, 2015). With the development of technology, the benefits of clove oil are much more varied. This is in accordance with research conducted by Putra et al. (2024) (Putra et al., 2024) in a bibliometric review to see the prospects for clove oil in the future. The prospect of clove oil needs has increased every year. In addition, Putra et al. (2024) also underlined the importance of oil extraction with environmentally friendly technology, free from solvent hazards, to support sustainable development goals. Therefore, the continuation of research on the benefits, extraction methods, and storage methods of clove oil is still very much needed.

TABLE 9.2
Clove oil components and benefits.

No	Compound	Benefits
(1)	Eugenol	Antibacterial, antifungal, antioxidant, and antineoplastic properties. Can serve as stabilizers and antioxidants in rubber and plastic manufacturing. Used as a component of food packaging materials.
(2)	Eugenol acetate	Antibacterial, anticancer, anti-inflammatory, and antioxidant properties. In addition, eugenol acetate can prevent skin cancer by inhibiting cancer cell proliferation.
(3)	β-Caryophyllene	Anti-inflammatory, antimicrobial, antibacterial, and antioxidant properties. It helps relieve pain, prevent osteoporosis, and relieve anxiety.
(4)	α-Humulene	Anti-inflammatory, antimicrobial, and anticancer properties. Can be developed as a food preservative.
(5)	Humulenol	Analgesic properties can help relieve pain. Good anti-inflammatory properties reduce inflammation in the airways.

9.7 CONCLUSIONS

There are various methods used to extract clove essential oil, ranging from traditional methods to modern methods. Each of the methods indicates its advantages and disadvantages. Microwave-assisted extraction is one of the most promising modern extraction methods because it is more effective and efficient, has good oil quality, and is energy efficient, so it is environmentally friendly. Cloves extracted by this method show recovery rates up to 4.5 times better compared to traditional methods. The ultrasound-assisted extraction method is an alternative clove extraction method that utilizes ultrasonic waves in its application. In addition to being environmentally friendly, ultrasound-assisted extraction also shows various advantages in extracting cloves. This method is considered to have a fast extraction time, use less energy, be easy to operate, be safe, and produce high-quality essential oils. To achieve better results, the ultrasound-assisted extraction method was developed by combining it with the hydrodistillation method or with the addition of supercritical CO_2 solvents. In addition to being combined with ultrasound-assisted extraction, supercritical CO_2 is also utilized for supercritical fluid extraction. The greater the pressure used for supercritical fluid extraction, the more clove oil is obtained. To maintain the quality of clove essential oil, storage is one of the most influential factors. Clove oil can be stored in a dark bottle in the refrigerator at a temperature of 4°C. If the temperature used to store clove oil is high, oxidation will occur, which will affect the content of clove essential oil. For this reason, clove oil is stored well to maintain the quality of its contents.

REFERENCES

Baliga, M. S., Mane, P. P., Timothy Nallemgera, J., Thilakchand, K. R., & Kalekhan, F. (2015). Dietary spices in the prevention of rheumatoid arthritis. In *Foods and Dietary Supplements in the Prevention and Treatment of Disease in Older Adults* (pp. 41–49). Elsevier. https://doi.org/10.1016/B978-0-12-418680-4.00005-1

Benmoussa, H., Béchohra, I., He, S., Elfalleh, W., & Chawech, R. (2023). Optimization of sonohydrodistillation and microwave assisted hydrodistillation by response surface methodology for extraction of essential oils from Cinnamomum cassia barks. *Industrial Crops and Products*, *192*, 115995. https://doi.org/10.1016/j.indcrop.2022.115995

Blackburn, C. de W. (2006). Managing microbial food spoilage: An overview. In *Food Spoilage Microorganisms* (pp. 147–170). Elsevier. https://doi.org/10.1533/9781845691417.2.147

Bogusz, M. J. (2008). *Chapter 1 Opioids: Methods of Forensic Analysis* (pp. 3–72). https://doi.org/10.1016/S1567-7192(06)06001-3

Bontzolis, C. D., Dimitrellou, D., Plioni, I., Kandylis, P., Soupioni, M., Koutinas, A. A., & Kanellaki, M. (2024). Effect of solvents on aniseed aerial plant extraction using soxhlet and ultrasound methods, regarding antimicrobial activity and total phenolic content. *Food Chemistry Advances*, *4*, 100609. https://doi.org/10.1016/j.focha.2024.100609

Boudraa, H., Kadri, N., Mouni, L., & Madani, K. (2021). Microwave-assisted hydrodistillation of essential oil from fennel seeds: Optimization using Plackett–Burman design and response surface methodology. *Journal of Applied Research on Medicinal and Aromatic Plants*, *23*, 100307. https://doi.org/10.1016/j.jarmap.2021.100307

Bustaman, S. (2011). Potensi pengembangan minyak daun cengkih sebagai komoditas ekspor untuk Maluku. *Jurnal Penelitian Dan Pengembangan Pertanian*, *30*(4), 132–139. https://doi.org/10.21082/jp3.v30n4.2011

Chemat, F., Rombaut, N., Sicaire, A.-G., Meullemiestre, A., Fabiano-Tixier, A.-S., & Abert-Vian, M. (2017). Ultrasound assisted extraction of food and natural products. Mechanisms, techniques, combinations, protocols and applications. A review. *Ultrasonics Sonochemistry*, *34*, 540–560. https://doi.org/10.1016/j.ultsonch.2016.06.035

Chen, Y., Xu, F., Pang, M., Jin, X., Lv, H., Li, Z., & Lee, M. (2022). Microwave-assisted hydrodistillation extraction based on microwave-assisted preparation of deep eutectic solvents coupled with GC-MS for analysis of essential oils from clove buds. *Sustainable Chemistry and Pharmacy*, *27*, 100695. https://doi.org/10.1016/j.scp.2022.100695

Cui, H., Zhao, C., & Lin, L. (2015). The specific antibacterial activity of liposome-encapsulated Clove oil and its application in tofu. *Food Control*, *56*, 128–134. https://doi.org/10.1016/j.foodcont.2015.03.026

Das, S., & Mukherjee, A. (2023). Nanotechnology as sustainable strategy for remediation of soil contaminants, air pollutants, and mitigation of food biodeterioration. In *Environmental Applications of Microbial Nanotechnology* (pp. 3–16). Elsevier. https://doi.org/10.1016/B978-0-323-91744-5.00009-6

Della Porta, G., Taddeo, R., D'Urso, E., & Reverchon, E. (1998). Isolation of clove bud and star anise essential oil by supercritical CO2Extraction. *LWT—Food Science and Technology*, *31*(5), 454–460. https://doi.org/10.1006/fstl.1998.0381

Fan, S., Yin, X., Liu, X., Wang, G., & Qiu, W. (2024). Enhancing bread preservation through non-contact application of starch-based composite film infused with clove essential oil nanoemulsion. *International Journal of Biological Macromolecules*, *263*, 130297. https://doi.org/10.1016/j.ijbiomac.2024.130297

Frohlich, P. C., Santos, K. A., Palú, F., Cardozo-Filho, L., da Silva, C., & da Silva, E. A. (2019). Evaluation of the effects of temperature and pressure on the extraction of eugenol from clove (Syzygium aromaticum) leaves using supercritical CO2. *The Journal of Supercritical Fluids*, *143*, 313–320. https://doi.org/10.1016/j.supflu.2018.09.009

Gonzalez-Rivera, J., Duce, C., Campanella, B., Bernazzani, L., Ferrari, C., Tanzini, E., Onor, M., Longo, I., Ruiz, J. C., Tinè, M. R., & Bramanti, E. (2021). In situ microwave assisted extraction of clove buds to isolate essential oil, polyphenols, and lignocellulosic compounds. *Industrial Crops and Products*, *161*, 113203. https://doi.org/10.1016/j.indcrop.2020.113203

Gopalakrishnan, N., Shanti, P. P. V., & Narayanan, C. S. (1990). Composition of clove (Syzygium aromaticum) bud oil extracted using carbon dioxide. *Journal of the Science of Food and Agriculture*, *50*(1), 111–117. https://doi.org/10.1002/jsfa.2740500113

Gowri, G., & Manimegalai, K. (2019). Phytochemical screening and GC-MS analysis of ethanol extract of Syzygium aromaticum L. *Journal of Pharmacognosy and Phytochemistry*, *8*(5), 746–750. www.phytojournal.com

Guan, W., Li, S., Yan, R., Tang, S., & Quan, C. (2007). Comparison of essential oils of clove buds extracted with supercritical carbon dioxide and other three traditional extraction methods. *Food Chemistry*, *101*(4), 1558–1564. https://doi.org/10.1016/j.foodchem.2006.04.009

Guldiken, B., Catalkaya, G., Ozkan, G., Ceylan, F. D., & Capanoglu, E. (2021). Toxicological effects of commonly used herbs and spices. In *Toxicology* (pp. 201–213). Elsevier. https://doi.org/10.1016/B978-0-12-819092-0.00021-2

Hasheminejad, N., Khodaiyan, F., & Safari, M. (2019). Improving the antifungal activity of clove essential oil encapsulated by chitosan nanoparticles. *Food Chemistry*, *275*, 113–122. https://doi.org/10.1016/j.foodchem.2018.09.085

Hastuti, L. T., Saepudin, E., Cahyana, A. H., Rahayu, D. U. C., Murni, V. W., & Haib, J. (2017). The influence of sun drying process and prolonged storage on composition of essential oil from clove buds (Syzygium aromaticum). *AIP Conference Proceedings*, *1862*(1), 030092. https://doi.org/10.1063/1.4991196

Hatami, T., Johner, J. C. F., Zabot, G. L., & Meireles, M. A. A. (2019). Supercritical fluid extraction assisted by cold pressing from clove buds: Extraction performance, volatile oil composition, and economic evaluation. *The Journal of Supercritical Fluids*, *144*, 39–47. https://doi.org/10.1016/j.supflu.2018.10.003

Ivanovic, J., Dimitrijevic-Brankovic, S., Misic, D., Ristic, M., & Zizovic, I. (2013). Evaluation and improvement of antioxidant and antibacterial activities of supercritical extracts from clove buds. *Journal of Functional Foods*, *5*(1), 416–423. https://doi.org/10.1016/j.jff.2012.11.014

Jacob Pulikkotil, S. (2015). Potential of clove of Syzygium aromaticum in development of a therapeutic agent for periodontal disease. A review. *SADJ: Journal of the South African Dental Association = Tydskrif van Die Suid-Afrikaanse Tandheelkundige Vereniging*, *70*, 108–115.

Jadhav, J. J., Jadeja, G. C., & Desai, M. A. (2023). Ultrasound-assisted hydrodistillation for extraction of essential oil from clove buds—A step towards process improvement and sustainable outcome. *Chemical Engineering and Processing—Process Intensification*, *189*, 109404. https://doi.org/10.1016/j.cep.2023.109404

Kapadiya, S., & Desai, M. (2017). Isolation of essential oil from buds of syzygium aromaticum using hydrodistillation: Multi-response optimization and predictive modelling. *International Journal of Advance Research in Science and Engineering*, *6*, 405–418.

Katekar, V. P., Rao, A. B., & Sardeshpande, V. R. (2023). A hydrodistillation-based essential oils extraction: A quest for the most effective and cleaner technology. *Sustainable Chemistry and Pharmacy*, *36*, 101270. https://doi.org/10.1016/j.scp.2023.101270

Kementerian Pertanian. (2018, August 15). Tanaman Cengkih (Syzigium aromaticum). *Kementerian Pertanian*.

Khajeh, M., Yamini, Y., Sefidkon, F., & Bahramifar, N. (2004). Comparison of essential oil composition of Carum copticum obtained by supercritical carbon dioxide extraction and hydrodistillation methods. *Food Chemistry*, *86*(4), 587–591. https://doi.org/10.1016/j.foodchem.2003.09.041

Kusuma, H. S., Lantip, G. I. Al, Mutiara, X., & Iqbal, M. (2023). Evaluation of mini bibliometric analysis, moisture ratio, drying kinetics, and effective moisture diffusivity in the drying process of clove leaves using microwave-assisted drying. *Applied Food Research*, *3*(1), 100304. https://doi.org/10.1016/j.afres.2023.100304

Lehotay, S. J., & Schenck, F. J. (2000). Multiresidue methods: Extraction. In *Encyclopedia of Separation Science* (pp. 3409–3415). Elsevier. https://doi.org/10.1016/B0-12-226770-2/05621-0

Lekjing, S. (2016). A chitosan-based coating with or without clove oil extends the shelf life of cooked pork sausages in refrigerated storage. *Meat Science*, *111*, 192–197. https://doi.org/10.1016/j.meatsci.2015.10.003

Leon-Méndez, G., Osorio-Fortich, M., Ortega-Toro, R., Pajaro-Castro, N., Torrenegra-Alarcón, M., & Herrera-Barros, A. (2018). Design of an emulgel-type cosmetic with antioxidant activity using active essential oil microcapsules of thyme (*Thymus vulgaris* L.), cinnamon (*Cinnamomum verum* J.), and clove (*Eugenia caryophyllata* T.). *International Journal of Polymer Science*, *2018*, 1–16. https://doi.org/10.1155/2018/2874391

Liu, X., Zhang, C., Liu, S., Gao, J., Cui, S. W., & Xia, W. (2020). Coating white shrimp (Litopenaeus vannamei) with edible fully deacetylated chitosan incorporated with clove essential oil and kojic acid improves preservation during cold storage. *International Journal of Biological Macromolecules*, *162*, 1276–1282. https://doi.org/10.1016/j.ijbiomac.2020.06.248

Lo, T. C., & Baird, M. H. I. (2003). Solvent extraction. In *Encyclopedia of Physical Science and Technology* (pp. 341–362). Elsevier. https://doi.org/10.1016/B0-12-227410-5/00713-4

Madhura, B., Lakshmipathy, K., D.V., C., Sunil, C. K., Baskaran, N., Rawson, A., & Dhivya, R. (2024). Evaluation of mint (Mentha spicata) hydrodistillation aqueous byproducts and its utilization in development of bioactive-rich functional drink. *Food and Humanity*, *2*, 100263. https://doi.org/10.1016/j.foohum.2024.100263

Mai, C. (2020). Effect of storage conditions on the chemical composition of Vietnamese clove (Syzygium aromaticum) essential oil. *Journal of Global Pharma Technology*, *12*, 447–454.

Nour, A. H., Modather, R. H., Yunus, R. M., Elnour, A. A. M., & Ismail, N. A. (2024). Characterization of bioactive compounds in patchouli oil using microwave-assisted and traditional hydrodistillation methods. *Industrial Crops and Products*, *208*, 117901. https://doi.org/10.1016/j.indcrop.2023.117901

Nurdjannah, N., & Bermawie, N. (2012). Cloves. In *Handbook of Herbs and Spices* (pp. 197–215). Elsevier. https://doi.org/10.1533/9780857095671.197

Özdikicierler, O., & Günç Ergönül, P. (2022). Encapsulation of clove (Syzygium aromaticum) essential oil. In *Clove (Syzygium Aromaticum)* (pp. 517–530). Elsevier. https://doi.org/10.1016/B978-0-323-85177-0.00028-8

Pandey, V. K., Srivastava, S., Ashish, Dash, K. K., Singh, R., Dar, A. H., Singh, T., Farooqui, A., Shaikh, A. M., & Kovacs, B. (2024). Bioactive properties of clove (Syzygium aromaticum) essential oil nanoemulsion: A comprehensive review. *Heliyon*, *10*(1), e22437. https://doi.org/10.1016/j.heliyon.2023.e22437

Pandey, V. K., Srivastava, S., Singh, R., Dar, A. H., & Dash, K. K. (2023). Effects of clove essential oil (*Caryophyllus aromaticus* L.) nanoemulsion incorporated edible coating on shelf-life of fresh cut apple pieces. *Journal of Agriculture and Food Research*, *14*, 100791. https://doi.org/10.1016/j.jafr.2023.100791

Patil, S. S., & Rathod, V. K. (2023). Extraction and purification of curcuminoids from Curcuma longa using microwave assisted deep eutectic solvent based system and cost estimation. *Process Biochemistry*, *126*, 61–71. https://doi.org/10.1016/j.procbio.2022.11.010

Putra, N. R., Rizkiyah, D. N., Che Yunus, M. A., Kusuma, H. S., & Darmokoesoemo, H. (2024). Mapping the landscape of clove oil as essential oil for health and wellness: A bibliometric review of advances, challenges, and future directions. *Journal of Essential Oil Bearing Plants*, *27*(2), 300–326. https://doi.org/10.1080/0972060X.2024.2325099

Rashd, J. A., Lalung, J., Kassim, M. A., Wijaya, D., Allzrag, A. M. M., & Shaah, M. A. (2024). Kinetics and thermodynamic studies on biodiesel synthesis via Soxhlet extraction of Scenedesmus parvus algae oil. *Energy Conversion and Management: X*, 100633. https://doi.org/10.1016/j.ecmx.2024.100633

Ruíz-González, N., López-Malo, A., Palou, E., Ramírez-Corona, N., & Jiménez-Munguía, M. T. (2019). Antimicrobial activity and physicochemical characterization of oregano, thyme and clove leave essential oils, nonencapsulated and nanoencapsulated, using emulsification. *Environmental Science, Chemistry*, 237–246. https://api.semanticscholar.org/CorpusID:208075940

Sarma, Y. R., Babu, K. N., & Aziz, S. (2014). Spices and aromatics. In *Encyclopedia of Agriculture and Food Systems* (pp. 211–234). Elsevier. https://doi.org/10.1016/B978-0-444-52512-3.00153-4

Shahavi, M. H., Hosseini, M., Jahanshahi, M., & Najafpour, G. (2015). Optimization of encapsulated clove oil particle size with biodegradable shell using design expert methodology. *Pakistan Journal of Biotechnology*, *12*, 149–160.

Sharma, A., Bhardwaj, G., Sohal, H. S., & Gohain, A. (2022). Eugenol. In *Nutraceuticals and Health Care* (pp. 177–198). Elsevier. https://doi.org/10.1016/B978-0-323-89779-2.00007-7

Sharma, S., Perera, K. Y., Jaiswal, A. K., & Jaiswal, S. (2024). Natural antimicrobials from fruits and plant extract for food packaging and preservation. In *Food Packaging and Preservation* (pp. 133–152). Elsevier. https://doi.org/10.1016/B978-0-323-90044-7.00008-2

Singh, P. A., & Baldi, A. (2023). Quality risk assessment on traditional immunity booster plants cultivators in Punjab, India. *Indian Journal of Traditional Knowledge*, *22*(2), 273–282.

Singh, P. A., Tushir, S., Hazra, S., Purewal, S. S., & Agarwal, G. (2024). Novel approaches for the extraction and identification of phytoconstituents from herbs and spices. *The Natural Products Journal*. http://doi.org/10.2174/0122103155293641240417072907

Suttiarporn, P., Seangwattana, T., Srisurat, T., Kongitthinon, K., Chumnanvej, N., & Luangkamin, S. (2024). Enhanced extraction of clove essential oil by ultrasound and microwave assisted hydrodistillation and their comparison in antioxidant activity. *Current Research in Green and Sustainable Chemistry*, *8*, 100411. https://doi.org/10.1016/j.crgsc.2024.100411

Talón, E., Lampi, A.-M., Vargas, M., Chiralt, A., Jouppila, K., & González-Martínez, C. (2019). Encapsulation of eugenol by spray-drying using whey protein isolate or lecithin: Release kinetics, antioxidant and antimicrobial properties. *Food Chemistry*, *295*, 588–598. https://doi.org/10.1016/j.foodchem.2019.05.115

Tekin, K., Akalın, M. K., & Şeker, M. G. (2015). Ultrasound bath-assisted extraction of essential oils from clove using central composite design. *Industrial Crops and Products*, *77*, 954–960. https://doi.org/10.1016/j.indcrop.2015.09.071

Temelli, F., Saldaña, M. D. A., & Comin, L. (2012). Application of supercritical fluid extraction in food processing. In *Comprehensive Sampling and Sample Preparation* (pp. 415–440). Elsevier. https://doi.org/10.1016/B978-0-12-381373-2.00142-3

Wei, M.-C., Lin, P.-H., Hong, S.-J., Chen, J.-M., & Yang, Y.-C. (2016). Development of a green alternative procedure for the simultaneous separation and quantification of clove oil and its major bioactive constituents. *ACS Sustainable Chemistry & Engineering*, *4*(12), 6491–6499. https://doi.org/10.1021/acssuschemeng.6b01263

Wei, M.-C., Xiao, J., & Yang, Y.-C. (2016). Extraction of α-humulene-enriched oil from clove using ultrasound-assisted supercritical carbon dioxide extraction and studies of its fictitious solubility. *Food Chemistry*, *210*, 172–181. https://doi.org/10.1016/j.foodchem.2016.04.076

Yamada, T., Taguchi, K., & Bamba, T. (2017). Separation of lipids. In *Supercritical Fluid Chromatography* (pp. 419–438). Elsevier. https://doi.org/10.1016/B978-0-12-809207-1.00014-8

Yazdani, F., Mafi, M., Farhadi, F., Tabar-Heidar, K., Aghapoor, K., Mohsenzadeh, F., & Darabi, H. R. (2005). Supercritical CO2 extraction of essential oil from clove bud: Effect of operation conditions on the selective isolation of eugenol and eugenyl acetate. *Zeitschrift Für Naturforschung B*, *60*(11), 1197–1201. https://doi.org/10.1515/znb-2005-1113

Yunusa, S., Um, Y., & Haruna, I. (2019). Comparison of essential oils of clove buds extracted using soxhlet and ultrasonic-assisted extraction methods (short communication). *Journal of Experimental Research*, *7*(1), 75–78.

10 Commercial Uses in Food and Cosmetic Industries

Gustavo Adolfo Castillo-Herrera, Hugo Espinosa-Andrews, Zaira Yunuen Garcia-Carvajal, Moisés Martínez Velázquez, and José Nabor Haro-González

10.1 FOOD INDUSTRY

Syzygium aromaticum L., belonging to the family Myrtaceae, includes more than 3,000 species and 130–150 genera, such as myrtle, eucalyptus, cloves, and guava. Cloves, aromatic flower buds, are grown in subtropical countries such as Madagascar, Sri Lanka, Indonesia, and China. Traditionally, they have been used as antimicrobial agents, insecticides, analgesics, and anesthetics. The FDA classifies them as generally recognized as safe (GRAS), making them suitable for use in perfumes, cosmetics, medical devices, drugs, and food. Cloves are widely used whole or powdered as a spice in world cuisine due to their strong, spicy, and woody aroma and their intense flavor that is sweet, bitter, and spicy. They are primarily used to flavor sauces, soups, marinades, and dressings, as well as in curries and chili dishes. Their sweet notes also allow them to be incorporated into desserts such as pan de muerto, pumpkin pie, gingerbread, and cookies (Haro-González et al., 2023, 2024).

The food business has encountered many difficulties lately in maintaining the nutritional value and sensory appeal of safe foods with longer shelf lives. Food deterioration generates hazardous secondary metabolites that endanger human health in addition to causing significant financial losses. Because of their antibacterial, antifungal, and antioxidant qualities, spices are positioned as a natural alternative to artificial food preservatives in response to the increased consumer desire for natural alternatives (Singh et al., 2018). The fundamental difficulty, though, is preserving the functional qualities of spices as food preservatives without affecting the meal's flavor or the hunger of the consumer. The main food applications are mentioned in Table 10.1. (Ahmed et al., 2021; Chaichi et al., 2021; Haro-González et al., 2021; Ju et al., 2018; Khaleque et al., 2016; Rajaei et al., 2017; Takahashi et al., 2021).

Generally, complex food matrices employ powders, extracts, or essential oils of spices, which require higher concentrations than those used in in vitro tests. For instance, high-protein foods can form complexes between proteins and phenolic

DOI: 10.1201/9781003521785-10

TABLE 10.1
Main food applications of clove buds.

Form	Food category	Food	Application form	Dose	Results	Reference
Powder	Packaging material	Wheat grain	Storage	30–50 g	Extended shelf life up to 30 day	(Qin et al., 2023)
	Baked foods	Cake	Fortification	1 g	Improved antioxidant and extended shelf life	(Dev et al., 2021)
	Meat, poultry, and seafood products	Pork patties	Fortification	1%	Improved antioxidant and extended shelf life	(Jin et al., 2021)
Hydro-ethanolic	Plain drinking water	Plain drinking water	Fortification	250 mg	Improved antioxidant	(Johannah et al., 2015)
	Dairy products	Yogurt	Fortification	250 mg	Improved antioxidant	(Johannah et al., 2015)
	Processed food	Fruit juice	Fortification	250 mg	Improved antioxidant	(Johannah et al., 2015)
	Natural food	Honey	Fortification	250 mg	Improved antioxidant	(Johannah et al., 2015)
Aqueous extract	Meat, poultry, and seafood products	Silver carp (Hypophthalmichthys molitrix) fillets	Coating	2%	Extended shelf life up to 3 days	(Shi et al., 2014)
Ethanol extract	Processed food	Chocolate beverage	Fortification	4%	Improved antioxidant	(Indiarto et al., 2024)
Clove Essential Oil	Baked foods	Cake, bread, green bean cake, and Buddha's hand citron cake	Storage	1%	Extended shelf life up to 2–12 days	(Ju et al., 2018)
		Bread	Storage	250 mg/g	Extended shelf life up to 15 days	(Ju et al., 2020)
		Refrigerated steamed buns	Coating	0–1.2%	Extended shelf life up to 10 days, but volatile components evaporate during the re-steaming process	(Li et al., 2020)
	Dairy products	Fresh soft cheese	Fortification	0.01%	Extended shelf life up to 3 weeks	(Li et al., 2020)
	Meat, poultry, and seafood products	Fresh rainbow trout	Coating		Extended shelf life up to 5–12 days	(Dehghani et al., 2018)
		Chicken breast meat	Coating		Extended shelf life up to 12 days	(Chaichi et al., 2021)
		Beef sucuk	Coating	1.50%	Extended shelf life up to 45 days	(Saricaoglu & Turhan, 2019)
		Beef cutlets	Coating	2 mg/g	Extended shelf life up to 12 days	(Rajaei et al., 2017)
		Sea bream	Storage	10–15 mg/kg	Extended shelf life up to 15 days	(Navarro-Segura et al., 2019)
		Bluefin tuna	Coating	0.5 mL	Extended shelf life up to 14 days	(Echeverría et al., 2018)
		Ground beef	Fortification	10%	Extended shelf life up to 7 days; at refrigeration and chilling temperatures 60 days Can produce unpleasant flavors	(Khaleque et al., 2016)

TABLE 10.1 *(Continued)*
Main food applications of clove buds.

Form	Food category	Food	Application form	Dose	Results	Reference
		Gelatin–chitosan film, cod fillets	Coating	15%	Extended shelf life up to 12 days; improved mechanical, structural, and barrier properties	(Gómez-Estaca et al., 2010)
		Raw grass carp fillets	Coating	0.1–1.0%	Reduced content of off-odor volatiles for 12 days	(Yu et al., 2018)
		White shrimp	Coating	0.25–0.5%	Extended shelf life up to 15 days and inhibited melanosis	(Liu et al., 2020)
		Salmon burgers	Fortification	0.005–0.01%	Extended shelf life up to 14 days and inhibited melanosis	(Jonušaite et al., 2021)
		Chicken patties	Coating	0.50%	Extended shelf life up to 35 days and inhibited melanosis	(Shukla et al., 2020)
		Chicken breast	Storage	0.2–0.5%	Extended shelf life up to 15 days and inhibited melanosis	(Hosseini et al., 2021)
	Packaging material	Mechanically deboned chicken meat protein film	Fortification	1%	Improved antioxidant and antimicrobial properties	(Sarıcaoglu & Turhan, 2020)
		Poly (lactic acid) biocomposite food packaging film	Fortification	3%	Improved antimicrobial properties	(Lu et al., 2021)
		Polylactide/poly(ε-caprolactone)/zinc oxide/CEO and scrambled eggs	Fortification	25%	Extended shelf life up to 21 days, improved mechanical, structural, and barrier properties	(J. Ahmed et al., 2019)
		Chitosan–gum Arabic film	Fortification	5%		(Xu et al., 2019)
		Citrus pectin film	Fortification	0.5–1.5%	Improved barrier, mechanical, antioxidant, and antimicrobial properties of pectin film	(Nisar et al., 2018)
		Chicken eggs	Storage	10–80 μg/g	Extended shelf life up to 30 days, less weight reduction	(Phothisuwan et al., 2020)
	Processed food	Ketchup	Fortification	500 ppm		(Omidbeygi et al., 2007)
		Sausages	Fortification	2000 mg/L	Prolonged shelf life for 14 days	(Takahashi et al., 2021)
	Vegetables	Mango (cv. Banganapalli and cv. Totapuri)	Storage	106 μL	Extended shelf life up to 20–21 days	(Perumal et al., 2021)
		Persimmon	Storage	1.56%	Inhibited mold growth on persimmon fruits for 28 days	(Vurmaz & Gündüz, 2020)
		Pak choi	Storage	0.02%	Extended shelf life up to 17 days	(Park et al., 2019)
		Avocado	Coating	0.20%	Extended shelf life up to 7 days	(Nilmini et al., 2021)

essential oils, thereby diminishing the effectiveness of the latter. Additionally, the lipid fraction of foods can absorb the antimicrobial agent, reducing its bactericidal action. Similarly, a reduction in the water content of foods could hinder the transfer of antimicrobial agents to the active site in the microbial cell. Furthermore, external factors such as storage temperature, packaging, initial concentration, method of application, and the type of microorganism can influence the efficacy of spices (Ahmed et al., 2021; Chaichi et al., 2021; Haro-González et al., 2021; Ju et al., 2018; Li et al., 2020; Omidbeygi et al., 2007).

10.1.1 Baked Food

The baked goods industry focuses on preventing mold growth and maintaining safety and nutritional value (Kumari et al., 2024). Preservation methods for baked goods include modifying storage atmosphere, irradiation, aseptic packaging, and the use of preservative acids. However, the use of organic acids (propionic, benzoic, and sorbic acids) has been restricted in several countries due to their adverse effects on human health. Clove essential oil has a broad-spectrum activity against spoilage and foodborne pathogenic microorganisms. Its addition to baked products can extend shelf life without altering the original taste, aroma, texture, appearance, or sensory acceptance (Ju et al., 2018, 2020; Li et al., 2020).

10.1.2 Dairy Products

Dairy products, such as cheese, are the cause of various outbreaks of food-borne diseases. Ahmed et al. (2021) applied approximately 1 kg of clove essential oil (CEO) per 200 liters of raw milk as an antimicrobial agent for the production of cheese. CEO demonstrated significant antimicrobial effects at 4°C for a month without affecting organoleptic properties, which showed potential for cost-effective use (L. I. Ahmed et al., 2021; Gómez-Estaca et al., 2010).

10.1.3 Processed Food

In recent years, there has been notable growth in the market for ready-to-cook processed foods, a trend propelled by shifts in consumer lifestyles and advancements in chilled distribution networks. Microbial spoilage of these processed items often leads to undesirable outcomes such as offensive smells, discoloration, increased sliminess, sediment formation, gas production, and a decline in pH levels, all of which can compromise the integrity of the food and pose health hazards for consumers. Research indicates that incorporating 5% (w/w) of clove essential oil into processed foods can adversely impact their sensory attributes. Consequently, its use has been mainly directed toward exploiting its properties as a flavor additive with antimicrobial and antioxidative benefits (Khaleque et al., 2016; Omidbeygi et al., 2007; Takahashi et al., 2021).

Also, in many food processes, the use of vegetable oils is required, for example in frying food and emulsions as flavor carriers, but the oxidation of these oils has been a target in research studies. For example, it has been reported that clove bud powder decreases the oxidation once it is combined with palm olein/sunflower oil blends during storage at high temperatures, but a better antioxidant effect results when ethanolic clove extracts are used, indicating that clove buds are a potential source of natural antioxidants that can be used in the preservation of vegetable oils (Djikeng et al., 2023).

10.1.4 Meat, Poultry, and Seafood Products

The application of clove essential oils to food products of animal origin mitigates undesirable reactions that involve the degradation of flavor, odor, color, texture, and sensory properties. Its antimicrobial activity reduces bacterial counts, decreases the deamination capacity of non-protein nitrogenous compounds, and lowers the formation of hydroperoxides due to its antioxidant properties. The mechanisms behind the antioxidant properties of clove essential oils include binding to transition metals, inhibition of chain reactions, decomposition of hydroperoxides, and interaction with free radicals. Clove essential oil has been applied to white shrimp, salmon burgers, fish fillets, ground meat, chicken patties, and chicken breasts during refrigerated or frozen storage. Films reinforced with clove essential oils can reduce weight loss, water activity, lipid oxidation, color changes, and microorganism growth in animal-origin foods for up to 45 days if treated with heat and up to 12 days if refrigerated (Chaichi et al., 2021; Dehghani et al., 2018; Echeverría et al., 2018; Gómez-Estaca et al., 2010; Jonušaite et al., 2021; Khaleque et al., 2016; Liu et al., 2020; Rajaei et al., 2017; Saricaoglu & Turhan, 2019; Shukla et al., 2020).

10.1.5 Vegetables

The deterioration of vegetables after harvest during transportation and storage generates significant economic losses in the supply chain. The antimicrobial properties of clove essential oils can prevent fungal spoilage in vegetables and adverse health effects, serving as a potential alternative to chemical fungicides. The antimicrobial activity can be enhanced when combined with UV-C light treatments or modified packaging. These processes allow for effective control of post-harvest decay and preserve the physicochemical quality of vegetables, extending their shelf life without affecting their organoleptic properties. Clove essential oils are added to the washing treatment of fresh-cut vegetables as an alternative to acetic acid, sodium bicarbonate, and chlorine-based disinfectants, reducing microbiological risks and extending shelf life. Furthermore, washing with clove essential oils does not impact the attributes of color, bioactive content, composition, or sensory properties. Therefore, the use of clove essential oils, along with cold storage, is an excellent eco-friendly alternative that could be further optimized for commercial applications, delivering vegetables to the market with better post-harvest quality, and greater consumer acceptance (Nilmini et al., 2021; Park et al., 2019; Perumal et al., 2021; Vurmaz & Gündüz, 2020)

10.1.6 Packaging Materials

In recent years, new biodegradable packaging materials (polysaccharides, lipids, proteins) from natural polymers have been developed. They can increase the antioxidant and antimicrobial properties of spices or their derivatives by increasing their shelf-life and reducing or inhibiting food-borne pathogens. The inclusion of spices and their derivatives in coating films is intended to change functional properties such as water vapor permeability and antimicrobial and antioxidant properties. Clove essential oil-based films have shown antibacterial properties and are deactivated for 21 days due to the penetration and destruction of cell structures by clove essential oil compounds (J. Ahmed et al., 2019). The addition of clove essential oil can change the moisture content of packaging materials, improve spatial distances in the film matrix, and produce thicker films (J. Ahmed et al., 2019; Gómez-Estaca et al., 2010; Lu et al., 2021; Nisar et al., 2018; Saricaoglu & Turhan, 2019; Sarıcaoglu & Turhan, 2020; Xu et al., 2019).

The optical properties of biodegradable films influence the appearance and quality of food, for example, biodegradable poly(lictide)-poly(butylene adipate-co-terephthalate) [PLA-PBAT] films containing clove oil, in addition to antimicrobial activity, reduces the transmission by extracting color, increasing the UV barrier. Application of coating also reduces the rate of lipid oxidation. In this respect, different coloring components of the essential oils of cloves can change the colour of the film. Its incorporation increases transparency due to the increase in light distribution caused by oil droplets in the film network. These factors reduce transparency, which is a benefit for photosensitive foods (Gengatharan & Rahim, 2023; Nisar et al., 2018; Sarıcaoglu & Turhan, 2020).

The introduction of clove essential oils into the film network partially replaces the interactions between stronger polymers and thinner polymers (polymer and oil). This results in a more heterogeneous network and a microstructure that does not change by rearranging the polymer. Similarly, the use of clove essential oil reduces the glass transition temperature and film elastic modulus. Sarıcaoglu and Turhan (2020) observed a decrease in elastic modulus and tensile strength when a coating of mechanically dissected chicken meat protein was added to the clove essential oil. However, the tensile strength is above 3.5 MPa, which is recommended for food film coating. These structural changes, which were made by incorporating clove essential oil, also produce rougher and more porous films (J. Ahmed et al., 2019; Lu et al., 2021; Nisar et al., 2018; Sarıcaoglu & Turhan, 2020; Xu et al., 2019).

Due of their antimicrobial activity, clove extracts should be considered as active ingredients for developing edible films and improving food safety, and, once the films are consumed, they still have antimicrobial effects, limiting their growth, for example, in *Helicobacter pylori* (Peng et al., 2024).

10.2 COSMETIC INDUSTRY

Cosmetics, which are used for cleaning and care, can be natural or chemical in nature to be beautiful and attractive and to change our appearance without affecting our health as represented in Figure 10.1. If the cosmetic is made up exclusively of natural sources, then we can call it natural cosmetics. The regular application of cosmetics helps to reduce skin and hair problems due to daily effects caused by unfriendly environmental conditions such as dust, sunlight, humidity, UV rays, climate changes, and personal habits. It is important to note that use of cosmetics on the external layer of the skin, enhancing its beauty and appearance and giving the appearance of healthy skin, does not cure disease or damage to the skin.

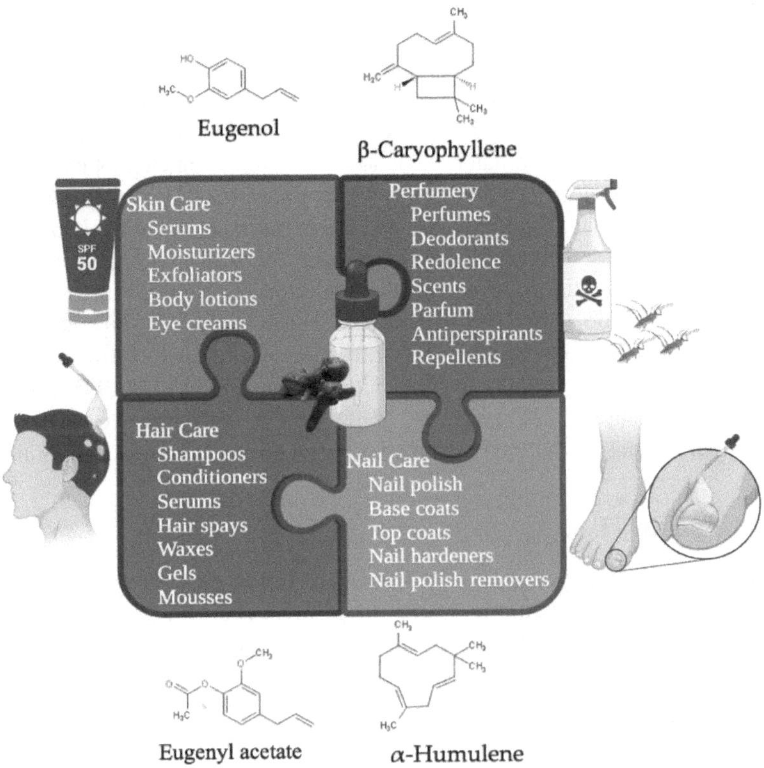

FIGURE 10.1 The different products used in the cosmetics sector.

Cosmeceuticals are products that contain active pharmaceutical or natural substances that can help heal the skin by attacking the root of the problem and improving the beauty of the skin (Shah et al., 2024). Due to their great acceptance throughout the centuries, several products, such as creams, lotions, deodorants, shampoos, conditioners, gels, sprays, pastes, powders, etc. have been treated (Warade & Bhosale, 2023), and commercial products containing clove essential oils or its main component eugenol are no exception. However, many products on the market are marketed without studies demonstrating their effectiveness and biological safety (Table 10.2).

Cutaneous fungal infections are usually caused by dermatophytes, non-dermatophyte fungi, and yeasts (*Candida spp.*) that superficially affect the skin, hair, and nails. For example, fungal infections caused by *Candida* species can cause many public health problems and even cause people to be hospitalized due to life-threatening infections, especially those with impaired immune systems. Thus, clove bud extracts are a natural source that helps in fungal treatment. Studies have shown that acetonic extracts have better antifungal activity than ethanol, ether, or dichloromethane extracts, and because of the nature of these extracts, they can easily be incorporated in some cosmeceutical formulas, such as balms, lotions, or serums, which help to treat the fungal infections (Mostafa et al., 2023; Shah et al., 2024).

10.2.1 Acne

Many people suffer from acne, which may be related to the presence of fungi and bacteria such as *Propionibacterium acnes, Staphylococcus epidermidis, Staphylococcus aureus*, and *Candida albicans*. For example, Tsai et al. (2017) reported that the bacterial activity of P. acnes can promote inflammatory acne lesions by releasing chemotactic substances and triggering the release of proinflammatory cytokines. In an in vitro study conducted by Melo et al. (2015) with different acne precursor bacteria, including *P. acnes*, it was observed that clove essential oil (*E. caryophyllata*) showed high antibacterial activity (600 and 800 ppm), as did the essential oils of thyme (*T. vulgaris*) and cinnamon (*C. verum*) (Melo et al., 2015). Similarly, Misar et al., (Misar et al., 2020) reported that cream formulations with clove essential oils (0.5–1.0) obtained by hydrodistillation presented dose-dependent inhibitory effects in an in vitro study. Tsai et al. (2017) reported that mouse ear edema caused by *P. acnes* improved with intradermal injection of eugenol or clove essential oil, suggesting the possibility of a reduction in cutaneous inflammation.

10.2.2 Healthy Hair

Due to its biological activities, clove essential oil has been used in different studies to improve alopecia. For example, Muhammad et al. reported the effect of eugenol emulsions (1%) on hair growth in shaved male Wistar albino rats for a period of 28 days. The hair length of the clove oil emulsion group was similar to that of the 5% minoxidil (alopecia treatment drug) group, but the hair length of the clove oil emulsion group was up to five times longer than that of the untreated group. The authors suggested that the emulsion with clove essential oil may help promote blood flow to the hair follicles (Shahtalebi et al., 2016).

Clove oil was tested as a permutation agent on the skin in finasteride, a hair loss and acne inhibitory agent. The authors observed in a Franz diffusion cell that 4 hours of pretreatment with clove oil resulted in the highest penetration rate for finasteride, which was almost 14.3 times greater than that of the control (water) and was more effective than urea or a freeze-dried grape seed extract powder (Salimi et al., 2022).

Because of its antifungal properties, clove oil has been used to combat dandruff. Khusnul et al. (2021) reported that dandruff can be caused by different types of fungi, including *P. ovale, C. albicans*, and *M. gypseum*, and that these can be inhibited by clove essential oil. The results showed that *M. gypseum* presented the greatest inhibition halo on the plaque, with inhibition halos similar to those of 2% ketoconazole at concentrations less than 10%, while for *C. albicans*, it was 50%, and for P. ovale it was 100% essential oil.

10.2.3 Nail Health

Dermatophytosis, tinea, or ringworm infections are names for the areas infected with dermatophytes (*Trichophyton, Epidermophyton* and *Microsporum*), which require keratin to grow and are limited to the stratum corneum, hair shafts, and nails. The irrational use of pharmacological treatment has led to a certain degree of resistance, so the option of using natural products alone or in conjunction with drugs can be used for the treatment of nail disorders. Shalayel et al. (2021) compared five antifungal drugs (fluconazole, ketoconazole, itraconazole, terbinafine and griseofulvin) and clove oil on samples from patients clinically suspected of having dermatophytosis. After an analysis of the samples, the authors concluded that 39% of the patients were positive for dermatophytes, (*T. verrucosum, T soudanense, T. mentagrophytes, M. Canis*) and non-dermatophytes, principally Aspergillus sp. Syzygium aromaticum oil, which displayed an affordable antifungal efficacy (MIC of approximately of 1.25–2 μg/mL), with better activity than that observed for fluconazole (from 2 to 8 μg/mL) but significantly greater activity than that observed for ketoconazole (from 0.25 to 0.5 μg/mL), itraconazole (from 0.05 to 0.5 μg/mL), terbinafine (from 0.002 to 0.5 μg/mL), and griseofulvin (from 0.5 to 1 μg/mL). These results were attributed to the eugenol content in clove essential oil, which can act on the cell membrane of T. rubrum by inhibiting the biosynthesis of ergosterol.

10.2.4 Oral Health

Some studies have shown the possible application of clove essential oil for oral care; however, it will be necessary to opt for formulations with other agents traditionally used to diminish the number of clove sensory notes to which consumers are not accustomed. Thosar et al. (Thosar et al., 2013) reported that eugenol oil required a low concentration to inhibit the four pathogens found in the oral cavity and that the minimum inhibitory concentration and minimum fungicidal concentration for *C. albicans* were 0.1 μl/ml, while the minimum inhibitory concentration and minimum bactericidal concentration for *S. aureus* were 0.4 μl/ml, and for *E. Coli* and *E. facelis* they were 1 μl/ml. de Oliveira Carvalho et al. (2020) explored the use of different essential oils alone or in combination for the inhibition of oral bacteria and film formation by the formation of a toothpaste (3–5%). The inhibitory effect on bacteria (*S. mutans* > 90%; *S. aureus* and *E. faecalis* > 60%) was greater at the lowest concentration tested and was more effective in pastes formulated with cinnamon and clove essential oil blends than in those formulated with individual compounds.

10.2.5 Repellent

Insect-borne diseases are an ongoing challenge to public health. Commonly used repellents cause significant health problems and have long-lasting adverse effects on the environment. In addition, an increase in resistance against repellents has been reported. Because of this, investigations have focused on developing natural repellents based on essential oils (Elzayyat et al., 2018; Neupane et al., 2020; Reuss et al., 2020).

The clove essential oil has shown high levels of repellency and toxicity against fleas, aphids, ants, mosquito, and cockroaches. Clove essential oil has shown repellent effects against mosquitoes that carry yellow fever, dengue, chikungunya, and Zika fever (Elzayyat et al., 2018; Lambert et al., 2020; Neupane et al., 2020; Reuss et al., 2020).

The primary targets of clove essential oils and other essential oils are octopamine and gamma-aminobutyric acid (GABA) receptors and transient receptor potential (TRP) channels (Toledo et al., 2020). The dose–response ratio of clove essential oil showed an increased mortality rate with increasing concentration (Elzayyat et al., 2018). The clove essential oil increased permeability activity on the cell membrane, disrupted the cytoplasmic membrane, and interacted with proteins, ATPase, histidine decarboxylase, amylase, and protease enzymes, which were also inhibited. The clove essential oils and its main constituents are products that have low toxicity to mammals and zero residual concentration, thus their application as insect repellents (Lambert et al., 2020). The common applications of clove buds are mentioned in Table 10.2.

TABLE 10.2
Main cosmetic applications of clove buds.

Form	Cosmetic category	Dose	Results	Reference
Clove bud powder	Repellent	30 mg/cm^2 killed 92%	Repellancy and contact toxicity against German cockroaches	(Neupane et al., 2020)
Clove essential oil	Oral health	0.019–0.156%	Combating bad breath against halitosis-associated bacteria	(Thosar et al., 2013)
		0.1–1 µl/ml	Antimicrobial activity against *Enterococcus fecalis, Escherischia coli, Staphylococcus aureus,* and *Candida albicans*	(Thosar et al., 2013)
	Acne	1%	Effective against microorganisms responsible for acne	(Misar et al., 2020)
	Healthy hair	2 mL	Increased the penetration of finasteride, an inhibitor used to treat hair loss and acne	(Salimi et al., 2022)
		10%	Inhibit the growth of some fungi that cause dandruff	(Khusnul et al., 2021)
	Repellent	4.00 ml/cm^2 killed 95%	Repellancy and contact toxicity against German cockroaches	(Neupane et al., 2020)
		LC_{50} of 17 mg/l	Oviposition deterrent and larvicide of Aedes japonicus japonicus	(Reuss et al., 2020)
		5.70 µg cm^{-2} in 24 h and 3.91 µg cm^{-2} in 48 h	Ecotoxicological against cat flea	(Lambert et al., 2020)
		0.08 µL of essential oil per cm^2	Ecotoxicological against aphids	(Toledo et al., 2020)
			Repellent effects against *Cimex lectularius L*	(Gomes et al., 2024)
		0.002 µg/cm^2	Repellancy against a common species of mosquito	(Deng et al., 2023)
Ethanolic clove extract	Acne	259 µg/ml	Skin mask hydrogel cytocompatibility and antimicrobial activity	(Akl et al., 2023)
		200 µg/ml	Suppress inflammatory responses elicited by *Propionibacterium acnes*	(Tsai et al., 2017)

10.3 CONCLUSIONS

Clove, widely used in world cuisine, has seen increased utilization in the food and cosmetic industries due to its antimicrobial and antioxidant properties. In the food industry, cloves are presented as a natural alternative to artificial preservatives, extending the shelf life of food products. However, it is necessary to optimize its application to avoid negative impacts on the sensory properties of the products without compromising their biological activity. Currently, cloves are used in meats, poultry, seafood, vegetables, dairy products, and edible coating films. In the cosmetics industry, the focus is on the development of skin, hair, nail and mouth care products, as well as repellent products. Although clove essential oil is the most commonly used form, there is significant research potential in other forms of cloves, such as powder, solvent extracts, tinctures, and aqueous extracts. Likewise, the application of encapsulation systems for compounds and clove products could alter solubility, absorption, bioavailability, and shelf life by avoiding degradation (photo, oxidative, or thermal), being an alternative for its application, improving its organoleptic preposition in food, and reducing irritability in cosmetics. Therefore, further studies are still required to define its biological functions and activities to promote its safe application in new product development.

REFERENCES

Ahmed, J., Mulla, M., Jacob, H., Luciano, G., T.B., B., & Almusallam, A. (2019). Polylactide/poly(ε-caprolactone)/zinc oxide/clove essential oil composite antimicrobial films for scrambled egg packaging. *Food Packaging and Shelf Life*, *21*, 100355. https://doi.org/10.1016/j.fpsl.2019.100355

Ahmed, L. I., Ibrahim, N., Abdel-Salam, A. B., & Fahim, K. M. (2021). Potential application of ginger, clove and thyme essential oils to improve soft cheese microbial safety and sensory characteristics. *Food Bioscience*, *42*, 101177. https://doi.org/10.1016/j.fbio.2021.101177

Akl, E. M., Hasanin, M. S., & Dacrory, S. (2023). Skin mask hydrogel-based natural sources: Characterization and biological properties evaluations. *Bioactive Carbohydrates and Dietary Fibre*, *29*, 100355. https://doi.org/10.1016/j.bcdf.2023.100355

Chaichi, M., Mohammadi, A., Badii, F., & Hashemi, M. (2021). Triple synergistic essential oils prevent pathogenic and spoilage bacteria growth in the refrigerated chicken breast meat. *Biocatalysis and Agricultural Biotechnology*, *32*, 101926. https://doi.org/10.1016/j.bcab.2021.101926

de Oliveira Carvalho, I., Purgato, G. A., Píccolo, M. S., Pizziolo, V. R., Coelho, R. R., Diaz-Muñoz, G., & Alves Nogueira Diaz, M. (2020). In vitro anticariogenic and antibiofilm activities of toothpastes formulated with essential oils. *Archives of Oral Biology*, *117*, 104834. https://doi.org/10.1016/j.archoralbio.2020.104834

Dehghani, P., Hosseini, S. M. H., Golmakani, M.-T., Majdinasab, M., & Esteghlal, S. (2018). Shelf-life extension of refrigerated rainbow trout fillets using total Farsi gum-based coatings containing clove and thyme essential oils emulsions. *Food Hydrocolloids*, *77*, 677–688. https://doi.org/10.1016/j.foodhyd.2017.11.009

Deng, W., Li, M., Liu, S., Logan, J. G., & Mo, J. (2023). Repellent screening of selected plant essential oils against dengue fever mosquitoes using behavior bioassays. *Neotropical Entomology*, *52*(3), 521–529. https://doi.org/10.1007/s13744-023-01039-z

Dev, M., Ghosh, M., & Bhattacharyya, D. kumar. (2021). Physico-chemical, antimicrobial, and organoleptic properties of roasted aromatic spice (clove bud) in baked product. *Applied Biochemistry and Biotechnology, 193*(6), 1813–1835. https://doi.org/10.1007/s12010-021-03504-0

Djikeng, F. T., Ewusi, K. E. K., Ngangoum, E. S., Sylvia, V.-T. Z. N., Nyemb, G. M., & Tiencheu, B. (2023). Stabilization of multi-source edible oils from palm olein and sunflower oil using clove (Syzygium aromaticum) extract and powder during accelerated storage. *Food and Humanity, 1*, 1584–1591. https://doi.org/10.1016/j.foohum.2023.11.008

Echeverría, I., López-Caballero, M. E., Gómez-Guillén, M. C., Mauri, A. N., & Montero, M. P. (2018). Active nanocomposite films based on soy proteins-montmorillonite- clove essential oil for the preservation of refrigerated bluefin tuna (Thunnus thynnus) fillets. *International Journal of Food Microbiology, 266*, 142–149. https://doi.org/10.1016/j.ijfoodmicro.2017.10.003

Elzayyat, E., Elleboudy, N., Moustafa, A., & Ammar, A. (2018). Insecticidal, oxidative, and genotoxic activities of syzygium aromaticum and Eucalyptus globulus on culex pipiens adults and larvae. *Turkish Journal of Parasitology, 42*(3), 213–222. https://doi.org/10.5152/tpd.2018.5626

Gengatharan, A., & Rahim, M. H. A. (2023). The application of clove extracts as a potential functional component in active food packaging materials and model food systems: A mini-review. *Applied Food Research, 3*(1), 100283. https://doi.org/10.1016/j.afres.2023.100283

Gomes, E. N., Caputi, C., Patel, H. K., Zorde, M., Vasilatis, A., Wu, Q., Wang, C., Wyenandt, C. A., & Simon, J. E. (2024). Chemical variability and insect repellent effects of lemon catnip essential oil and related phytochemicals against Cimex lectularius L. *Journal of Natural Pesticide Research, 8*, 100074. https://doi.org/10.1016/j.napere.2024.100074

Gómez-Estaca, J., López de Lacey, A., López-Caballero, M. E., Gómez-Guillén, M. C., & Montero, P. (2010). Biodegradable gelatin–chitosan films incorporated with essential oils as antimicrobial agents for fish preservation. *Food Microbiology, 27*(7), 889–896. https://doi.org/10.1016/j.fm.2010.05.012

Haro-González, J. N., Castillo-Herrera, G. A., Martínez-Velázquez, M., & Espinosa-Andrews, H. (2021). Clove essential oil (Syzygium aromaticum L. Myrtaceae): Extraction, chemical composition, food applications, and essential bioactivity for human health. *Molecules, 26*(21), 6387. https://doi.org/10.3390/molecules26216387

Haro-González, J. N., Martínez-Velázquez, M., Castillo-Herrera, G. A., & Espinosa-Andrews, H. (2024). Clove essential oil nanoemulsions: Development, physical characterization, and anticancer activity evaluation. *Journal of Dispersion Science and Technology*, 1–9. https://doi.org/10.1080/01932691.2024.2302067

Haro-González, J. N., Schlienger de Alba, B. N., Martínez-Velázquez, M., Castillo-Herrera, G. A., & Espinosa-Andrews, H. (2023). Optimization of clove oil nanoemulsions: Evaluation of antioxidant, antimicrobial, and anticancer properties. *Colloids and Interfaces, 7*(4), 64. https://doi.org/10.3390/colloids7040064

Hosseini, M., Jamshidi, A., Raeisi, M., & Azizzadeh, M. (2021). Effect of sodium alginate coating containing clove (*Syzygium Aromaticum*) and lemon verbena (*Aloysia Citriodora*) essential oils and different packaging treatments on shelf life extension of refrigerated chicken breast. *Journal of Food Processing and Preservation, 45*(3). https://doi.org/10.1111/jfpp.14946

Indiarto, R., Herwanto, J. A., Filianty, F., Lembong, E., Subroto, E., & Muhammad, D. R. A. (2024). Total phenolic and flavonoid content, antioxidant activity and characteristics of a chocolate beverage incorporated with encapsulated clove bud extract. *CyTA—Journal of Food, 22*(1). https://doi.org/10.1080/19476337.2024.2329144

Jin, S.-K., Kim, G.-D., & Jeong, J.-Y. (2021). Evaluation of the effect of inhibiting lipid oxidation of natural plant sources in a meat model system. *Journal of Food Quality, 2021*, 1–8. https://doi.org/10.1155/2021/6636335

Johannah, N., Renny, R., Gopakumar, G., Balu, M., Sureshkumar, D., & Krishnakumar, I. (2015). Beyond the flavour: A de-flavoured polyphenol rich extract of clove buds (Syzygium aromaticum L) as a novel dietary antioxidant ingredient. *Food & Function*, *6*(10), 3373–3382. https://doi.org/10.1039/C5FO00682A

Jonušaite, K., Venskutonis, P. R., Martínez-Hernández, G. B., Taboada-Rodríguez, A., Nieto, G., López-Gómez, A., & Marín-Iniesta, F. (2021). Antioxidant and antimicrobial effect of plant essential oils and sambucus nigra extract in salmon burgers. *Foods*, *10*(4), 776. https://doi.org/10.3390/foods10040776

Ju, J., Xie, Y., Yu, H., Guo, Y., Cheng, Y., Qian, H., & Yao, W. (2020). A novel method to prolong bread shelf life: Sachets containing essential oils components. *LWT*, *131*, 109744. https://doi.org/10.1016/j.lwt.2020.109744

Ju, J., Xu, X., Xie, Y., Guo, Y., Cheng, Y., Qian, H., & Yao, W. (2018). Inhibitory effects of cinnamon and clove essential oils on mold growth on baked foods. *Food Chemistry*, *240*, 850–855. https://doi.org/10.1016/j.foodchem.2017.07.120

Khaleque, M. A., Keya, C. A., Hasan, K. N., Hoque, M. M., Inatsu, Y., & Bari, M. L. (2016). Use of cloves and cinnamon essential oil to inactivate Listeria monocytogenes in ground beef at freezing and refrigeration temperatures. *LWT*, *74*, 219–223. https://doi.org/10.1016/j.lwt.2016.07.042

Khusnul, K., Anisa, P. P., & Virgianti, D. P. (2021). Effect of Clove essential oil (Syzygium aromaticum) against the growth of dandruff scalps-causing fungal pathogen using Kirby- Bauer method in vitro. *Journal of Microbial Systematics and Biotechnology*, *3*(1), 41–48. https://doi.org/10.37604/jmsb.v3i1.78

Kumari, S., Singh, P. A., Hazra, S., Sindhwani, R., & Singh, S. (2024). Ocimum sanctum: The journey from sacred herb to functional food. *Recent Advances in Food Nutrition & Agriculture*, *15*(2), 83–102.

Lambert, M. M., Campos, D. R., Borges, D. A., de Avelar, B. R., Ferreira, T. P., Cid, Y. P., Boylan, F., Scott, F. B., de Almeida Chaves, D. S., & Coumendouros, K. (2020). Activity of Syzygium aromaticum essential oil and its main constituent eugenol in the inhibition of the development of Ctenocephalides felis felis and the control of adults. *Veterinary Parasitology*, *282*, 109126. https://doi.org/10.1016/j.vetpar.2020.109126

Li, K., Zhang, M., Bhandari, B., Xu, J., & Yang, C. (2020). Improving storage quality of refrigerated steamed buns by mung bean starch composite coating enriched with nano-emulsified essential oils. *Journal of Food Process Engineering*, *43*(9). https://doi.org/10.1111/jfpe.13475

Liu, X., Zhang, C., Liu, S., Gao, J., Cui, S. W., & Xia, W. (2020). Coating white shrimp (Litopenaeus vannamei) with edible fully deacetylated chitosan incorporated with clove essential oil and kojic acid improves preservation during cold storage. *International Journal of Biological Macromolecules*, *162*, 1276–1282. https://doi.org/10.1016/j.ijbiomac.2020.06.248

Lu, W., Cui, R., Zhu, B., Qin, Y., Cheng, G., Li, L., & Yuan, M. (2021). Influence of clove essential oil immobilized in mesoporous silica nanoparticles on the functional properties of poly(lactic acid) biocomposite food packaging film. *Journal of Materials Research and Technology*, *11*, 1152–1161. https://doi.org/10.1016/j.jmrt.2021.01.098

Melo, G. E. M., Méndez, G. L., & Fortich, M. del R. O. (2015). In vitro antibacterial activity of nineteen essential oils against acne-associated bacteria. *Revista Cubana de Farmacia*, *49*(1), 103–116. http://scielo.sld.cu/scielo.php?script=sci_arttext&pid=S0034-75152015000100011&lng=en&nrm=iso&tlng=es

Misar, K. S., Kulkarni, S. B., & Gurnule, W. B. (2020). Formulation and evaluation of anti-acne cream by using Clove oil. *Materials Today: Proceedings*, *29*, 1251–1258. https://doi.org/10.1016/j.matpr.2020.06.106

Mostafa, A. A.-F., Yassin, M. T., Al–Askar, A. A., & Al-Otibi, F. O. (2023). Phytochemical analysis, antiproliferative and antifungal activities of different Syzygium aromaticum solvent extracts. *Journal of King Saud University—Science*, *35*(1), 102362. https://doi.org/10.1016/j.jksus.2022.102362

Navarro-Segura, L., Ros-Chumillas, M., López-Cánovas, A. E., García-Ayala, A., & López-Gómez, A. (2019). Nanoencapsulated essential oils embedded in ice improve the quality and shelf life of fresh whole seabream stored on ice. *Heliyon*, *5*(6), e01804. https://doi.org/10.1016/j.heliyon.2019.e01804

Neupane, A. C., Sapakuka, S., Tao, P., & Kafle, L. (2020). Repellancy and contact toxicity of clove bud oil and its constituents against German cockroaches, *Blatella germanica* (Dictyoptera: Blattellidae), under laboratory conditions. *International Journal of Pest Management*, *66*(4), 289–297. https://doi.org/10.1080/09670874.2019.1641250

Nilmini, R. K., Kodituwakku, T. D., Abeywickrama, K., & Kuruppu, M. (2021). In vitro and in vivo application of eco-friendly treatments to control postharvest stem-end rot of naturally infected avocado (cv. Pollock). *Journal of Agricultural Sciences—Sri Lanka*, *16*(2), 283–299. https://doi.org/10.4038/jas.v16i2.9335

Nisar, T., Wang, Z.-C., Yang, X., Tian, Y., Iqbal, M., & Guo, Y. (2018). Characterization of citrus pectin films integrated with clove bud essential oil: Physical, thermal, barrier, antioxidant and antibacterial properties. *International Journal of Biological Macromolecules*, *106*, 670–680. https://doi.org/10.1016/j.ijbiomac.2017.08.068

Omidbeygi, M., Barzegar, M., Hamidi, Z., & Naghdibadi, H. (2007). Antifungal activity of thyme, summer savory and clove essential oils against Aspergillus flavus in liquid medium and tomato paste. *Food Control*, *18*(12), 1518–1523. https://doi.org/10.1016/j.foodcont.2006.12.003

Park, J.-B., Kang, J.-H., & Song, K. Bin. (2019). Clove bud essential oil emulsion containing benzethonium chloride inactivates Salmonella Typhimurium and Listeria monocytogenes on fresh-cut pak choi during modified atmosphere storage. *Food Control*, *100*, 17–23. https://doi.org/10.1016/j.foodcont.2019.01.001

Peng, C., Feng, Z., Zou, Y., Ou, L., Lai, Y., Su, B., Chen, M., Zhang, C., Zhu, W., Gan, G., Zhang, G., & Yao, M. (2024). Studies on the mechanisms of Helicobacter pylori inhibition by Syzygium aromaticum aqueous extract. *Phytomedicine*, *130*, 155540. https://doi.org/10.1016/j.phymed.2024.155540

Perumal, A. B., Nambiar, R. B., Sellamuthu, P. S., & Emmanuel, R. S. (2021). Use of modified atmosphere packaging combined with essential oils for prolonging post-harvest shelf life of mango (cv. Banganapalli and cv. Totapuri). *LWT*, *148*, 111662. https://doi.org/10.1016/j.lwt.2021.111662

Phothisuwan, S., Preechatiwong, W., & Matan, N. (2020). Enhancement of antibacterial activity of essential oil vapor released from a paper egg tray in combination with UV-C radiation against pathogenic bacteria on chicken eggs. *Journal of Food Processing and Preservation*, *44*(10). https://doi.org/10.1111/jfpp.14794

Qin, Y.-L., Zhang, S.-B., Ding, W.-Z., Lv, Y.-Y., Zhai, H.-C., Wei, S., Ma, P.-A., & Hu, Y.-S. (2023). The effect of volatile compounds of Syzygium aromaticum flower buds against Aspergillus flavus growth on wheat grain at postharvest stage. *Food Control*, *145*, 109450. https://doi.org/10.1016/j.foodcont.2022.109450

Rajaei, A., Hadian, M., Mohsenifar, A., Rahmani-Cherati, T., & Tabatabaei, M. (2017). A coating based on clove essential oils encapsulated by chitosan-myristic acid nanogel efficiently enhanced the shelf-life of beef cutlets. *Food Packaging and Shelf Life*, *14*, 137–145. https://doi.org/10.1016/j.fpsl.2017.10.005

Reuss, F., Kreß, A., Braun, M., Magdeburg, A., Pfenninger, M., Müller, R., & Mehring, M. (2020). Knowledge on exotic mosquitoes in Germany, and public acceptance and effectiveness of Bti and two self-prepared insecticides against Aedes japonicus japonicus. *Scientific Reports*, *10*(1), 18901. https://doi.org/10.1038/s41598-020-75780-5

Salimi, A., Mohammad Soleymani, H., & Mohammad Soleymani, S. (2022). Altered skin permeation of finasteride using clove oil, urea, and lyophilized powder of grape seed extract. *Advanced Pharmaceutical Bulletin*. https://doi.org/10.34172/apb.2023.010

Saricaoglu, F. T., & Turhan, S. (2019). Performance of mechanically deboned chicken meat protein coatings containing thyme or clove essential oil for storage quality improvement of beef sucuks. *Meat Science*, *158*, 107912. https://doi.org/10.1016/j.meatsci.2019.107912

Sarıcaoglu, F. T., & Turhan, S. (2020). Physicochemical, antioxidant and antimicrobial properties of mechanically deboned chicken meat protein films enriched with various essential oils. *Food Packaging and Shelf Life*, *25*, 100527. https://doi.org/10.1016/j.fpsl.2020.100527

Shah, K. C., Verma, K., Trivedi, J. D., Mehta, S. H., Kumar, S., Dhameliya, T. M., & Kathuria, D. (2024). Natural products for cosmeceuticals. In *Bioprospecting of Natural Sources for Cosmeceuticals* (pp. 1–17). Royal Society of Chemistry. https://doi.org/10.1039/9781837672288-00001

Shahtalebi, M. A., Sadat-Hosseini, A., & Safaeian, L. (2016). Preparation and evaluation of clove oil in emu oil self-emulsion for hair conditioning and hair loss prevention. *Journal of HerbMed Pharmacology*, *5*(2), 72–77.

Shalayel, M. H. F., Nour, S., Al-Mazaideh, G. M., Mahmoud, S. Y., & Farrag, E. S. (2021). Antifungal efficacy of clove flower buds (Syzygium aromaticum) oil and aqueous extract against dermatophytes clinical isolates. *Medical Science*, *25*(109), 744–750.

Shi, C., Cui, J., Yin, X., Luo, Y., & Zhou, Z. (2014). Grape seed and clove bud extracts as natural antioxidants in silver carp (Hypophthalmichthys molitrix) fillets during chilled storage: Effect on lipid and protein oxidation. *Food Control*, *40*, 134–139. https://doi.org/10.1016/j.foodcont.2013.12.001

Shukla, V., Mendiratta, S. K., Zende, R. J., Agrawal, R. K., & Kumar Jaiswal, R. (2020). Effects of chitosan coating enriched with *Syzygium aromaticum* essential oil on quality and shelf-life of chicken patties. *Journal of Food Processing and Preservation*, *44*(11). https://doi.org/10.1111/jfpp.14870

Singh, P. A., Desai, S. D., & Singh, J. (2018). A review on plant antimicrobials of past decade. *Current Topics in Medicinal Chemistry*, *18*(10), 812–833.

Takahashi, H., Nakamura, A., Fujino, N., Sawaguchi, Y., Sato, M., Kuda, T., & Kimura, B. (2021). Evaluation of the antibacterial activity of allyl isothiocyanate, clove oil, eugenol and carvacrol against spoilage lactic acid bacteria. *LWT*, *145*, 111263. https://doi.org/10.1016/j.lwt.2021.111263

Thosar, N., Basak, S., Bahadure, R. N., & Rajurkar, M. (2013). Antimicrobial efficacy of five essential oils against oral pathogens: An in vitro study. *European Journal of Dentistry*, *7*(S01), S071–S077. https://doi.org/10.4103/1305-7456.119078

Toledo, P. F. S., Viteri Jumbo, L. O., Rezende, S. M., Haddi, K., Silva, B. A., Mello, T. S., Della Lucia, T. M. C., Aguiar, R. W. S., Smagghe, G., & Oliveira, E. E. (2020). Disentangling the ecotoxicological selectivity of clove essential oil against aphids and non-target ladybeetles. *Science of the Total Environment*, *718*, 137328. https://doi.org/10.1016/j.scitotenv.2020.137328

Tsai, T.-H., Huang, W.-C., Lien, T.-J., Huang, Y.-H., Chang, H., Yu, C.-H., & Tsai, P.-J. (2017). Clove extract and eugenol suppress inflammatory responses elicited by *Propionibacterium acnes in vitro* and *in vivo*. *Food and Agricultural Immunology*, *28*(5), 916–931. https://doi.org/10.1080/09540105.2017.1320357

Vurmaz, A. K., & Gündüz, G. T. (2020). Inhibition of mold growth on the surface of dried persimmons using combined treatments of UV-C light and clove oil. *Innovative Food Science & Emerging Technologies*, *61*, 102336. https://doi.org/10.1016/j.ifset.2020.102336

Warade, D. D., & Bhosale, P. M. (2023). A review on herbal cosmetics used in skin and hair care. *International Research Journal of Modernization in Engineering Technology and Science*. https://doi.org/10.56726/IRJMETS37305

Xu, T., Gao, C., Feng, X., Huang, M., Yang, Y., Shen, X., & Tang, X. (2019). Cinnamon and clove essential oils to improve physical, thermal and antimicrobial properties of chitosan-gum arabic polyelectrolyte complexed films. *Carbohydrate Polymers*, *217*, 116–125. https://doi.org/10.1016/j.carbpol.2019.03.084

Yu, D., Xu, Y., Regenstein, J. M., Xia, W., Yang, F., Jiang, Q., & Wang, B. (2018). The effects of edible chitosan-based coatings on flavor quality of raw grass carp (*Ctenopharyngodon idellus*) fillets during refrigerated storage. *Food Chemistry*, *242*, 412–420. https://doi.org/10.1016/j.foodchem.2017.09.037

11 Uses in Pharmaceutical Industry

Zaira Yunuen Garcia-Carvajal, Moisés Martínez Velázquez, Gustavo Adolfo Castillo-Herrera, Hugo Espinosa-Andrews, and José Nabor Haro-González

11.1 GLOBAL CLOVE MARKET

The market for clove (*Syzygium aromaticum*) a medicinal agent comprises the global trade and use of clove due to its therapeutic properties. The medicinal application of clove is due to its anti-inflammatory, antioxidant, and antimicrobial properties (RBI, 2024). The clove market size was valued at USD 5.9 billion in 2023 and is projected to reach USD 6.9 billion by 2030, growing at a compound annual growth rate (CAGR) of 3.70% during the forecast period 2024–2030 (VMR, 2024).

In terms of applications, the medicinal clove market serves various industries, including hospitals, medical centers, laboratories, and various other sectors. Primarily, clove is used in the pharmaceutical and healthcare industries to treat various ailments and promote general well-being (VMR, 2024).

Growing consumer awareness of the health benefits of cloves, along with the growing preference for natural (herbal) remedies, has fueled the expansion of the clove-based pharmaceutical and nutraceutical products market (see Table 11.1).

Aura Cacia, Nature's Alchemy, Rocky Mountain Oils, and Olive Nation are considered market leaders in the medicinal clove industry. New entrants like Aroma Aromatics & Flavours and Saipro Biotech Private Limited offer industry innovation and novel perspectives in the medicinal clove industry (VMR, 2024).

11.2 BIOLOGICAL ACTIVITIES OF CLOVE

According to the *Handbook of Medicinal Herbs*, the potential uses for clove include Alzheimer's, anorexia, arthrosis, *Aspergillus*, athlete's foot (15% tincture in 70% alcohol), bacteria, bronchosis, bug bite, bunion, callus, cancer, *Candida*, caries, childbirth, cholera, cold, colic, convulsion, cough, cramp, dermatosis, diarrhea, dyspepsia, enterosis, *Escherichia*, fever, fungus, gas, gastrosis, gingivosis, halitosis, headache, heart, hernia, herpes, hiccup, infection, infertility, inflammation, maculosis, mucososis, myalgia, mycosis, nausea, nervousness, ophthalmia, pain, pharyngosis, phthisis, polyp, retinosis, rhinosis, sore, sore throat, spasm, *Staphylococcus*,

DOI: 10.1201/9781003521785-11

TABLE 11.1
Medicinal clove market in various regions worldwide.

Region	Countries	Comments
North America (NA)	United States, Canada	The growing focus on natural remedies and healthy lifestyle choices drives the demand for medicinal cloves.
Asia Pacific (APAC)	China, Japan, South India, Australia, Indonesia, Thailand, Malaysia	The traditional use of cloves in various cuisines and medicinal practices contributes to the market growth.
Europe	Germany, France, U.K., Italy, Russia	The increasing awareness about the health benefits of cloves propels the market forward.
Latin America Middle East Africa Turkey Saudi Arabia UAE Korea	Mexico, Brazil, Argentina, Colombia	These regions are experiencing significant growth in medicinal cloves due to their rich cultural heritage and growing preference for natural remedies.

stomatosis, teething, toothache, *Trichomonas*, ulcer, uterosis, vaginosis, virus, vomiting, wart, worm, wounds, and yeast (Duke, 2002).

Moreover, the reported dosages are 120–320 mg clove, 100–300 mg powdered clove, 0.05–0.2 ml clove oil, mouthwashes with 1–5% essential oil, and 2–4 ml concentrated clove infusion (Duke, 2002).

Clove has been shown to have different health benefits, mainly due to the eugenol content. However, compounds such as eugenyl acetate, β-caryophyllene and α-humulene also contribute to the bioactivity of clove. Clove essential oil (CEO) has been particularly studied in more depth. Table 11.2 provides examples of biological activities of CEO.

11.2.1 Antimicrobial

CEO has shown inhibitory activity against diverse pathogens. The antibacterial mechanism is based on the interaction of functional groups (such as –OH groups) of CEO bioactive compounds with the cytoplasmic membrane of microbial cells (Behbahani et al., 2019; El-Darier et al., 2018; Marmouzi et al., 2019; Rajkowska et al., 2016; Shahbazi, 2019; Singh et al., 2018). Due to its lipophilic nature, CEO can interact with cell membrane components, such as polysaccharides, fatty acids, and phospholipids. The interaction causes loss of cell membrane integrity, leakage of cellular contents, and interference with proton pump activity, resulting in cell death (Behbahani et al., 2019; El-Darier et al., 2018; Morsy, 2017; Rajkowska et al., 2016; Shahbazi, 2019). CEO inhibits both gram-negative (*E. coli*, *Salmonella*,

TABLE 11.2
Examples of biological activities of CEO.

Pharmaceutical form	Bioactivity	Mechanism of action	Model	Dose	References
CEO	Analgesic	Mediated by opioidergic and cholinergic receptors Inhibits voltage-gated Na^+ channels and activates TRPV1	Adult male Wistar rats Yellowtail clownfish *Amphiprion clarkia*	40–500 µl/l	Correia et al., 2017; Khalilzadeh et al., 2016
	Anesthetic	Inhibits voltage-gated Na^+ channels and activates TRPV1 Reduces dorsal muscle contraction	Wistar rats Cardinal tetra Paracheirodon axelrodi Angelfish *Pterophyllum scalare* *Cherax quadricarinatus* Adult male Tilapia del Nilo *Oreochromis niloticus*	50–500 µl/l	da Silva et al., 2021; Fujimoto et al., 2018; Ghanawi et al., 2019; Khalilzadeh et al., 2016
	Anticancer	Decreases inflammatory biomarkers levels Inhibits tissue remodeling proteins Inhibits pro-inflammatory proteins Induces apoptosis Changes polarization of cancer cells Inhibits proton pumps and ATP production	Human dermal fibroblasts, cancer cell lines (cervical, liver, breast, prostate, colon, erythroleukemia, lung)	13–127 µg/ml	Behbahani et al., 2019; Das et al., 2018; El-Darier et al., 2018; Han & Parker, 2017; Najar et al., 2020; Rajkowska et al., 2016; Ramadan et al., 2015
	Anticoagulant	Delays blood coagulation time	Male Swiss mice (*Mus musculus*)	0.0625–4 mg/ml	Uchôa Lopes et al., 2020
	Antidiarrheal	Balances gut microbiota Helps intestinal motility Potentiates digestive process Regulates neurotransmitters such as histamine and dopamine	Male Swiss mice (*M. musculus*)	50–100 mg/kg	Uchôa Lopes et al., 2020

(Continued)

TABLE 11.2 *(Continued)*

Examples of biological activities of CEO.

Pharmaceutical form	Bioactivity	Mechanism of action	Model	Dose	References
	Anti-inflammatory	Inhibits release or synthesis of inflammation mediators Decreases inflammatory biomarkers levels Inhibits tissue remodeling proteins Down-regulates pro-inflammatory proteins Inhibits prostaglandin synthesis and neutrophil chemotaxis Inhibits NF-kB in activation of TNF-α Inhibits expression of COX-2	Rats Human dermal fibroblasts BALB/c mice	100–250 mg/kg	Han & Parker, 2017; Marmouzi et al., 2019; Sugihartini et al., 2019
	Antimicrobial	Inhibits growth Destabilizes membrane permeability and integrity Disintegrates phospholipid membrane	*Candida albicans, Klebsiella spp., E. coli, Proteus spp., Pseudomonas aeruginosa., Agrobacterium tumefaciens, Erwinia spp., S. aureus, Listeria innocua, Bacillus subtilis, B. cereus, L. monocytogenes, Salmonella typhimurium, Lactobacillus acidophilus, L. reuteri, L. casei, L. rhamnosus, Aspergillus niger, Tetrahymena pyriformis*	1.25–6.25 mg/ml	Behbahani et al., 2019; El-Darier et al., 2018; Marmouzi et al., 2019; Rajkowska et al., 2016; Shahbazi, 2019
	Antinociceptive	Inhibits COX-2 and TRPV in primary neurons	Female Wistar rats	100 µg/kg	Beltrán-Villalobos et al., 2017
	Antioxidant	Shows radical scavenging activity Inhibits lipid peroxidation Neutralizes free radicals and blocks oxidative processes Shows protective effect on ROS-induced biochemical changes and histopathological damage	DPPH, β-carotene-linoleate, ABTS, FRAP, Folin–Ciocalteu, flavones and flavonols, flavonoids, TAC Wistar rats/blood, histopathological study	30–600 µg/ml	Bakaour et al., 2018; B. A. Behbahani et al., 2019; Marmouzi et al., 2019; Ramadan et al., 2015; Shahbazi, 2019

TABLE 11.2 *(Continued)*
Examples of biological activities of CEO.

Pharmaceutical form	Bioactivity	Mechanism of action	Model	Dose	References
	Antipyretic	Reduces chemotaxis Inhibits COX-1 and COX-2	Male Swiss mice (*M. musculus*)	50–100 mg/kg	Uchôa Lopes et al., 2020
	Hemolytic	Interacts with the cell membrane	Male Swiss mice (*M. musculus*)	0.625–2.5 mg/ml	Uchôa Lopes et al., 2020
CEO microemulsion	Anti-inflammatory	Promotes re-epithelialization and formation of dermis and epidermis Increases collagen synthesis	m5S cell line Male Wistar rats	0.2 g	Banerjee et al., 2020
CEO nanoemulsion	Antimicrobial	Destabilizes membrane permeability	*S. aureus*	19–24 μg/ml	Nirmala et al., 2019
	Anticancer	Shows antiproliferative effect Exhibits cytotoxic activity Induces necrosis	Thyroid cancer cell line	19–24 μg/ml	Nirmala et al., 2019
CEO nanoemulsion	Wound healing	Reduces wound epithelialization period Increases collagen content Induces neovascularization	Female albino Wistar rats	0.61 mg/g	Alam et al., 2017

Klebsiella pneumoniae, *Erwinia carotovora*, *Agrobacterium*, and *Pseudomonas aeruginosa*) and gram-positive bacteria (*S. aureus*, *Streptococcus*, and *L. monocytogenes*), *Aspergillus* (*A. flavus*, *A. parasiticus*, and *A. ochraceus*), *Penicillium C. albicans*, and yeast (Behbahani et al., 2019; El-Darier et al., 2018; Rajkowska et al., 2016; Shahbazi, 2019). CEO has greater activity against gram-positive than gram-negative bacteria, which is attributed to differences in the cell membrane composition (Behbahani et al., 2019). Similarly, food-related pathogens are more sensitive to CEO than probiotics and fungi (Shahbazi, 2019).

11.2.2 Antioxidant

High levels of reactive oxygen species (ROS) have been related to the promotion of diseases such as cancer, arteriosclerosis, Alzheimer's, and Parkinson's (Morsy, 2017). CEO has shown radical scavenging activity and inhibition of lipid peroxidation (Dahham et al., 2015; Ramadan et al., 2015). Antioxidant compounds contained in CEO, such as eugenol, eugenyl acetate, β-caryophyllene, and α-humulene, protect cells from free radical oxidation. The aromatic ring of eugenol contains a hydroxyl group, which is responsible for the antioxidant activity (Marmouzi et al., 2019). Phenolic compounds transfer electrons or hydrogen atoms and neutralize free radicals, resulting in the blocking of the oxidative process (Shahbazi, 2019).

A protective effect of CEO on the biochemical changes and histopathological injuries induced by ROS in the kidney, liver, and brain has been demonstrated (Bakaour et al., 2018).

11.2.3 Antiviral

Antiviral activity of CEO against influenza A (Dai et al., 2013), Ebola, and herpes simplex virus types 1 and 2 (Lane et al., 2019) has been demonstrated. de Oliveira et al. (2019) showed that eugenol derivatives inhibited the West Nile Virus activity, evidencing promising compounds against flaviviruses such as dengue, Zika, and yellow fever virus. Eugenol also acted as an inhibitor of the initial stage of HIV-1 infection, reducing virus replication and increasing lymphocyte production (Behbahani et al., 2013).

CEO is active against feline calicivirus, which is used as a surrogate for human norovirus. Thus, CEO applied in the process of washing fruits and vegetables eliminates any viral load that may exist. Furthermore, CEO applied to cleaning wipes allows the decontamination of surfaces (Aboubakr et al., 2016). Moreover, CEO increases the resistance of tomato plants to tomato yellow leaf curl virus to a greater extent than moroxydine hydrochloride (Sun et al., 2016).

11.2.4 Antinociceptive

Inflammatory nociceptive pain is largely treated by using nonsteroidal anti-inflammatory drugs (NSAIDs). They inhibit cyclooxygenase (COX) and decrease prostaglandin levels that cause nociceptive pain. Eugenol has shown antinociceptive and anti-inflammatory properties by inhibiting COX-2 and transient receptor potential vanilloid (TRPV) channel in primary afferent neurons (Beltrán-Villalobos et al.,

2017). The antinociceptive activity is more related to opioid, cholinergic, and $\alpha 2$-adrenergic receptors than serotoninergic receptors. Eugenol administration also inhibits gamma-aminobutyric acid (GABA) receptor currents in trigeminal ganglion neurons and inhibits GABA $\alpha 1\beta 2\gamma 2$ expressed in these neurons. GABA modulation could also explain the antinociceptive activity of eugenol (Beltrán-Villalobos et al., 2017; Khalilzadeh et al., 2016).

11.2.5 Anti-inflammatory and Wound Healing

It is known that pathophysiological conditions such as diabetes, hypertension, and cardiovascular and neurodegenerative diseases are related to inflammation and oxidative stress processes (Marmouzi et al., 2019). Banerjee et al. (2020) reported the anti-inflammatory properties of CEO and eugenol. They found that activities were comparable to diclofenac gel, significantly reducing the inflammation after 3 hours. Similarly, inflicted wounds in rats treated with CEO showed a significant contraction. These results showed that CEO-treated animals underwent a similar wound healing to those that were neomycin treated. CEO inhibited the expression of antiproliferative biomarkers, decreased the levels of inflammatory biomarkers, and inhibited the expression of tissue remodeling proteins (Alam et al., 2017; Han & Parker, 2017; Sugihartini et al., 2019). Otherwise, eugenol inhibits the expression of COX-2 and reduces the production of mediators of inflammation (Han & Parker, 2017; Sugihartini et al., 2019). Moreover, Fernandes et al. (2007) reported anti-inflammatory effects of compounds alpha-humulene and (–)-trans-caryophyllene.

11.2.6 Analgesic

CEO has been used to treat headaches, joint pains, and toothaches. As analgesics, CEO and eugenol are safe, effective, and inexpensive. The analgesic activity of eugenol in different pain models is well documented. Khalilzadeh et al. (2016) reported that the opioidergic and cholinergic systems mediate the analgesic activity of CEO in acute corneal pain. Likewise, eugenol modulates its analgesic activity by inhibiting the voltage-gated channels (Na^+ and Ca^{2+}) and activating TRPV1 channel. Both CEO and eugenol show analgesic activities similar to those of lidocaine. Furthermore, Correia et al. (2017) evidenced an analgesic effect of CEO in fish and Sueksakit et al. (2013) in rats.

11.2.7 Anesthetic

At low concentrations, CEO exhibits anesthetic properties in vertebrates and invertebrates, showing no side effects. CEO quickly induces anesthesia, has a brief period of reflex recovery, and shows a low mortality rate, while it does not affect the external stimuli response (Correia et al., 2017; da Silva et al., 2021; Ghanawi et al., 2019). Khalilzadeh et al. (2016) demonstrated that CEO and eugenol, when topically applied, reduce corneal sensitivity in rats to a similar extent to lidocaine. Additionally, CEO shows anesthetic effects in Nile tilapia, cardinal tetra, ringed cichlid, and angelfish, which affect swimming ability and balance and decrease the external stimuli response, until complete immobilization. Based on the concentration and time of exposure, side effects of CEO were not found (da Silva et al., 2021; Fujimoto et al., 2018; Khalilzadeh et al., 2016). CEO also induced anesthesia efficiently in red claw crayfish and other crustaceans, such as *Nephrops norvegicus* and grass shrimp (Ghanawi et al., 2019).

11.2.8 Anticancer

CEO and its components, eugenol, α-humulene, and β-caryophyllene, have shown cytotoxic and antitumor effects, so they have been proposed as surrogates in the prevention and co-treatment of cancer (Sung et al., 2012; Lesgards et al., 2014; Fadilah et al., 2017; Ambrož et al., 2019). Several studies show that EOs can reduce the undesirable effects of chemotherapy, such as nausea, vomiting, loss of appetite, and weight loss (Bayala et al., 2014; Lee et al., 2018; Morsy, 2017). The anticancer effect is associated with the antioxidant and anti-inflammatory properties, since the elevation of ROS levels activates specific signaling pathways and promotes the development of tumors, favoring cell proliferation, angiogenesis, and metastasis (Bayala et al., 2014; Morsy, 2017). CEO has been active against different cancer types, such as colon, lung, breast, pancreatic, leukemia, cervical, and prostate (Dahham et al., 2015; Ramadan et al., 2015; Das et al., 2018; Najar et al., 2020).

Mechanisms such as cell cycle arrest or apoptosis, activation of detoxifying enzymes, DNA damage by oxidative stress, reduction of Akt and MMP-2 expression levels, among others explain the anticancer properties of CEO and eugenol (Das et al., 2018; Han & Parker, 2017; Nirmala et al., 2019; Rajkowska et al., 2016; Aranha et al., 2019).

11.2.9 Other Bioactivities

It has been reported that CEO has antiseptic, natural stimulant, carminative, anticoagulant, anthelminthic, antiemetic, antidiarrheal, antispasmodic, hepatoprotective, spasmolytic, antimutagenic, anticonvulsant, antidepressant, renal reinforcement, antipyretic, neuroprotective, antistress, antiallergic, antidiabetic, and hypocholesterolemic effects (Guan et al., 2007; Din et al., 2015; Liu et al., 2015; Kennouche et al., 2015; Khalilzadeh et al., 2016; Wei et al., 2016; Khalil et al., 2017; Behbahani et al., 2019; Shahbazi, 2019; Thapa et al., 2019; Banerjee et al., 2020; Uchôa Lopes et al., 2020).

11.3 CLOVE PRODUCTS IN THE MARKET

Clove has been part of traditional medicine worldwide. Up to 85% of clove essential oil comprises eugenol, the primary bioactive component (Batiha et al., 2020; Pandey et al., 2024). The medicinal clove market offers various products (Figure 11.1), including powder, oil, extracts, and gels (VMR, 2024; Batiha et al., 2020).

An exhaustive search has been carried out through the drug bank platform for commercial products with clove oil or extract offered on the market. Table 11.3 offers an overview of the clove market products in various regions worldwide. The clove oil and eugenol can be used as a natural antibacterial agent against cariogenic and periodontopathogenic bacteria. Furthermore, combining clove oil or eugenol

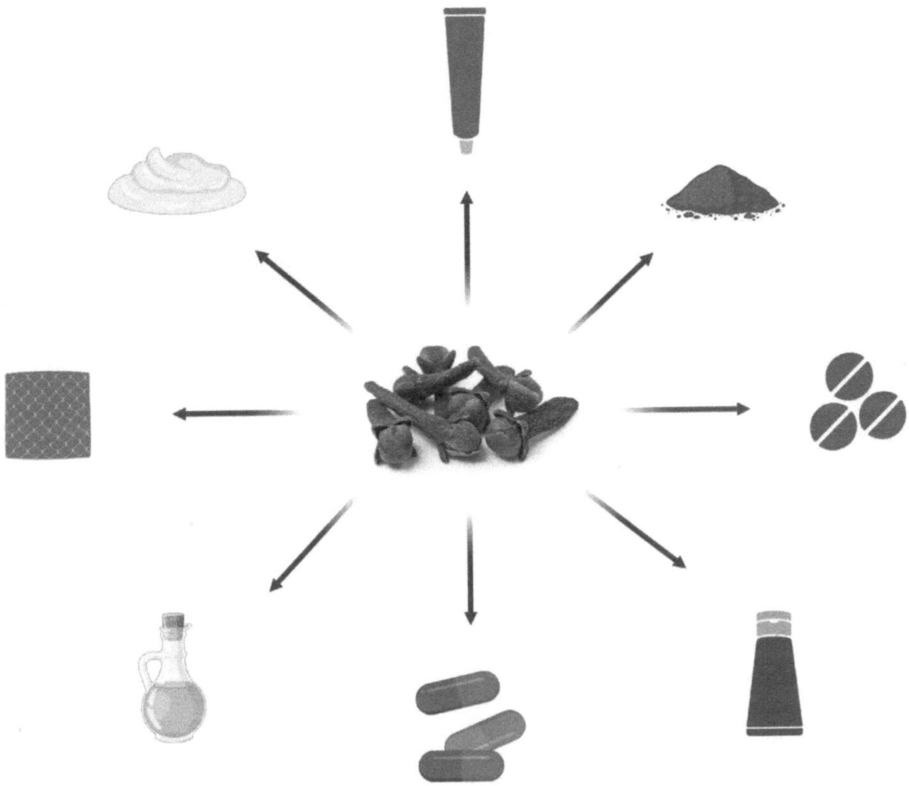

FIGURE 11.1 A variety of clove-based medicinal products are offered on the market.

with antibiotics could reduce the antibacterial activity (Moon et al., 2011). It is well known that most of the products are over-the-counter (OTC), referring to medicine products that can be bought without a doctor's prescription.

The U.S. Food and Drug Administration (FDA) has approved clove oil for use in food as a flavoring agent, in dentistry as an analgesic, in dental cement, as a fragrance in personal care products, and in aromatherapy oils, and clove oil is generally recognized as safe (GRAS) by the FDA (Pandey et al., 2022). Several OTC products are available in the U.S. market. In the OTC active ingredients published in 2010 by the FDA, clove oil is approved for skin protectants like astringent. Likewise, eugenol is used as an external analgesic and anesthetic and for the relief of oral discomfort (FDA, 2010).

A similar situation occurs in Europe. The only country with marketing authorization in Europe is the United Kingdom (UK; EMA, 2011). Thornton & Ross Ltd sells a CLOVE OIL BP dental solution containing clove oil 100% v/v to relieve toothache temporarily (www.vademecum.es/reino-unido/medicamento/31015087/clove-oil-bp-dental-solution). In Belgium and Latvia, products containing cloves are registered as food supplements (Pandey et al., 2022).

In Canada, the Arthrimed-Plus® spray products are offered on the market as topical analgesics for human OTC drugs labeled for temporary relief of aches and

TABLE 11.3
An overview of the clove market products with therapeutic effects in various regions worldwide.

Product name	Qualitative and quantitative composition	Pharmaceutical form	Therapeutic indications	Company	Web link
Clove Oil BP	Clove oil 100% v/v	Dental solution	For the temporary relief of toothache	Thornton & Ross Ltd (UK)	www.medicines.org.uk/emc/product/4817
Wilisan clove oil preparation	Clove oil 87% v/v	Dental solution	For the temporary relief of throbbing, persistent toothache due to a cavity until a dentist can be seen	Chong Park Pharmaceutical SDN BHD (Malaysia)	https://cheesengtong.com/product/wilisan-clove-oil-preparation/
Tiger balm soft	Clove oil (1.5%) + Cajuput oil (13%) + Camphor (10%) + Levomenthol (12%)	Ointment (topical) OTC	Provides symptomatic relief of muscular aches, pains, and strains	Tiger Balm (M) SDN BHD (Malaysia)	www.npra.gov.my/index.php/en/consumer-medication-information-leaflets-rimup/item/tiger-balm-soft.html
Letomint	18 mg/1	Tablet (oral) OTC Unapproved by FDA	Vaginal discharge, leukorrhea, period pain, and menstrual irregularity	H&C 21 America Inc (USA)	https://dailymed.nlm.nih.gov/dailymed/fda/fdaDrugXsl.cfm?setid=fe10b461–11a8–4d0b-9774-dcdbd8517953&type=display
Jamu Sexty	18 mg/1	Tablet (vaginal) OTC Unapproved by FDA	Vaginal discharge, leukorrhea, period pain, and menstrual irregularity	JMBio Co., Ltd. (USA)	https://dailymed.nlm.nih.gov/dailymed/drugInfo.cfm?setid=c5327c6a-30f6–478d-95e2-c4229b7c2453&audience=consumer
Nose-clove aerosol	0.75 mg/100ml	Nasal aerosol OTC Unapproved by FDA	Nasal moisturizer	US EXCY-LINE INC. (USA)	https://fda.report/DailyMed/EF870793–3FD5–487A-AFE7-C10C3545A950.pdf
Soma Circum	0.0275 g/g	Pill (oral) OTC Unapproved by FDA	Menopausal disorder, menstrual pain, menstrual irregularity, infertility, cold hands and feet syndrome, insomnia, depression, alopecia, and aging	Soma Biotechnology Co., Ltd (USA)	https://fda.report/NDC/58370–1010 https://dailymed.nlm.nih.gov/dailymed/fda/fdaDrugXsl.cfm?setid=3cbb8727-f2e4–42a8–952c-08c41c446a89&type=display

TABLE 11.3 *(Continued)*
An overview of the clove market products with therapeutic effects in various regions worldwide.

Product name	Qualitative and quantitative composition	Pharmaceutical form	Therapeutic indications	Company	Web link
Soma Prouri	0.0275 g/1g	Pill (oral) OTC Unapproved by FDA	To help the treatment of prostate, sexual dysfunction, urinary incontinence, and muscular dystrophy	Soma Biotechnology Co., Ltd (USA)	https://dailymed.nlm.nih.gov/dailymed/lookup.cfm?setid=2a36e43e-2c55–4d26-adfe-2406c77c9652 https://fda.report/NDC/58370–1011
Nosmo King smoking deterrent	Clove (0.11875 g/1) + Lemon (0.001 g/1) + Menthol (0.0015 g/1)	Tablet, soluble (oral) OTC Unapproved by FDA	To help smokers abstain from smoking by greatly reducing withdrawal symptoms causing cravings to smoke	Nosmo Co., Ltd (USA)	https://dailymed.nlm.nih.gov/dailymed/drugInfo.cfm?setid=365f85b3–96b7–4a02-bc4f-48a205c0edbe https://ndclist.com/ndc/34605–5001/label
Nosmotea smoking deterrent	Clove (20 mg/1g) + Lemon (1 mg/1g) + Menthol (1.5 mg/1g)	Powder (oral) OTC Unapproved by FDA	Smoking deterrent, stop smoking aid, remove nicotine	Nosmo Co., Ltd (USA)	https://dailymed.nlm.nih.gov/dailymed/drugInfo.cfm?setid=8f65a3a1-f921–4a2a-b93d-ebf10a0ef1a1 file:///C:/Users/ZAIRA/Downloads/20100306_8f65a3a1-f921–4a2a-b93d-ebf10a0ef1a1%20(1).pdf
Arthrimed-Plus ® spray	Menthol (synthetic) at 2.5%, camphor (*Cinnamomum camphora*) (wood) at 3%, eucalyptus essential oil (*Eucalyptus globulus*) (leaves) at 3%, and clove oil (*Eugenia caryophyllata*) flower at 0.2%	External spray (topical) OTC	Temporary relief of aches and pains of muscles and joints associated with back pain, lumbago, bruises, sprains, strains, arthritic and rheumatic pains, sore tendons, and ligaments as well as for tired and sore muscles	ArthriMed Plus (Canada)	www.arthrimed-plus.ca/

(Continued)

TABLE 11.3 *(Continued)*

An overview of the clove market products with therapeutic effects in various regions worldwide.

Product name	Qualitative and quantitative composition	Pharmaceutical form	Therapeutic indications	Company	Web link
Lee Buan Soa Pill (Fishing Brand)	Clove (3 mg) + belladonna (3 mg) + cinnamon (1 mg) + glycyrrhiza glabra (69 mg) + levomenthol (2 mg) + peppermint (30 mg) + wood creosote (20 mg)	Pill (oral) OTC	For stomachache, indigestion, flatulence, abdominal pain and discomfort, diarrhea due to food or water poisoning, spasm, nausea, vomiting, and any digestive or stomach-related symptoms	Lee Buan Soa, Fishing Brand (Thailand)	www.leebuansoa.com/index.php/en/producten/42-lee-buan-soa-pill-fishing-brand
South moon patch	Turmeric (2mg/10mg), ascorbic acid (2 mg/10mg), ginger (2mg/10mg), *Pueraria montana* var. *lobata* root (2mg/10mg), clove oil (2mg/10mg)	Patch (cutaneous) OTC	Breast enhancement patch	Shantou Youjia E-Commerce Co., Ltd. (China)	https://dailymed.nlm.nih.gov/dailymed/fda/fdaDrugXsl.cfm?setid=13370f34–9e77–3bb5-e063–6294a90a2f97&type=display
Eelhoe fresh breath essence	*Eucalyptus radiata* leaf (3mg/30 mg), lemon (7.5mg/30mg), clove oil (4.5mg/30 mg), peppermint oil (6mg/30mg), water	Liquid (dental) OTC	Antibacterial	Shantou Youjia E-Commerce Co., Ltd. (China)	https://dailymed.nlm.nih.gov/dailymed/drugInfo.cfm?setid=131ea0d6-b0d4–7f40-e063–6294a90a6373

**OTC*: over the counter.

pains of muscles and joints associated with back pain, lumbago, strains, bruises, sprains, strains, arthritic and rheumatic pains, sore tendons, and ligaments as well as for tired and sore muscles (https://dailymed.nlm.nih.gov/dailymed/drugInfo.cfm?setid=e531848c-43e2-29f1-e053-2995a90a2087).

For the Asian market, the situation was very different. Several products have been found on the drug bank platform. Most products are natural remedies, herbal preparations, and ayurvedic products with different therapeutic effects.

Published information on the indications for clove-based products is limited. However, based on existing documentation, the European Medicines Agency (EMA) proposes that pharmacological activities support the traditional use of clove to temporarily relieve toothache and the symptomatic treatment of minor inflammations in the mouth or throat.

The Committee on Herbal Medicinal Products (HMPC) compiles and assesses scientific data on herbal substances, preparations, and combinations, focusing on safety and efficacy. This activity supports the harmonization of the European market, given that the national competent authorities can refer to one unique set of information on a herbal substance or preparation when evaluating marketing applications. The herbal monographs contain the HMPC's scientific opinion on safety and efficacy data about a herbal substance and its preparations intended for medicinal use. The HMPC evaluates all available information—including non-clinical and clinical data—and documents long-standing use and experience in Europe.

The herbal monographs are divided into two sections:

(1) Well-established use (marketing authorization): demonstrated sufficient safety and efficacy data.
(2) Traditional use (simplified registration): accepted based on sufficient safety data and plausible efficacy.

A final EU monograph can be used in application reference material by a marketing authorization applicant (well-established use part) and by a traditional-use registration applicant (traditional-use part).

The "Community herbal monograph on *Syzygium aromaticum* (L.) Merill et L.M. Perry, floris aetheroleum" was published with technical information about well-established and traditional use. Its use is focused on the symptomatic treatment of minor inflammations in the mouth or the throat. The temporary relief of toothache due to a dental cavity was reported (EMA, 2011). However, safety is compromised due to eugenol, as it is known to be carcinogenic and genotoxic. Figure 11.2 shows the main information reported in the herbal clove monograph.

11.4 CLOVE FORMULATIONS IN CLINICAL TRIALS

Table 11.4 lists a number of clove formulations currently under evaluation in clinical trials, developed for the treatment of different conditions such as pain, dental caries, and gingivitis, as well as for the detection of olfactory disorders.

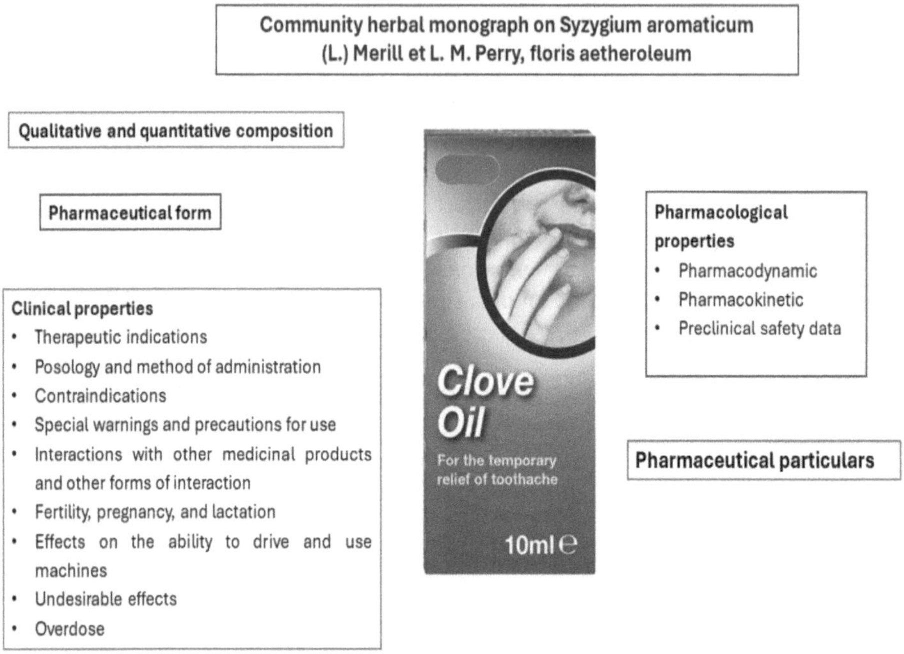

FIGURE 11.2 Technical information of herbal clove monograph.

11.5 CONTRAINDICATIONS, INTERACTIONS, AND SIDE EFFECTS OF CLOVE

Clove (*S. aromaticum*) is a highly appreciated spice, widely used for many medicinal purposes and as a food preservative. Clove contains approximately 15–20% wt. of essential oil, eugenol being the major bioactive compound found both in clove spice and clove essential oil, accounting for up to 90% (Haro-González et al., 2021). Eugenol is thought to be responsible for the aromatic as well as both beneficial and harmful effects of clove. Side effects of eugenol have been reported, its cytotoxicity and genotoxicity being the major factors that limit the consumption of both clove and eugenol by humans. A number of mechanisms of action have recently been proposed to explain the cytotoxicity of eugenol, which include formation of hepatotoxic eugenol intermediates, plasma membrane damage due to highly lipophilic eugenol nature, uncoupling of oxidative phosphorylation in mitochondria, formation of oxygen-free radicals, and protein deactivation due to eugenol binding (Özbek & Ergönül, 2022). Neither eugenol nor other clove extracts have been approved for use in any medical condition in the United States (LiverTox, 2019). Clove oil is reported to have an oral LD_{50} of 2650 mg/kg body weight in rats (Duke & duCellier, 1993). Accumulated evidence indicates that eugenol, when used in low doses, has shown side effects such as local irritation, allergic reactions, and contact dermatitis, while ingestion of high doses (as with overdoses) can result in tissue damage, with acute onset of seizures, coma, and severe liver and kidney injury (LiverTox, 2019).

TABLE 11.4
Clove formulations in ongoing clinical trials.

NCT number	Study title	Study status	Conditions	Interventions	Study type
NCT02865980	Effect of washing with clove extract on inflammation and infection place logging shaldon needle	Unknown	Renal failure	Clove extract/Betadine	Interventional
NCT05507359	Effect of topical application of clove gel versus benzocaine gel on pain reduction during application of simple dental procedures in children	Not yet recruiting	Pain	Clove gel/Benzocaine gel	Interventional
NCT04390256	Remineralization of early carious lesion using natural agents versus bioadhesive polymers	Completed	Dental caries	Clove water extract/Carbopol/ Carboxymethyle cellulose	Interventional
NCT03553628	Using a mouthwash containing propolis, clove oil, and chlorhexidine to improve the caries risk of high risk patients	Completed	Caries, dental/risk reduction	Chlorhexidine, propolis and clove oil mouthwash/Chlorhexidine mouthwash	Interventional
NCT05130957	The effect of clove oil on labor pain	Completed	Labor pain	Massage application with clove oil/ massage application without any oil	Interventional
NCT01385891	Clove in the treatment of relapsed or resistant acute leukemia in children	Completed	Acute leukemia	Drug: Clofarabine VP 16 ciclophosphamide	Interventional
NCT06382701	Effect of aromatherapy on pain, anxiety, and satisfaction in cannulation	Active not recruiting	Hemodialysis patients	Aromatherapy/placebo	Interventional
NCT06327633	Effects of olfactory training on the brain function in T2DM patients with mild cognitive impairment	Recruiting	Type 2 Diabetes mellitus/mild cognitive impairment	Olfactory training	Interventional
NCT05908318	Effect of preoperative olfactory training on postoperative delirium in the elderly undergoing orthopedic surgeries	Not yet recruiting	Preoperative olfactory training/ postoperative delirium	Behavioral: olfactory training	Interventional

(Continued)

TABLE 11.4 ***(Continued)***
Clove formulations in ongoing clinical trials.

NCT number	Study title	Study status	Conditions	Interventions	Study type
NCT04710394	Visual-olfactory training in participants with COVID-19 resultant loss of smell	Completed	Anosmia/COVID-19/ageusia/ hyposmia/ hypogeusia/ SARS-CoV-2 infection	Behavioral: smell training	Interventional
NCT04598763	Evaluation of two methods of olfactory rehabilitation in post-viral loss of smell: classic and intensive	Unknown	Olfaction disorders	Smell the odors of the olfactory rehabilitation kit according to the classic or intensive method	Interventional
NCT05912881	Chemosensory dysfunction and recovery in endoscopic endonasal skull base surgery	Recruiting	Anosmia/skull base neoplasms	Olfactory training/placebo training	Interventional
NCT03049358	Olfactory training in improving sense of smell after radiation therapy in patients with paranasal sinus or nasopharyngeal cancer	Terminated	Paranasal sinus cancer/ nasopharyngeal carcinoma	Physiologic testing/quality of life assessment/procedure: sham intervention/procedure: therapeutic procedure/rose essential oil/lemon essential oil/clove essential oil/ eucalyptus essential oil/canola oil placebo	Interventional
NCT04973748	Conditioned open-label placebo (COLP) for peri-operative pain management in patients with head and neck cancer	Recruiting	Postoperative pain/ opioid use	Behavioral: conditioned open-label placebo (COLP)/behavioral: surveys about pain, opioid use, and depression symptoms	Interventional
NCT06290219	The effect of platelet-rich plasma nasal injection in the treatment of traumatic olfactory dysfunction	Not yet recruiting	Traumatic olfactory nerve injury with anosmia (diagnosis)/ effect of drug	Combination product: platelet-rich plasma combined with hyaluronic acid/ drug: zinc gluconate 10 mg oral tablet/ device: 4 bottles of phenyl ethyl alcohol, lemon, eucalyptus, and clove oil	Interventional

TABLE 11.4 *(Continued)*
Clove formulations in ongoing clinical trials.

NCT number	Study title	Study status	Conditions	Interventions	Study type
NCT04952389	Acupuncture therapy for Covid-related olfactory loss	Completed	Olfactory dysfunction/ COVID-19	Device: acupuncture therapy/drug: budesonide/olfactory training	Interventional
NCT05467631	Topical phytoterpenes for plantar fasciitis	Completed	Plantar fasciitis	Topical oil application	Interventional
NCT06295822	The effectiveness of Chinese medicine Nuan-gong-ye on primary dysmenorrhea	Recruiting	Chinese medicine/ primary dysmenorrhea	Drug: Nuan-gong-ye/placebo	Interventional
NCT05518019	An exploratory investigation of a CBD supplement's impact on sleep, stress, and focus	Completed	Anxiety/fear/stress/ low mood/sleep	Dietary supplement: the daily supplement	Interventional
NCT03287011	Efficacy of a plaque disclosing toothpaste on home oral hygiene procedures	Unknown	Chronic gingivitis, plaque	Shoplaq toothpaste/Colgate toothpaste	Interventional
NCT05815654	Research and development of WPAPS based on the modern technology of traditional Chinese medicine	Recruiting	Dysmenorrhea	Device: warm palace analgesic point sticker	Interventional
NCT05199623	The effect of Chinese herbal medicine as an adjuvant therapy in the treatment of traumatic anosmia	Completed	Traumatic anosmia	Drug: tailored Guizhi decoction/ behavioral: olfactory training	Interventional
NCT05539560	Association between smell training and quality of life in patients with impaired sense of smell following COVID-19	Recruiting	COVID-19	Olfactory training with essential oils/ olfactory training with fragrance-free oils	Interventional
NCT05054023	Topical piroxicam vs Soulagel in the treatment of acute extremity pain after emergency department discharge	Completed	Acute pain	Drug: Soulagel®/drug: Arthrosyl®	Interventional

(Continued)

TABLE 11.4 ***(Continued)***

Clove formulations in ongoing clinical trials.

NCT number	Study title	Study status	Conditions	Interventions	Study type
NCT03372109	Evaluate the effects of repeated periods of modified fasting to support healthy natural weight management and prevention of weight gain	Completed	Body weight changes	Dietary supplement: meal replacement shake/dietary supplement: multivitamin and mineral capsules/ dietary supplement: fish oil with sesame lignans and olive extract softgel/dietary supplement: prebiotic chewable tablet/dietary supplement: clove extract and maqui berry extract capsule/dietary supplement: whole food blend capsule/dietary supplement: turmeric extract capsule/ dietary supplement: European white kidney bean capsule/dietary supplement: saffron extract capsule/ dietary supplement: ubiquinol with fulvic acid complex softgel/dietary supplement: calcium with hesperidin and gynostemma extract tablet/dietary supplement: multivitamin and mineral supplement	Interventional
NCT00954902	Effects of antioxidants on cardiovascular risk measures (spice study)	Completed	Cardiovascular/ psychological stress	Dietary supplement: high antioxidant spice blend/behavioral: trier psychological stressor/placebo antioxidant concentrate	Interventional
NCT03420976	Novel supplement-based therapy for the treatment of small intestinal bacterial overgrowth	Withdrawn	Small intestinal bacterial overgrowth	Dietary supplement: novel supplement-based therapy	Interventional

TABLE 11.4 *(Continued)*
Clove formulations in ongoing clinical trials.

NCT number	Study title	Study status	Conditions	Interventions	Study type
NCT04374474	Anosmia rehabilitation in patients post coronavirus disease (COVID 19)	Withdrawn	Olfactory disorder	Olfactory retraining/drug: corticosteroid nasal irrigation/smell household items/nasal irrigation	Interventional
NCT06107595	Conditioned open label placebo for postoperative pain management	Recruiting	Pain, postoperative	Behavioral: placebo	Interventional
NCT04764981	Olfactory training for olfactory dysfunction after coronavirus disease—19 (COVID-19)	Unknown	COVID-19/SARS-CoV infection/olfactory disorder/anosmia	Essence oils	Interventional
NCT04606680	Herb-partitioned moxibustion plus lifestyle modification treatment for simple obesity	Unknown	Obesity	Device: heat application at acupoints/device: medicated plaster at acupoints/device: herb-partitioned moxibustion at acupoints	Interventional
NCT03738657	Gustatory function following radiotherapy to the head and neck	Unknown	Head and neck cancer	Olfactory screening/quantitative gustatory testing using taste strips	Observational

11.6 CONCLUSIONS

Clove exhibits a widely diverse range of biological activities, and its use has extended to the pharmaceutical industry. Medicinal clove products are used in hospitals, medical centers, and laboratories, and they are offered to the users in a variety of pharmaceutical forms. Moreover, several experimental clove formulations for pain and oral health are currently under evaluation in clinical trials, ensuring that new products are on the way. However, a warning regarding the use of clove products is important due to the side effects reported for this spice.

REFERENCES

Aboubakr, H. A., Nauertz, A., Luong, N. T., Agrawal, S., El-Sohaimy, S. A., Youssef, M. M., & Goyal, S. M. (2016). *In vitro* antiviral activity of clove and ginger aqueous extracts against feline calicivirus, a surrogate for human norovirus. *Journal of Food Protection*, *79*(6), 1001–1012. https://doi.org/10.4315/0362-028X.JFP-15-593

Alam, P., Ansari, M. J., Anwer, Md. K., Raish, M., Kamal, Y. K. T., & Shakeel, F. (2017). Wound healing effects of nanoemulsion containing clove essential oil. *Artificial Cells, Nanomedicine, and Biotechnology*, *45*(3), 591–597. https://doi.org/10.3109/21691401.2016.1163716

Ambrož, M., Šmatová, M., Šadibolová, M., Pospíšilová, E., Hadravská, P., Kašparová, M., Skarková, V. H., Králová, V., & Skálová, L. (2019). Sesquiterpenes α-Humulene and *β*-Caryophyllene oxide enhance the efficacy of 5-Fluorouracil and oxaliplatin in colon cancer cells. *Acta Pharmaceutica*, *69*(1), 121–128. https://doi.org/10.2478/acph-2019-0003

Aranha, E. S. P., de Azevedo, S. G., dos Reis, G. G., Silva Lima, E., Machado, M. B., & de Vasconcellos, M. C. (2019). Essential oils from *Eugenia* spp.: *In vitro* antiproliferative potential with inhibitory action of metalloproteinases. *Industrial Crops and Products*, *141*, 111736. https://doi.org/10.1016/j.indcrop.2019.111736

Bakaour, M., Soulo, N., Hammas, N., Fatemi, H., Aboulghazi, A., Taroq, A., Abdellaoui, A., Al-Qaili, N., & Lyoussi, B. (2018). The antioxidant content and protective effect of argan oil and *Syzygium aromaticum* essential oil in hydrogen peroxide-induced biochemical and histological changes. *International Journal of Molecular Sciences*, *19*(2), 610. https://doi.org/10.3390/ijms19020610

Banerjee, K., Madhyastha, H., Sandur V., R., N.T., M., N., T., & Thiagarajan, P. (2020). Anti-inflammatory and wound healing potential of a clove oil emulsion. *Colloids and Surfaces B: Biointerfaces*, *193*, 111102. https://doi.org/10.1016/j.colsurfb.2020.111102

Batiha, G. El-S., Alkazmi, L. M., Wasef, L. G., Beshbishy, A. M., Nadwa, E. H., & Rashwan, E. K. (2020). *Syzygium aromaticum* L. (Myrtaceae): Traditional uses, bioactive chemical constituents, pharmacological and toxicological activities. *Biomolecules*, *10*(2), 202. https://doi.org/10.3390/biom10020202

Bayala, B., Bassole, I. H. N., Scifo, R., Gnoula, C., Morel, L., Lobaccaro, J. M. A., & Simpore, J. (2014). Anticancer activity of essential oils and their chemical components—A review. *American Journal of Cancer Research*, *4*(6), 591–607.

Behbahani, B. A., Noshad, M., & Falah, F. (2019). Study of chemical structure, antimicrobial, cytotoxic and mechanism of action of *Syzygium aromaticum* essential oil on foodborne pathogens. *Potravinarstvo Slovak Journal of Food Sciences*, *13*(1), 875–883. https://doi.org/10.5219/1226

Behbahani, M., Mohabatkar, H., & Soltani, M. (2013). Anti-HIV-1 activities of aerial parts of *Ocimum basilicum* and its parasite *Cuscuta campestris*. *Journal of Antivirals & Antiretrovirals*, *5*(3), 057–061. https://doi.org/10.4172/jaa.1000064

Beltrán-Villalobos, K. L., Déciga-Campos, M., Aguilar-Mariscal, H., González-Trujano, M. E., Martínez-Salazar, M. F., Ramírez-Cisneros, M. de los Á., Rios, M. Y., & López-Muñoz, F. J. (2017). Synergistic antinociceptive interaction of *Syzygium aromaticum* or *Rosmarinus officinalis* coadministered with ketorolac in rats. *Biomedicine & Pharmacotherapy, 94*, 858–864. https://doi.org/10.1016/j.biopha.2017.07.166

Correia, A. M., Pedrazzani, A. S., Mendonça, R. C., Massucatto, A., Ozório, R. A., & Tsuzuki, M. Y. (2017). Basil, tea tree and clove essential oils as analgesics and anaesthetics in *Amphiprion clarkii* (Bennett, 1830). *Brazilian Journal of Biology, 78*(3), 436–442. https://doi.org/10.1590/1519-6984.166695

Dahham, S., Tabana, Y., Iqbal, M., Ahamed, M., Ezzat, M., Majid, A. A., & Majid, A. A. (2015). The anticancer, antioxidant and antimicrobial properties of the sesquiterpene *β*-Caryophyllene from the essential oil of *Aquilaria crassna. Molecules, 20*(7), 11808–11829. https://doi.org/10.3390/molecules200711808

Dai, J.-P., Zhao, X.-F., Zeng, J., Wan, Q.-Y., Yang, J.-C., Li, W.-Z., Chen, X.-X., Wang, G.-F., & Li, K.-S. (2013). Drug screening for autophagy inhibitors based on the dissociation of Beclin1-Bcl2 complex using BiFC technique and mechanism of eugenol on anti-influenza a virus activity. *PLoS ONE, 8*(4), e61026. https://doi.org/10.1371/journal.pone.0061026

Das, A., K, H., S K, D. K., K, H. R., & Jayaprakash, B. (2018). Evaluation of therapeutic potential of eugenol a natural derivative of *Syzygium aromaticum* on Cervical Cancer. *Asian Pacific Journal of Cancer Prevention: APJCP, 19*(7), 1977–1985. https://doi.org/10.22034/APJCP.2018.19.7.1977

da Silva, D. R., Arvigo, A. L., Giaquinto, P. C., Delicio, H. C., Barcellos, L. J. G., & Barreto, R. E. (2021). Effects of clove oil on behavioral reactivity and motivation in Nile tilapia. *Aquaculture, 532*, 736045. https://doi.org/10.1016/j.aquaculture.2020.736045

de Oliveira, A. S., Gazolla, P. A. R., Oliveira, A. F. C. da S., Pereira, W. L., de S. Viol, L. C., Maia, A. F. da S., Santos, E. G., da Silva, Í. E. P., Mendes, T. A. de O., da Silva, A. M., Dias, R. S., da Silva, C. C., Polêto, M. D., Teixeira, R. R., & de Paula, S. O. (2019). Discovery of novel West Nile Virus protease inhibitor based on isobenzonafuranone and triazolic derivatives of eugenol and indan-1,3-dione scaffolds. *PLoS ONE, 14*(9), e0223017. https://doi.org/10.1371/journal.pone.0223017

Din, M. ud, Sarfraz, R. A., & Shahid, T. (2015). Biological activity-based assessment of essential oil emulsions. *Journal of Food Processing and Preservation, 39*(6), 1452–1456. https://doi.org/10.1111/jfpp.12364

Duke, J. A. (2002). *Handbook of Medicinal Herbs*. Second Edition. CRC Press, Boca Raton, FL. 893 pp.

Duke, J. A., & duCellier, J. L. (1993). *Handbook of Alternative Cash Crops*. CRC Press, Boca Raton, FL. 544 pp.

El-Darier, S. M., El-Ahwany, A. M. D., Elkenany, E. T., & Abdeldaim, A. A. (2018). An *in vitro* study on antimicrobial and anticancer potentiality of thyme and clove oils. *Rendiconti Lincei. Scienze Fisiche e Naturali, 29*(1), 131–139. https://doi.org/10.1007/s12210-018-0672-0

EMA, European Medicines Agency (2011). Assessment report on *Syzygium aromaticum* (L.) Merill et L.M. Perry, flos and *Syzygium aromaticum* (L.) Merill et L.M. Perry, floris aetheroleum. EMA/HMPC/534946/2010.

Fadilah, F., Yanuar, A., Arsianti, A., & Andrajati, R. (2017). Phenylpropanoids, eugenol scaffold, and its derivatives as anticancer. *Asian Journal of Pharmaceutical and Clinical Research, 10*(3), 41–46. https://doi.org/10.22159/ajpcr.2017.v10i3.16071

FDA, Food and Drug Administration (2010). *OTC Active Ingredients*. Available at www.fda.gov/media/75750/download.

Fernandes, E. S., Passos, G. F., Medeiros, R., da Cunha, F. M., Ferreira, J., Campos, M. M., Pianowski, L. F., & Calixto, J. B. (2007). Anti-inflammatory effects of compounds alpha-Humulene and (–)-trans-Caryophyllene isolated from the essential oil of *Cordia verbenacea*. *European Journal of Pharmacology*, *569*(3), 228–236. https://doi.org/10.1016/j.ejphar.2007.04.059

Fujimoto, R. Y., Pereira, D. M., Silva, J. C. S., de Oliveira, L. C. A., Inoue, L. A. K. A., Hamoy, M., de Mello, V. J., Torres, M. F., & Barbas, L. A. L. (2018). Clove oil induces anaesthesia and blunts muscle contraction power in three Amazon fish species. *Fish Physiology and Biochemistry*, *44*(1), 245–256. https://doi.org/10.1007/s10695-017-0430-8

Ghanawi, J., Saoud, G., Zakher, C., Monzer, S., & Saoud, I. P. (2019). Clove oil as an anaesthetic for Australian redclaw crayfish *Cherax quadricarinatus*. *Aquaculture Research*, *50*(12), 3628–3632. https://doi.org/10.1111/are.14319

Guan, W., Li, S., Yan, R., Tang, S., & Quan, C. (2007). Comparison of essential oils of clove buds extracted with supercritical carbon dioxide and other three traditional extraction methods. *Food Chemistry*, *101*(4), 1558–1564. https://doi.org/10.1016/j.foodchem.2006.04.009

Han, X., & Parker, T. L. (2017). Anti-inflammatory activity of clove (*Eugenia caryophyllata*) essential oil in human dermal fibroblasts. *Pharmaceutical Biology*, *55*(1), 1619–1622. https://doi.org/10.1080/13880209.2017.1314513

Haro-González, J. N., Castillo-Herrera, G. A., Martínez-Velázquez, M., & Espinosa-Andrews, H. (2021). Clove essential oil (*Syzygium aromaticum* L. Myrtaceae): Extraction, chemical composition, food applications, and essential bioactivity for human health. *Molecules*, *26*, 6387. https://doi.org/10.3390/molecules26216387.

Kennouche, A., Benkaci-Ali, F., Scholl, G., & Eppe, G. (2015). Chemical composition and antimicrobial activity of the essential oil of *Eugenia caryophyllata* Cloves extracted by conventional and microwave techniques. *Journal of Biologically Active Products from Nature*, *5*(1), 1–11. https://doi.org/10.1080/22311866.2014.961100

Khalil, A. A., Rahman, U. ur, Khan, M. R., Sahar, A., Mehmood, T., & Khan, M. (2017). Essential oil eugenol: Sources, extraction techniques and nutraceutical perspectives. *RSC Advances*, *7*(52), 32669–32681. https://doi.org/10.1039/C7RA04803C

Khalilzadeh, E., Hazrati, R., & Saiah, G. (2016). Effects of topical and systemic administration of *Eugenia caryophyllata* buds essential oil on corneal anesthesia and analgesia. *Research in Pharmaceutical Sciences*, *11*(4), 293–302. https://doi.org/10.4103/1735-5362.189297

Lane, T., Anantpadma, M., Freundlich, J. S., Davey, R. A., Madrid, P. B., & Ekins, S. (2019). The natural product eugenol is an inhibitor of the ebola virus in vitro. *Pharmaceutical Research*, *36*(7), 1–6. https://doi.org/10.1007/s11095-019-2629-0

Lee, G., Yang, E., Kim, S., Park, C.-S., Park, Y. S., & Jin, Y.-H. (2018). Parapheromones suppress chemotherapy side effects. *Journal of Pharmacology and Experimental Therapeutics*, *367*(2), 215–221. https://doi.org/10.1124/jpet.118.251363

Lesgards, J.-F., Baldovini, N., Vidal, N., & Pietri, S. (2014). Anticancer activities of essential oils constituents and synergy with conventional therapies: A review. *Phytotherapy Research*, *28*(10), 1423–1446. https://doi.org/10.1002/ptr.5165

Liu, B.-B., Luo, L., Liu, X.-L., Geng, D., Li, C.-F., Chen, S.-M., Chen, X.-M., Yi, L.-T., & Liu, Q. (2015). Essential oil of *Syzygium aromaticum* reverses the deficits of stress-induced behaviors and hippocampal p-ERK/p-CREB/brain-derived neurotrophic factor expression. *Planta Medica*, *81*(3), 185–192. https://doi.org/10.1055/s-0034-1396150

LiverTox (2019). *Clinical and Research Information on Drug-Induced Liver Injury [Internet]. Bethesda (MD): National Institute of Diabetes and Digestive and Kidney Diseases; 2012–. Eugenol (Clove Oil).* [Updated 2019 Oct 28]. Available at www.ncbi.nlm.nih.gov/books/

Marmouzi, I., Karym, E. M., Alami, R., El Jemli, M., Kharbach, M., Mamouch, F., Attar, A., Faridi, B., Cherrah, Y., & Faouzi, M. E. A. (2019). Modulatory effect of *Syzygium aromaticum* and *Pelargonium graveolens* on oxidative and sodium nitroprusside stress and inflammation. *Oriental Pharmacy and Experimental Medicine, 19*(2), 201–210. https://doi.org/10.1007/s13596-018-0335-9

Moon, S.-E., Kim, H.-Y., & Cha, J.-D. (2011). Synergistic effect between clove oil and its major compounds and antibiotics against oral bacteria. *Archives of Oral Biology, 56*, 907–916. https://doi.org/10.1016/j.archoralbio.2011.02.005

Morsy, N. F. S. (2017). Chemical structure, quality indices and bioactivity of essential oil constituents. In *Active Ingredients from Aromatic and Medicinal Plants* (pp. 175–206). InTech. https://doi.org/10.5772/66231

Najar, B., Shortrede, J. E., Pistelli, L., & Buhagiar, J. (2020). Chemical composition and *in vitro* cytotoxic screening of sixteen commercial essential oils on five cancer cell lines. *Chemistry & Biodiversity, 17*(1), e1900478. https://doi.org/10.1002/cbdv.201900478

Nirmala, M. J., Durai, L., Gopakumar, V., & Nagarajan, R. (2019). Anticancer and antibacterial effects of a clove bud essential oil-based nanoscale emulsion system. *International Journal of Nanomedicine, 14*, 6439–6450. https://doi.org/10.2147/IJN.S211047

Özbek, Z. A., & Ergönül, P. G. (2022). Chapter 17-Clove (*Syzygium aromaticum*) and eugenol toxicity. In *Clove (Syzygium aromaticum) Chemistry, Functionality and Applications.* Ed. Mohamed Fawzy Ramadan (pp. 267–314). Academic Press. https://doi.org/10.1016/B978-0-323-85177-0.00029-X.

Pandey, V. K., Shams, R., Singh, R., Dar, A. H., Pandiselvam, R., Rusu, A. V., & Trif, M. (2022). A comprehensive review on clove (*Caryophyllus aromaticus* L.) essential oil and its significance in the formulation of edible coatings for potential food applications. *Frontiers in Nutrition, 9*, 987674. https://doi.org/10.3389/fnut.2022.987674.

Pandey, V. K., Srivastava, S., Ashish, Dash, K. K., Singh, R., Dar, A. H., Singh, T., Farooqui, A., Shaikh, A. M., & Kovacs, B. (2024). Bioactive properties of clove (*Syzygium aromaticum*) essential oil nanoemulsion: A comprehensive review. *Heliyon, 10*, e22437. https://doi.org/10.1016/j.heliyon.2023.e22437.

Rajkowska, K., Nowak, A., Kunicka-Styczyńska, A., & Siadura, A. (2016). Biological effects of various chemically characterized essential oils: Investigation of the mode of action against *Candida albicans* and HeLa cells. *RSC Advances, 6*(99), 97199–97207. https://doi.org/10.1039/C6RA21108A

Ramadan, M. M., Ali, M. M., Ghanem, K. Z., & El-Ghorabe, A. H. (2015). Essential oils from Egyptian aromatic plants as antioxidant and novel anticancer agents in human cancer cell lines. *Grasas y Aceites, 66*(2), e080. https://doi.org/10.3989/gya.0955142

RBI, Reliable Business Insights (2024). Medicinal clove market trends and market analysis forecasted for period 2024–2031. *Reliable Business Insights.* Available at www.reliablebusinessinsights.com/enquiry/request-sample/1230721?utm_campaign=4&utm_medium=cp_1967&utm_source=Medium&utm_content=ia&utm_term=medicinal-clove&utm_id=free

Shahbazi, Y. (2019). Antioxidant, antibacterial, and antifungal properties of nanoemulsion of clove essential oil. *Nanomedicine Research Journal, 4*(4), 204–208. https://doi.org/10.22034/nmrj.2019.04.001

Singh, P. A., Desai, S. D., & Singh, J. (2018). A review on plant antimicrobials of past decade. *Current Topics in Medicinal Chemistry, 18*(10), 812–833.

Sueksakit, K., Thisayakorn, K., Khueynok, V., Sriyam, K., Pahusee, D., & Buddhakala, N. (2013). Preliminary study of Syzygium aromaticum L. on analgesic activity in rats. *Thai Journal of Pharmaceutical Sciences, 38*, 63–65.

Sugihartini, N., Prabandari, R., Yuwono, T., & Rahmawati, D. R. (2019). The anti-inflammatory activity of essential oil of clove (*Syzygium aromaticum*) in absorption base ointment with addition of oleic acid and propylene glycol as enhancer. *International Journal of Applied Pharmaceutics*, *11*, 106–109. https://doi.org/10.22159/ijap.2019.v11s5.T0081

Sun, W.-J., Lv, W.-J., Li, L.-N., Yin, G., Hang, X., Xue, Y., Chen, J., & Shi, Z. (2016). Eugenol confers resistance to *Tomato yellow leaf curl virus* (TYLCV) by regulating the expression of *SlPer1* in tomato plants. *New Biotechnology*, *33*(3), 345–354. https://doi.org/10.1016/j.nbt.2016.01.001

Sung, B., Prasad, S., Yadav, V. R., & Aggarwal, B. B. (2012). Cancer cell signaling pathways targeted by spice-derived nutraceuticals. *Nutrition and Cancer*, *64*(2), 173–197. https://doi.org/10.1080/01635581.2012.630551

Thapa, D., Richardson, A. J., Zweifel, B., Wallace, R. J., & Gratz, S. W. (2019). Genoprotective effects of essential oil compounds against oxidative and methylated DNA damage in human colon cancer cells. *Journal of Food Science*, *84*(7), 1979–1985. https://doi.org/10.1111/1750-3841.14665

Uchôa Lopes, C., Saturnino de Oliveira, J., Holanda, V., Rodrigues, A., Martins da Fonseca, C., Galvão Rodrigues, F., Camilo, C., Lima, V., Coutinho, H., Kowalski, R., & da Costa, J. (2020). GC-MS analysis and hemolytic, antipyretic and antidiarrheal potential of *Syzygium aromaticum* (Clove) essential oil. *Separations*, *7*(2), 35. https://doi.org/10.3390/separations7020035

VMR, Verified Market Research (2024). Global clove market size by product type, by distribution channels, by end-use application, by geographic scope and forecast. *Verifiedmarketresearch.com*. Available at www.verifiedmarketresearch.com/product/clove-market/#:~:text=Clove%20Market%20Size%20And%20Forecast,the%20forecast%20period%202024%2D2030.

Wei, M.-C., Xiao, J., & Yang, Y.-C. (2016). Extraction of α-Humulene enriched oil from clove using ultrasound-assisted supercritical carbon dioxide extraction and studies of its fictitious solubility. *Food Chemistry*, *210*, 172–181. https://doi.org/10.1016/j.foodchem.2016.04.076

12 Traditional Formulations and Uses

Sneha Kumari, Preet Amol Singh, Ritika Sindhwani, Subhajit Hazra, Neha Bajwa, and Ankur Suri

12.1 INTRODUCTION

Spices like clove are an integral part of traditional healthcare systems due to their alimentary antimicrobial, carminative, and stomachic properties. They are also used as flavoring and preservative agents and have a wide range of industrial applications. When considering uses of clove, its rich in micronutrients are essential for the human body, such as sodium (6.01 mg in ~2.1g), potassium (21 mg in ~2.1g), calcium (10 mg in ~2.1g), and magnesium (3.99 mg in ~2.1g) (Cortés-Rojas et al., 2014; Cock et al., 2018; Food data Central, 2021). Besides, in relation to its traditional formulations, clove water is used for the alleviation of extreme thirst and nausea experienced in cholera, and consumption of extract of lavanga (clove) relieves indigestion and cholic. Some of the well-known *Ayurvedic* formulations of clove include, *Lavangadi churna*, *Lavangchatussama*, *Lavangadi vati*, *Avipattikara churna*, and *Suparipaka* (Pamnani et al., 2019). In Unani system of medicine, some of the formulations of clove include *Habb-i-Ambar*, *Habb-i-Munaish*, *Dawa-i-Hindi*, *Roghane-Surkh*, *Araq-i-Ambar*, *Araq-i-Chobchini*, etc. (Ara et al., 2022). As for homeopathy, *Eugenia caryophyllata* CH is a dilution made from clove, which is evidenced to have beneficial effects in toothache, fever, premature ejaculation, diabetes, etc. (Chaieb et al., 2007). Therefore, knowledge of the aforementioned nutritional and medicinal value of *Syzygium aromaticum* is what makes it a precious spice with many uses, such as in digestive, dental, and respiratory disorders. Surprisingly, the ground form of clove is more popular than its other forms. Variation of ground paste with pure or mixed spices forms a base for cooking in many types of cuisine in different parts of the world. As for its other traditional uses, it finds its use in preparation like mouthwashes, toothpaste, and tooth powders. Besides, infusions of cloves have also been known to control cough and asthma (Bhowmik et al., 2012).

12.2 HISTORICAL PERSPECTIVE OF CLOVE

Clove has been utilized for millennia due to its well-known therapeutic properties. Essentially, the key ingredient used is the dried, fragrant flower buds of Myrtaceae trees (Sachan et al., 2018). Milind and co-worker investigated the origin of clove and the reports of the study illustrated that the origin dates back to before the time of Christ, in the 1st century, during the Han dynasty in ancient

DOI: 10.1201/9781003521785-12

China, which dated from 207 BC to 220 AD At that time, clove was known as "chicken-tongue spice" or *hi-sho-hiang* ("bird's tongue") and was used as a breath freshener. Chinese physicians reported that court visitors to the Emperor had to keep a clove in their mouth (Milind & Deepa, 2011). During the 8th century, cloves were one of the most traded spices in the European market. In terms of cultural importance, in the Moluccas Islands (where clove was first discovered) parents planted a clove tree with the birth of a child (McCormick Science Institute, 2021). The places of origin of clove is depicted in Figure 12.1 (Hancock, J. World History, 2021; Royal Botanic Gardens, 2024; Mbaveng and Kuete, 2017).

In India, the plant appeared to have arrived in ~1700 BC and in southern Europe by the 1st century AD. Similar to Chinese practice, it was observed that the ancient Sanskrit text, *Charaka Saṃhita* (Ayurvedic textbook), mentioned clove as "a material for someone who wants clean, fresh, fragrant breath". For Romans, clove was

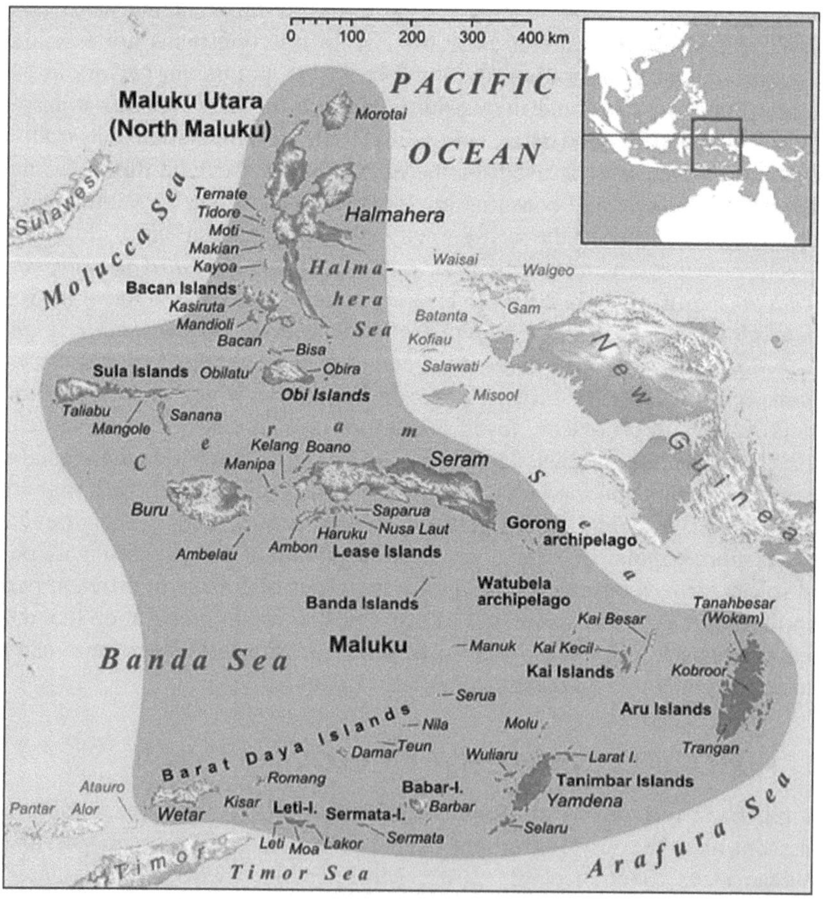

FIGURE 12.1 Maluka: native location of clove.

first described in their traditional book *Natural History* in 70 CE by Pliny the Elder (23–79 CE) which talked about a grain resembling that of pepper (but larger and more fragile) in India, called caryophyllom (i.e. *Syzygium aromaticum*) (Ecostore, 2022), which was reported to grow on the Indian lotus tree, which was imported to Rome due to its aroma. Also, clove has its mention in the circa 5th-century traditional manuscripts by the Greek physician, Paul of Aegina, and read as, "It is of the nature of a flower of some tree, woody, black, almost as thick as a finger; reputed aromatic, sour, bitterish, hot and dry in the third degree; excellent in relishes and other prescriptions". Consecutively in the 6th century CE, Byzantine (now Istanbul) physician Alexander of Tralles, in his 12 traditional textbooks of medicine, endorsed cloves for seasickness, gout, and appetite stimulation (Hancock, 2021). Interestingly, in the 16th century, it was the Portuguese who discovered the origin of clove plants in the Molucca Islands (De & De, 2022). Therefore, clove has been mentioned in various ancient traditional textbooks throughout the world.

12.3 MARKETED VALUE AND GLOBAL TRADE OF CLOVE

Clove, which is used for commercial purposes, consists of dried, unopened flower buds of the evergreen tree, *Syzygium aromaticum*, of the family Myrtaceae. As per the Harmonized System (HS) Codes, the product code includes clove (HS 40907); cloves [whole fruit, cloves and stems], neither crushed nor ground (HS 090710); cloves [whole fruit, cloves and stems], crushed or ground (HS 090720); cloves [whole fruit, cloves and stems] (HS 090700), and clove stems (HS 09071030). Moreover, in terms of processed product, it can be divided into clove extract (HS 09071010), clove oil (clove leaf or stem oil, HS 33012921; clove bud oil, HS 33012928), and clove oleoresins (HS 33019021) (Spices Board India, 2022). Their import and export data has been depicted in Table 12.1, signifying the economic importance or demand of clove throughout the world.

Countries producing the maximum output of clove include Indonesia, Zanzibar, and Madagascar. The major importing nations are India, the USA, Germany, France, and Singapore. The global production of clove is estimated to be 63,700 tons, where Indonesia accounts for 66% of the produce alone. In India, the two morphological varieties of clove are king clove (with bolder flower buds) and dwarf clove (smaller plant) (Sarma et al., 2014).

As far as the trade statistic of HS 09071010 (clove extract) is concerned, globally its import shipments stood at 9K, which was a conglomeration of 1,474 worldwide importers, from a collective of ~1,223 suppliers. Moreover, as of 7 January 2024, most of the demand for the import of HS 09071010 is fulfilled by Madagascar, India, and Indonesia, while the top ten importers include Madagascar (2,876 shipments), India (2,588 shipments), Indonesia (1,046 shipments), Comoros (1,013 shipments), Sri Lanka (880 shipments), Tanzania (565 shipments), Brazil (5 shipments), Croatia (1 shipment), Germany (1 shipment), and Singapore (1 shipment) (Volza, 2024a). In terms of its exports, shipment value stood at 9K, which was a conglomeration of 1,223 worldwide importers, from an aggregate of 1,474 suppliers. Moreover, as of 7 January 2024, most of the demand for the export of HS 09071010 is fulfilled by India, Canada, and the United States, while, the top ten importers include India (6,392

TABLE 12.1
Trade analysis of clove and its forms or products.

HS codes	World trade [year: 2022]	Top exporter [year: 2022] (country)	Top importer [year: 2022]	Export growth (CAGR)	Share of world trade [year: 2022]
HS 40907	$603M	$309M (Madagascar)	$309M (Indonesia)	23.2%	0.0025%
HS 090710	$663M	$371M (Madagascar)	$197M (Indonesia)	36.8%	0.0027%
HS 090700	$603M	$309M (Madagascar)	$143M (Indonesia)	23.2%	0.0025%
HS 090720	$19.9M	$2.81M (Indonesia)	$4.23M (U.S.)	15.3%	< 0.01%
HS 09071030	–	–	–	–	–
HS 09071010	–	–	–	–	–
HS 33012921	–	Clove leaf oil: $550M in 2023 [market size] Clove stem oil: $250M in 2022 [market size]		5% [2023–2033] ~12.99% [2022–2032]	–
HS 33012928	–	$250M in 2022 [market size]		13% [2023–2032]	–
HS 33019021	–	$1.63Bn in 2023 [market size]		6.36% [2023–2032]	–

Note: HS, harmonized system; CAGR, compound annual growth rate [2021–2022]; $, US Dollars; Bn, Billion; M, million; k, thousand; US, United States.

Source: Observatory of Economic Complexity (OEC) and various market research groups.

shipments), Canada (307 shipments), the United States (298 shipments), Germany (186 shipments), United Arab Emirates (181 shipments), United Kingdom (128 shipments), Kuwait (97 shipments), Australia (94 shipments), Singapore (94 shipments), and South Africa (88 shipments) (Volza, 2024b).

For HS 33012921 (clove oil), Indonesia is the largest producer (producing more than 70% of the world's cloves) and consumer of cloves (~90% of produce, for the Kretek cigarette industry, etc.). Therefore, it can export only ~10–15% of the country's clove produce. This is even more shocking when data from FAO show that clove production in the country has increased to 134.7 thousand metric tons in 2019 (under a total cultivation area of 569.4 thousand hectares). However, there was a decline in these numbers (133.6 thousand metric tons cultivated in 552.8 thousand hectares of land) in 2020 owing to bad weather. In the year 2021, clove was majorly exported to India (29.1% or 5,738 tons and was valued at $28,051k), Singapore, Bangladesh, the United Arab Emirates, the United States, and Pakistan (Mordor Intelligence, 2024). Moreover, as of 13 December 2023, the top three importers of clove oil were the

Netherlands (6,468 shipments), France (6,213 shipments), and Germany (6,054 shipments), while the top three exporters were India (10,752 shipments), France (10,331 shipments), and Germany (8,911 shipments) (Volza, 2024c).

12.4 ROLE OF CLOVE IN TRADITIONAL SYSTEMS OF MEDICINE

Clove has played a significant role in various traditional medicine systems across the globe, each utilizing its unique therapeutic attributes to address different health conditions (Tripathi et al., 2020; Ugbogu et al., 2021). The following section outlines the specific uses and principles of clove in various traditional medicine systems (Table 12.2).

TABLE 12.2
Utilization of clove in traditional systems of medicine.

Traditional medicine system	Primary uses of clove	Form of use	Additional benefits	Notable formulations	References
Chinese medicine	Treats yang deficiency, digestive and respiratory issues	Whole clove, clove oil	Warms internal organs, antiseptic for fungal infections	Herbal teas, topical ointments	(Li et al., 2020)
Ayurvedic medicine	Digestive issues, respiratory ailments, dental pain	Whole clove, clove oil	Balances doshas, antiseptic for wounds	Herbal pastes, medicinal oils	(Ravindran et al., 2021; Singh et al., 2021)
Homeopathic medicine	Dental issues, digestive disturbances, respiratory conditions	Highly diluted tinctures	Stimulates the body's healing process	Homeopathic tinctures	(Barve et al., 2021; Singh & Singh, 2019)
Unani medicine	Balances humors, digestive health, respiratory ailments	Whole clove, clove powder	Enhances mental function, aphrodisiac properties	Compound formulations, herbal mixtures	(Ahmad et al., 2018; Rahman et al., 2020)
Naturopathy	Digestive health, antimicrobial, anti-inflammatory	Whole clove, clove oil	Natural pain reliever and boosts immune function	Topical applications, herbal infusions	(Smith & Brown, 2020).
Sowa-Rigpa	Digestive disorders, respiratory conditions	Whole clove, clove powder	Improves circulation, reduces inflammation	Herbal pills, decoctions	(Tsering et al., 2020; Wangchuk et al., 2019)

(Continued)

TABLE 12.2 (Continued)
Utilization of clove in traditional systems of medicine.

Traditional medicine system	Primary uses of clove	Form of use	Additional benefits	Notable formulations	References
Bhutanese medicine	Digestive and respiratory issues	Whole clove, clove oil	Stimulates digestion and enhances appetite	Herbal teas, medicinal concoctions	(Tsering et al., 2020; Wangchuk et al., 2019)
Indonesian medicine	Digestion, colds, and flu, antiseptic for wounds	Whole clove, clove oil	Relieves muscle pain, enhances circulation	Jamu drinks, massage oils	(Utami et al., 2021)
African traditional medicine	Dental pain, infections, digestive issues	Whole clove, clove oil	Aromatic properties, used in rituals	Herbal mixtures, topical applications	(Mothibe & Sibanda, 2018; Oladunmoye & Kehinde, 2020)
European traditional medicine	Digestive issues, respiratory conditions, dental pain	Whole clove, clove oil	Preserves food, improves mood and mental clarity	Herbal mixtures, essential oils	(Harris, 2018; Johnson et al., 2020)
North American medicine	Dental pain, digestive discomfort, insect repellent	Whole clove, clove oil	Used in rituals, antiseptic properties	Herbal mixtures, topical applications	(Moerman, 2020; Turner & Bell, 2019)

12.4.1 Chinese System of Medicine

In traditional Chinese medicine (TCM), clove is known as "*ding xiang*". It is classified as a warm herb that is particularly effective in treating conditions related to yang deficiency. This classification aligns with TCM's principle of balancing yin and yang to maintain health. Clove is used to warm the middle burner (spleen and stomach) and lower burner (kidneys), helping to dispel internal cold. This makes it an effective remedy for ailments such as stomach colds, hiccups, vomiting, and diarrhea (Cortés-Rojas et al., 2014). Additionally, clove oil is used externally for its potent antiseptic properties to treat fungal infections. The aromatic nature of clove is also believed to invigorate the mind and enhance overall vitality, making it a holistic remedy in TCM practices (Li et al., 2020).

12.4.2 Ayurvedic System of Medicine

In Ayurveda, clove is referred to as "*Lavanga*" and is highly esteemed for its broad range of medicinal properties. It is categorized under "*Deepana*" (digestive), "*Pachana*" (carminative), and "*Krimighna*" (antimicrobial) therapies. Clove is extensively used to address digestive issues such as flatulence, nausea, and vomiting. It is also used to treat respiratory ailments like bronchitis and asthma (Ayushi et al., 2020; Ravindran et al., 2021). The principle of balancing the body's *doshas*

(*Vata*, *Pitta*, and *Kapha*) is central to Ayurveda, and clove plays a vital role in this process by promoting digestive fire and clearing mucus from the respiratory tract. Clove oil is also applied topically to alleviate dental pain and as an antiseptic for wounds (Ravindran et al., 2021).

12.4.3 Homeopathic System of Medicine

In homeopathy, clove is utilized in the form of *Caryophyllus aromaticus*. It is commonly used for dental issues, such as toothache and gum infections, reflecting homeopathy's principle of treating like with like. Clove is also employed to manage digestive disturbances, particularly nausea and vomiting, and respiratory conditions such as coughs and colds. Homeopathic preparations involve the use of highly diluted substances to stimulate the body's self-healing mechanisms (Barve et al., 2021; Singh & Singh, 2019).

12.4.4 Unani System of Medicine

In the Unani system, clove is known as "Qaranfal" and is valued for its ability to balance the four humors: phlegm, blood, bile, and black bile. This balancing act is crucial in Unani medicine, which aims to maintain harmony among these humors to achieve health. Clove is beneficial for digestive health, especially in relieving flatulence, colic, and diarrhea. It is also used to treat respiratory ailments and is known for its aphrodisiac properties. Clove is included in various formulations designed to enhance mental function and relieve stress (Ahmad et al., 2018; Rahman et al., 2020).

12.4.5 Naturopathy System of Medicine

Naturopathy promotes the use of natural remedies, and clove is a staple due to its natural healing properties. It is used to enhance digestive health, providing relief from bloating and indigestion, and its antimicrobial properties are leveraged to combat infections. Clove oil, known for its anti-inflammatory effects, is often recommended in naturopathy for topical application to relieve muscle pain and arthritis. The principle of supporting the body's inherent healing processes is central to naturopathy, and clove's diverse therapeutic properties make it an essential remedy in this system. Its use in preventive medicine highlights its role in maintaining health and preventing disease naturally (Smith & Brown, 2020).

12.4.6 Other Traditional Systems

In *Sowa-Rigpa* and *Bhutanese* medicine, clove is used for its warming and digestive properties, treating stomach pain, diarrhea, colds, and coughs and improving circulation (Tsering et al., 2020; Wangchuk et al., 2019). Indonesian Jamu incorporates clove for digestion, colds, flu, and as an antiseptic, as well as in massage oils for muscle pain relief (Utami et al., 2021), while, African traditional medicine employs clove for dental pain, infections, and digestive issues, and it is used in rituals and as an insect repellent (Mothibe & Sibanda, 2018; Oladunmoye & Kehinde, 2020). European traditional medicine, especially during the Middle Ages, used clove to preserve food, treat digestive and respiratory conditions, and relieve pain (Harris,

2018; Johnson et al., 2020. Native American medicine utilized clove for toothaches, gum infections, and digestive discomfort and as a natural insect repellent in various rituals (Moerman, 2020; Turner & Bell, 2019). These diverse applications highlight clove's importance in health and well-being across different cultures.

12.5 ROLE OF CLOVE DURING THE COVID PANDEMIC

Novel coronavirus (severe acute respiratory syndrome coronavirus 2 or SARS-CoV-2), first identified in China, was the cause of a global pandemic that wreaked havoc for almost 3 years. What's most concerning about the scenario was the fact that, for most of the time during this pandemic, therapy or drugs were absent for treating the virus (apart from vaccines). Nowadays herbal medicines are widely used to treat or manage various diseases; as indicated by WHO ~80% of the global population depends on herbal medicine to treat diseases (Yang et al., 2020).

In the context of COVID-19, herbal compounds were thought to have prophylactic and/or therapeutic importance. Such examples come from research conducted by Yu and colleagues, who evaluated 64 pure natural chemicals and discovered that scutellarein, derived from *Scutellaria lateriflora*, was a powerful inhibitor of SARS-CoV helicase protein and so could be utilized to combat the pandemic. Van Doremalen et al. discovered that tea-derived supplements (tannic acid and theaflavin-3-gallate from black tea) might help prevent SARS-CoV infection (Yu et al., 2012; van Doremalen et al., 2020). Most of this preliminary research is carried out via molecular docking and molecular dynamics simulations studies (Shanmugarajan et al., 2022) including compounds derived from *Withania somnifera* and *Zingiber officinale* that have been suggested as potential inhibitors of SARS-CoV-2 (Kumar et al., 2022). Other potential inhibitors with similar probable actions include griffithsin, silvestrol, emodin, escin, daidzin, 10-hydroxyusambarensine, quercetin, naringenin, oleuropein, resveratrol, ellagic acid, benzoic acid, and gallic acid (Huang et al., 2020). A similar study was conducted by Yunus docking studies with 17 herbal compounds for testing its efficacy in SARS-CoV-2 M^{pro}, 6 of them being from *Syzygium aromaticum*, such as acetyl eugenol, beta-caryophyllene, vanillin, crategolic acid, kaempferol, and rhamnetin. Results of the study showed that crategolic acid from *Syzygium aromaticum* (clove) had the highest binding affinity with SARS-CoV-2 M^{pro} protease, followed by sennoside (A, B, C, and D) compounds from *Cassia acutifolia* (Yunus, 2021).

As for the mechanism of action of SARS-CoV-2, to enter the cell of the host, the spike S1 binds to the ACE2 (Jackson et al., 2022). Therefore modern therapeutics focused on competitively inhibiting the binding of ACE2 to S1. *Ocimum tenuiflorum* and clove have been known to contain eugenol (Kumari et al., 2024), and what's more encouraging is the fact that eugenol inhibits the interaction between spike S1 and ACE2 in an AlphaScreen-based assay as described in a study conducted by Paidi et al. where it was evidenced that eugenol strongly suppressed the entry of pseudotyped SARS-CoV-2, into human ACE2-expressing HEK293 cells. Additionally, it also reduced SARS-CoV-2 spike S1-induced activation of NF-κB and the expression of IL-6, IL-1β, and TNFα in human A549 lung cells (Figure 12.2). This effect was valid when eugenol was administered orally as it seemed to reduce lung

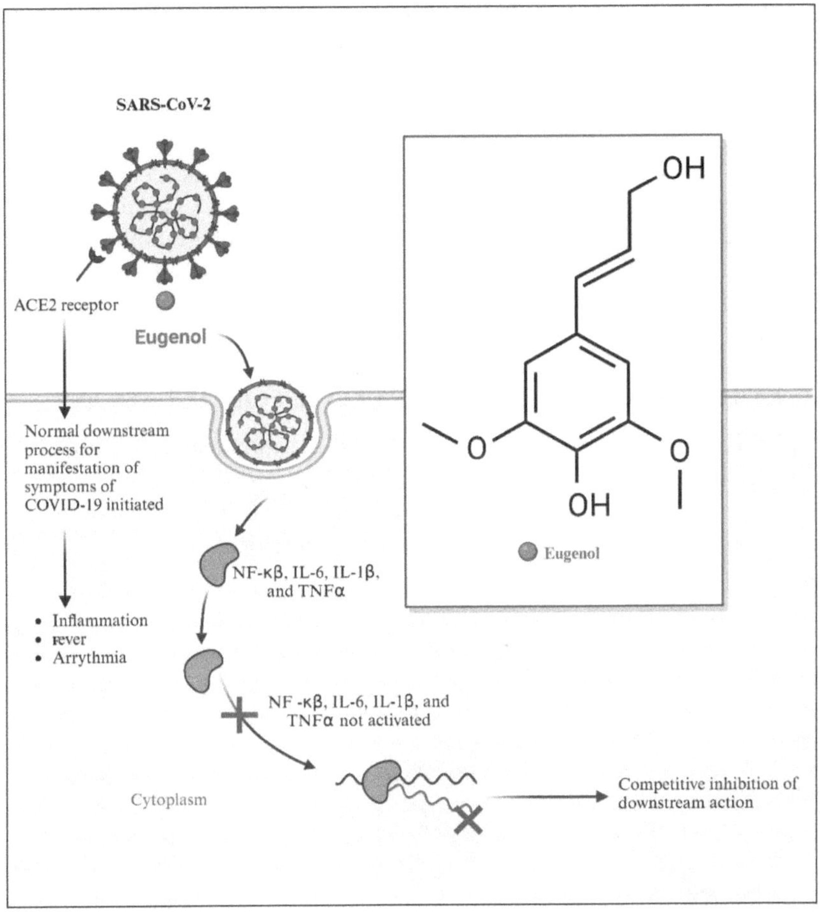

FIGURE 12.2 Mechanism of action of eugenol as an anti-Covid agent.

inflammation, decrease fever, improve the functioning of the cardiovascular system, and enhance locomotor activities in mice models of SARS-CoV-2. Therefore, selective targeting of SARS-CoV-2 spike S1 by eugenol was deemed to be effective for the treatment of COVID-19 (Paidi et al., 2021).

Regarding the particular solvent suitable for extraction of phytoconstituents for anti-Covid activity, it was the water and ethanol extract (in a dose-dependent manner) that suppressed SARS-CoV-2 spike protein binding to ACE2 and inhibited ACE2 activity, as seen in a study by Li and associates. The water extract demonstrated a stronger inhibitory effect than the ethanol extract on a dry-weight basis, besides being a more potent free radical scavenger (compared ethanolic extract, 58.44 and 2,298.01 μmol TE/g, respectively) as tested in DPPH and ABTS (536.9 and 3,525.06 μmol TE/g, respectively) assay (Li et al., 2022). Consecutively, a study by Hawkins et al. aimed at studying the potential for inhaled essential oils, containing an aromatherapy blend of thyme, orange, clove bud, and frankincense, in improving energy levels among female survivors of acute COVID-19 who experienced a lack of energy > 5 months following recovery. Results of the study showed that individuals who inhaled the essential oil blend for 2 weeks had a significant (p = 0.0201) lowering of fatigue scores, employment status, body mass index (BMI), and olfactory function. Subscale analysis identified subscales of vigor, as well as global, behavioral, general, and mental fatigue as benefiting from the intervention. Therefore, preliminary evidence suggests the role of clove in improving energy levels among women who are on the road to recovery from COVID-19 (Hawkins et al., 2022).

12.6 CLINICAL APPLICATIONS OF CLOVE FORMULATIONS

Clove has been the subject of numerous clinical trials and human case studies due to its extensive medicinal properties. These studies have explored its efficacy in treating a variety of diseases, its mechanisms of action, and its pharmacokinetics and pharmacodynamics (Figure 12.3). The following is a critical overview of the clinical applications of clove formulations:

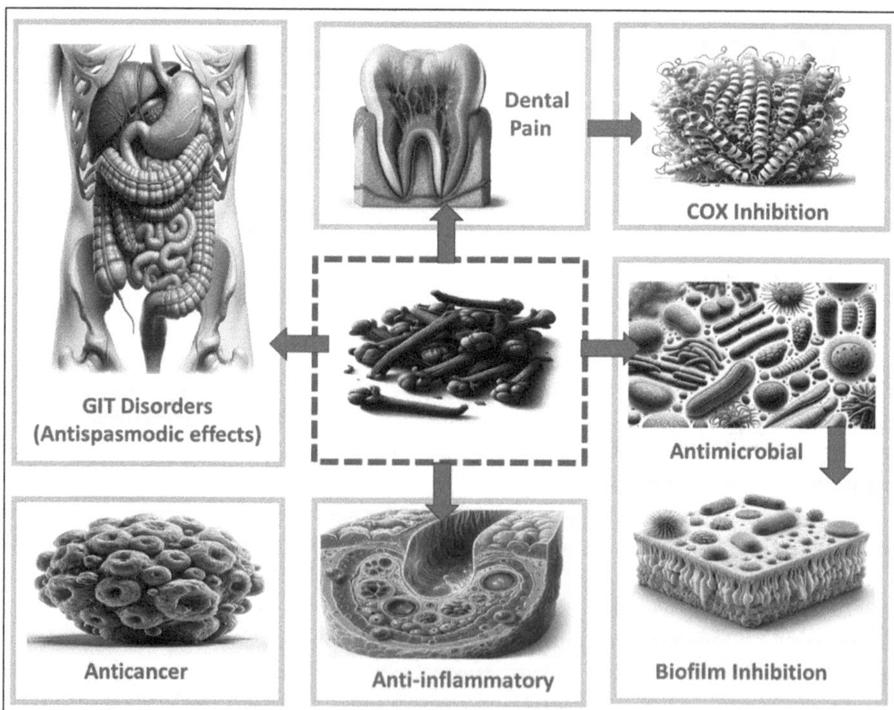

FIGURE 12.3 Efficacy of clove in treating various diseases.

12.6.1 Dental Pain and Inflammation

Clove oil is well-known for its analgesic properties, particularly in the context of dental pain. Reddy et al., demonstrated that eugenol, the active compound in clove oil, significantly reduces pain and inflammation in patients with dental issues. The study involved 120 participants who reported a reduction in pain intensity and inflammation after applying clove oil topically. Eugenol works by inhibiting cyclooxygenase (COX) enzymes, which are involved in the inflammatory process (Reddy et al., 2019). In a study involving 80 participants, clove oil was compared to benzocaine gel for its effectiveness in reducing pain during dental procedures. The results indicated that clove oil provided comparable pain relief to benzocaine, suggesting its potential as a natural alternative for dental analgesia (Ganaie et al., 2016), while Ertas et al. reported that clove oil significantly reduced gum inflammation and improved oral health over 4 weeks (Ertas et al., 2020).

12.6.2 Antimicrobial Properties

Clove has demonstrated significant antimicrobial activity, particularly against oral pathogens. Chaieb et al. found that clove oil effectively reduces microbial load by disrupting cell membranes and inhibiting biofilm formation, especially against *Streptococcus mutans* and *Candida albicans* (Chaieb et al., 2020). Nzeako et al. showed that clove oil can inhibit the growth of multidrug-resistant *Staphylococcus aureus* (MRSA), highlighting its potential against resistant infections (Nzeako et al., 2014). Additionally, Marchese et al. confirmed clove oil's broad-spectrum antimicrobial properties, making it valuable in antifungal and antibacterial treatments (Marchese et al., 2017).

12.6.3 Gastrointestinal Disorders

Singh found that clove oil alleviates symptoms of irritable bowel syndrome (IBS) in 75% of participants by exerting antispasmodic effects on the smooth muscles of the gastrointestinal tract, reducing spasms and discomfort (Singh & Singh, 2018). Tadesse et al. conducted a clinical trial involving patients with peptic ulcers, showing that clove oil significantly reduced ulcer symptoms and improved gastric mucosal healing due to its antioxidant and anti-inflammatory properties (Tadesse et al., 2018). Similarly, Zhou et al. demonstrated clove extract's efficacy in treating gastroesophageal reflux disease (GERD), reporting decreased acid reflux symptoms and esophageal irritation among participants using clove extract (Zhou et al., 2019).

12.6.4 Anticancer Properties

Research suggests that clove oil may have potential anticancer properties. A study on patients with breast cancer found that clove extract inhibited cancer cell proliferation and induced apoptosis through the activation of the p 53 pathway (Gupta et al., 2021). Similarly, Prasad et al. demonstrated the anti-tumor effects of clove oil in liver cancer patients, showing it could induce apoptosis and inhibit tumor growth by modulating the expression of Bcl-2 and Bax proteins (Prasad et al., 2015). Furthermore, clove oil has demonstrated chemopreventive properties in colorectal cancer. Studies have revealed that it could reduce the incidence of colon cancer through its anti-inflammatory and antioxidant activities (Petrocelli et al., 2021; Salah et al., 2022).

12.6.5 Anti-inflammatory Properties

From an ethnobotanical perspective, buds of clove contain flavonoids such as β-caryophyllene, kaempferol, and rhamnetin, which are known to contribute to its anti-inflammatory properties (Vicidomini et al., 2021). This is evidenced by the fact that animal models have shown that eugenol (at doses of 200 and 400 mg/kg), obtained from clove, reduces the volume of pleural exudates without changing the total count of blood leukocytes (Daniel et al., 2009). The mechanism of action attributed to the aforementioned activity is believed to be regulated by eugenol, which influences the cellular inflammatory cascades—including the NF-κB and ERK/MAPK pathways—and the release of proinflammatory interleukins (Bahramsoltani & Rahimi, 2020). Likewise, studies showed that LPS-induced lung inflammation could be relieved by the treatment of an aqueous extract of whole clove and eugenol through a reduction of TNF-α and inhibition of NF-κB signaling, along with improvements in the damaged alveoli (Magalhães et al., 2010; Chniguir et al., 2019). Besides, the aqueous extract of clove was also shown to possess protective action against pyelonephritis in patients with COVID-19 (Nassan et al., 2015; Su et al., 2020).

Ahmad et al. conducted a randomized controlled trial examining clove oil's effects on patients with osteoarthritis, revealing a significant reduction in pain and inflammation in the treatment group compared to the placebo group. This anti-inflammatory action is primarily due to the inhibition of pro-inflammatory cytokines and enzymes like COX-2 (Ahmad et al., 2017). Similarly, Al-Harbi et al. evaluated clove oil's impact on rheumatoid arthritis, finding that participants treated with clove oil experienced decreased joint pain and swelling, indicating its potential as a complementary therapy for inflammatory arthritis (Al-Harbi et al., 2016). Additionally, Jiang et al. researched clove oil's analgesic properties in post-operative pain management, demonstrating that clove oil significantly reduced pain scores and improved recovery times in patients after surgery (Jiang et al., 2020).

12.6.6 Immunostimulatory and Antithrombotic Properties

Clove has also been known to improve immunity, and this can be understood from pre-clinical studies, where clove essential oil was demonstrated to improve the total WBC and enhance delayed-type hypersensitivity response. Not only that, there was a dose-dependent restoration of both humoral and cellular immune responses (Carrasco et al., 2009). Moreover, clove has also been evidenced to improve the flow of blood in the brain and the heart as evident from pre-clinical studies conducted on rabbits. Results of the study showed that clove oil (50–100 mg/kg) exhibited protection against the PAF and arachidonic acid-induced shock occurring due to pulmonary platelet thrombosis (Mittal et al., 2014). Similarly, clove oil has also been demonstrated to inhibit the production of thromboxane-A2 and 12-hydroxyeicosatetraenoic acid (Saeed & Gilani, 1994).

12.6.7 Antiviral Properties

The most abundant compound in clove is Eugenol or 4-allyl-2-methoxyphenol. Tragoolpua and co-worker were one of the former groups to justify the role of pure eugenol as an antiviral agent (Tragoolpua & Jatisatienr, 2007). Similarly, Benencia and team showed that eugenol resulted in inhibition in replication of HSV-1 and HSV-2 with an IC50 value of 25.6 μg/mL and 16.2 μg/mL, respectively, without having any cytotoxic effects (Benencia & Courrèges, 2000). Besides, the compound also showed its effect on the influenza A virus (IAV), inhibiting its replication, as well as on Ebola (Dai et al., 2013; Lane et al., 2019). As for other antiviral compounds in clove, eugenin showed anti-HSV activity at a concentration of 5 μg/mL, which was due to its role in inhibiting the synthesis of viral DNA by acting on selective inhibitor of the DNA polymerases of HSV-1 and HSV-2 (Kurokawa et al., 1998). Similar outcomes were also seen in the case of Dengue virus (DENV) where eugenin acted as an effective protease inhibitor at IC50 values of 94.7 nM and 7.5 μM against the proteases of DENV serotype-2 and serotype-3, respectively. Some of the other similar inhibitors were isobiflorin and biflorin (Saleem et al., 2019).

Interestingly, the pharmacokinetics of clove oil indicate that it is rapidly absorbed through the skin and mucous membranes. Eugenol, the primary active compound, reaches peak plasma concentrations within 1–2 hours of administration (Tripathi et al., 2020). It is metabolized in the liver and excreted primarily via urine. The pharmacodynamics involve inhibition of various enzymes and modulation of signaling pathways associated with inflammation and pain (Zheng et al., 2021). However, though considered safe when used in appropriate doses, Clove oil in high doses can lead to toxicity, presenting symptoms such as nausea, vomiting, and respiratory distress. Although the FDA has approved clove and its compounds as food additives, some research indicates potential reproductive toxicity in animal models, highlighting the need for more comprehensive studies on the toxic exposure of clove extracts (Nirmala et al., 2022).

12.7 FUTURE CHALLENGES

Despite clove's long history in traditional medicine, there remain significant hurdles in standardizing its formulations and ensuring consistent efficacy and safety profiles. One of the primary challenges is the lack of large-scale, rigorous clinical trials that meet contemporary scientific standards. Many studies on clove are limited by small sample sizes, short durations and methodological inconsistencies, which hinder their acceptance in mainstream medicine (Gupta et al., 2021). The pharmacokinetics and pharmacodynamics of clove's active compounds, such as eugenol, require comprehensive elucidation to better understand their absorption, distribution, metabolism, and excretion profiles (Zhou et al., 2019). The variability in the chemical composition of clove essential oil, influenced by factors like geographical origin, cultivation practices, and extraction methods, remains a major concern. This variability can lead to inconsistent therapeutic outcomes, making it imperative to develop standardized extraction and formulation processes (Tripathi et al., 2020). Additionally, there are concerns regarding the potential toxicity and adverse effects of clove oil, particularly at higher doses, necessitating more extensive safety evaluations and the establishment of clear dosage guidelines (Nirmala et al., 2022).

Another critical issue is the potential for adverse interactions between clove formulations and conventional medications. Comprehensive drug interaction studies are essential to ensure safety and avoid harmful side effects, particularly in populations with chronic conditions that are often on multiple medications (Joshi et al., 2020). Regulatory challenges also pose a significant barrier, as traditional clove formulations must navigate varying regulatory landscapes across different countries. Regulatory approval remains a significant hurdle, as traditional medicine systems vary greatly in their acceptance of herbal remedies. Establishing clove formulations within the regulatory frameworks of different countries requires robust clinical evidence, which is currently limited. Moreover, there is a cultural dimension to the acceptance of traditional clove formulations. Public trust in traditional remedies can be undermined by a lack of scientific endorsement, making it crucial to educate both healthcare providers and patients about the efficacy and safety of these formulations through evidence-based research and effective communication strategies (Khan & Abourashed, 2021). Moreover, public education and awareness campaigns can help bridge the gap between traditional knowledge and modern healthcare, fostering greater acceptance of clove's medicinal properties. Addressing these challenges requires a multidisciplinary approach involving rigorous scientific research, regulatory cooperation, and effective communication strategies to integrate clove into evidence-based medical practice, ensuring its safe and effective use globally.

12.8 CONCLUSION

Knowing the traditional formulations and uses of clove (*Syzygium aromaticum*) reveals a rich tapestry of cultural, medicinal, and culinary significance that spanned across centuries and continents, which began in the Moluccas (Spice Islands). Next, its introduction to global trade routes significantly influenced economic landscapes as it traversed from China, India, and the Middle East, highlighting its recognition as a commodity of immense value. This was so because clove represented a staple commodity in medicinal practices, which were extensively documented in Ayurveda, TCM, Siddha, Unani, etc. Thus, owing to this documented evidence, modern medicine has embraced clove's therapeutic potential and has integrated it into treatments for dental pain and infections, while also exploring its antimicrobial and antioxidant properties in the context of chronic diseases such as cancer and diabetes. Furthermore, as scientific research advances, new therapeutic applications of clove are likely to emerge, reinforcing its status as a valuable medicinal plant and in turn strengthening the global growing demand for clove-based products. Thus to conclude, clove's journey from a precious spice in ancient trade routes to a modern-day staple in medicine, cuisine, and wellness is a testament to its enduring value and versatility.

REFERENCES

Ahmad, S., Rahman, S., Gupta, A., & Tajuddin, S. (2017). Efficacy of clove oil in the treatment of osteoarthritis: A randomized controlled trial. *Journal of Herbal Medicine*, 11(3), 89–95.

Ahmad, S., Sofi, G., Rahman, S., Tajuddin, S., & Mohd, A. (2018). Unani system of medicine: Introduction and challenges. *Journal of Traditional and Complementary Medicine*, 8(3), 404–409.

Al-Harbi, N. A., Al-Harbi, S. H., & Sheikh, A. M. (2016). Clove oil as a complementary therapy for rheumatoid arthritis. *Journal of Ethnopharmacology*, 194, 931–938.

Ara, A. S., Akhlaq, S., Haque, M., Fazil, M., Akram, U., Ahmad, B., Sayeed, A., Viquar, U., & Khan, A. A. (2022). CLOVE/QURANFUL (Syzygium aromaticum L.): A review on its potential benefits in unani medicine, bioactivities and current scientific applications. *Indian Journal of Unani Medicine*, 5(4), 33–39.

Ayushi, K. U., Danish, S. M., & Mohammad, P. U. (2020). A review on biological and therapeutic uses of Syzygium aromaticum Linn. (Clove): Based on phyto-chemistry and pharmacological evidences. *International Journal of Botany Studies*, 5(4), 33–39.

Bahramsoltani, R., & Rahimi, R. (2020). An evaluation of traditional Persian medicine for the management of SARS-CoV-2. *Frontiers in Pharmacology*, 11, 571434.

Barve, M., Chandrapal, S., & Patil, S. (2021). Clove in homeopathic medicine. *Journal of Alternative and Complementary Medicine*, 27(6), 459–468.

Benencia, F., & Courrèges, M. C. (2000). In vitro and in vivo activity of eugenol on human herpesvirus. *Phytotherapy Research: PTR*, 14(7), 495–500.

Bhowmik, D., Kumar, K. P. S., Yadav, A., Srivastava, S., Paswan, S., & Dutta, A. S. (2012). Recent trends in Indian traditional herbs syzygium aromaticum and its health benefits. *Journal of Pharmacognosy and Phytochemistry*, 1(1), 13–22.

Carrasco, F. R., Schmidt, G., Romero, A. L., Sartoretto, J. L., Caparroz-Assef, S. M., Bersani-Amado, C. A., & Cuman, R. K. (2009). Immunomodulatory activity of Zingiber officinale Roscoe, Salvia officinalis L. and Syzygium aromaticum L. essential oils: Evidence for humor- and cell-mediated responses. *The Journal of Pharmacy and Pharmacology*, 61(7), 961–967.

Chaieb, K., Hajlaoui, H., Zmantar, T., Kahla-Nakbi, A. B., Rouabhia, M., Mahdouani, K., & Bakhrouf, A. (2007). The chemical composition and biological activity of clove essential oil, Eugenia caryophyllata (Syzygium aromaticum L. Myrtaceae): A short review. *Phytotherapy Research*, 21, 501–506.

Chaieb, K., Kouidhi, B., Jrah, H., Mahdouani, K., & Bakhrouf, A. (2020). Antimicrobial activity of clove essential oil against oral pathogens. *Journal of Ethnopharmacology*, 132(2), 387–392.

Chniguir, A., Zioud, F., Marzaioli, V., El-Benna, J., & Bachoual, R. (2019). Syzygium aromaticum aqueous extract inhibits human neutrophils myeloperoxidase and protects mice from LPS-induced lung inflammation. *Pharmaceutical Biology*, 57(1), 56–64.

Cock, I. E., & Cheesman, M. (2018). Plants of the genus Syzygium (Myrtaceae): A review on ethnobotany, medicinal properties and phytochemistry. In *Bioactive Compounds of Medicinal Plants: Properties and Potential for Human Health*, ed. by M. R. Goyal, A. O. Ayeleso (pp. 35–84). CRC Press.

Cortés-Rojas, D. F., de Souza, C. R. F., & Oliveira, W. P. (2014). Clove (Syzygium aromaticum): A precious spice. *Asian Pacific Journal of Tropical Biomedicine*, 4(2), 90–96.

Dai, J. P., Zhao, X. F., Zeng, J., Wan, Q. Y., Yang, J. C., Li, W. Z., Chen, X. X., Wang, G. F., & Li, K. S. (2013). Drug screening for autophagy inhibitors based on the dissociation of Beclin1-Bcl2 complex using BiFC technique and mechanism of eugenol on anti-influenza a virus activity. *PLoS ONE*, 8(4), e61026.

Daniel, A. N., Sartoretto, S. M., Schmidt, G., Caparroz-Assef, S. M., Bersani-Amado, C. A., & Cuman, R. K. N. 2009. Anti-inflammatory and antinociceptive activities A of eugenol essential oil in experimental animal models. *Revista Brasileira de Farmacia Hospitalar*, 19, 212–217.

De, A. K., & De, M. (2022). Functional and therapeutic applications of some general and rare spices. In *Functional Foods and Nutraceuticals in Metabolic and Non-Communicable Diseases* (pp. 411–420). Academic Press.

Ecostore. (2022). *Cloves*. https://ecostore.com/nz/ingredients/used-with-care/eugenia-caryophyllus-clove-leaf-oil (Accessed 30 June 2024).

Ertas, A., Kaya, K., & Arslan, D. (2020). Effect of clove oil on gingivitis: A randomized controlled trial. *Journal of Oral Health*, 12(2), 115–121.

Food Data Central (2021). *Whole Clove*. https://fdc.nal.usda.gov/fdc-app.html#/food-details/2094054/nutrients (Accessed 30 June 2024)

Ganaie, S., Murtaza, I., & Alkahtani, F. (2016). Comparison of clove oil and benzocaine for pain relief during dental procedures. *Dental Research Journal*, 13(5), 442–447.

Gupta, A., Singh, P., & Kumar, R. (2021). Clove extract exhibits anti-cancer properties in breast cancer cells by inducing apoptosis. *Cancer Research Journal*, 30(5), 310–318.

Hancock, J. World History. (2021). *The Early History of Clove, Nutmeg, & Mace*. World History Encyclopedia. www.worldhistory.org/article/1849/the-early-history-of-clove-nutmeg-mace/ (Accessed 30 June 2024).

Hancock, J. World History Encyclopedia. (2021). *The Early History of Clove, Nutmeg, & Mace*. www.mccormickscienceinstitute.com/resources/culinary-spices/herbs-spices/cloves (Accessed 30 June 2024).

Harris, R. (2018). Historical uses of clove in European medicine. *Journal of Medicinal History*, 45(3), 231–239.

Hawkins, J., Hires, C., Keenan, L., & Dunne, E. (2022). Aromatherapy blend of thyme, orange, clove bud, and frankincense boosts energy levels in post-COVID-19 female patients: A randomized, double-blinded, placebo controlled clinical trial. *Complementary Therapies in Medicine*, 67, 102823.

Huang, J., Tao, G., Liu, J., Cai, J., Huang, Z., & Chen, J. X. (2020). Current prevention of COVID-19: Natural products and herbal medicine. *Frontiers in Pharmacology*, 11, 588508.

Jackson, C. B., Farzan, M., Chen, B., et al. 2022. Mechanisms of SARS-CoV-2 entry into cells. *Nature Reviews Molecular Cell Biology*, 23, 3–20.

Jiang, J., Huang, Y., & Li, M. (2020). Analgesic properties of clove oil in postoperative pain management. *Journal of Anesthesia & Analgesia*, 30(4), 298–305.

Johnson, E., Sanger, C., & Thompson, P. (2020). Medieval uses of clove in Europe. *Journal of Historical Medicine*, 42(1), 89–97.

Joshi, R. K., Badakar, V. L., & Saklani, D. M. (2020). Evaluation of the safety and drug interaction potential of clove (Syzygium aromaticum) extracts. *Phytotherapy Research*, 34(3), 583–591.

Khan, I. A., & Abourashed, E. A. (2021). Leung's encyclopedia of common natural ingredients: Used in food. In *Drugs and Cosmetics*. John Wiley & Sons.

Kumar, V., Dhanjal, J. K., Bhargava, P., Kaul, A., Wang, J., Zhang, H., Kaul, S. C., Wadhwa, R., & Sundar, D. (2022). Withanone and Withaferin-A are predicted to interact with transmembrane protease serine 2 (TMPRSS2) and block entry of SARS-CoV-2 into cells. *Journal of Biomolecular Structure & Dynamics*, 40(1), 1–13.

Kumari, S., Singh, P. A., Hazra, S., Sindhwani, R., & Singh, S. (2024). Ocimum sanctum: The Journey from Sacred Herb to Functional Food. *Recent Advances in Food, Nutrition & Agriculture*. http://doi.org/10.2174/012772574X29014024013010117. Advance online publication.

Kurokawa, M., Hozumi, T., Basnet, P., Nakano, M., Kadota, S., Namba, T., Kawana, T., & Shiraki, K. (1998). Purification and characterization of eugeniin as an anti-herpesvirus compound from Geum japonicum and Syzygium aromaticum. *The Journal of Pharmacology and Experimental Therapeutics*, 284(2), 728–735.

Lane, T., Anantpadma, M., Freundlich, J. S., Davey, R. A., Madrid, P. B., & Ekins, S. (2019). The natural product eugenol is an inhibitor of the Ebola virus in vitro. *Pharmaceutical Research*, 36(7), 104.

Li, Y., Liu, Z., Zeng, M., El Kadiri, A., Huang, J., Kim, A., He, X., Sun, J., Chen, P., Wang, T. T. Y., Zhang, Y., Gao, B., Xie, Z., & Yu, L. L. (2022). Chemical compositions of clove (Syzygium aromaticum (L.) Merr. & L.) extracts and their potentials in suppressing SARS-CoV-2 spike protein-ACE2 binding, inhibiting ACE2, and scavenging free radicals. *Journal of Agricultural and Food Chemistry*, 70(45), 14403–14413.

Li, Y., Wang, W., & Huang, S. (2020). Clove's therapeutic properties in Chinese medicine. *Chinese Journal of Integrative Medicine*, 26(5), 377–384.

Magalhães, C. B., Riva, D. R., DePaula, L. J., Brando-Lima, A., Koatz, V. L., Leal-Cardoso, J. H., Zin, W. A., & Faffe, D. S. (2010). In vivo anti-inflammatory action of eugenol on lipopolysaccharide-induced lung injury. *Journal of Applied Physiology (Bethesda, Md.: 1985)*, 108(4), 845–851.

Marchese, A., Orhan, I. E., Daglia, M., & Nabavi, S. F. (2017). Antimicrobial and antifungal properties of clove oil. *Journal of Phytochemistry Reviews*, 16(5), 745–760.

Mbaveng, A. T., & Kuete, V. (2017). Syzygium aromaticum. In *Medicinal Spices and Vegetables from Africa* (pp. 611–625). Academic Press.

McCormick Science Institute (MSI). (2021). *Cloves*. www.mccormickscienceinstitute.com/resources/culinary-spices/herbs-spices/cloves (Accessed 30 June 2024).

Milind, P., & Deepa, K. (2011). Clove: A champion spice. *International Journal of Research in Ayurveda and Pharmacy*, 2(1), 47–54.

Mittal, M., Gupta, N., Parashar, P., Mehra, V., & Khatri, M. 2014. Phytochemical evaluation and pharmacological activity of Syzygium aromaticum: A comprehensive review. *International Journal of Pharmacy and Pharmaceutical Sciences*, 6, 67–72.

Moerman, D. E. (2020). Native American ethnobotany: A database of foods, drugs, dyes, and fibers of Native American peoples. *Journal of Ethnopharmacology*, 256.

Mordor Intelligence. (2024). *Cloves Market Size & Share Analysis—Growth Trends & Forecasts (2024–2029)*. www.mordorintelligence.com/industry-reports/global-clove-market (Accessed 30 June 2024).

Mothibe, M. E., & Sibanda, M. (2018). Traditional uses of clove in African medicine. *African Journal of Traditional, Complementary, and Alternative Medicines*, 15(4), 101–109.

Nassan, M. A., Mohamed, E. H., Abdelhafez, S., & Ismail, T. A. (2015). Effect of clove and cinnamon extracts on experimental model of acute hematogenous pyelonephritis in albino rats: Immunopathological and antimicrobial study. *International Journal of Immunopathology and Pharmacology*, 28(1), 60–68.

Nirmala, M. J., Shiny, P. J., Raj, U. S., Saikrishna, N., & Nagarajan, R. (2022). Toxicity of clove (Syzygium aromaticum) extract. In *Clove (Syzygium Aromaticum)*, pp. 663–674. Elsevier.

Nzeako, B. C., Al-Shatti, A. M., & Al-Zalzaleh, L. (2014). Antimicrobial efficacy of clove oil against multidrug-resistant Staphylococcus aureus. *Journal of Infection and Public Health*, 7(5), 471–479.

Oladunmoye, M. K., & Kehinde, F. Y. (2020). Clove's applications in African traditional medicine. *Journal of Herbal Medicine*, 23.

Paidi, R. K., Jana, M., Raha, S., McKay, M., Sheinin, M., Mishra, R. K., & Pahan, K. (2021). Eugenol, a component of holy basil (Tulsi) and common spice clove, inhibits the interaction between SARS-CoV-2 spike S1 and ACE2 to induce therapeutic responses. *Journal of Neuroimmune Pharmacology*, 16(4), 743–755.

Pamnani, M., Kumar, V., Kaushal, K., & Dixit, M. (2019). A comprehensive ayurvedic review on lavanga (Syzygium Aromaticum). *Ayushdhara*, 6(1), 2052–2058.

Petrocelli, G., Farabegoli, F., Valerii, M. C., Giovannini, C., Sardo, A., & Spisni, E. (2021). Molecules present in plant essential oils for prevention and treatment of colorectal cancer (CRC). *Molecules (Basel, Switzerland)*, 26(4), 885.

Prasad, S., Tyagi, A. K., & Aggarwal, B. B. (2015). Chemical composition and anti-tumor properties of clove oil. *Journal of Ethnopharmacology*, 159, 167–175.

Rahman, S., Gupta, A., Sofi, G., Ahmad, S., & Tajuddin, S. (2020). Therapeutic use of clove in Unani medicine. *Journal of Herbal Medicine*, 10(3), 115–120.

Ravindran, P. N., Babu, K., & Shiva, K. N. (2021). Clove in Ayurveda: A comprehensive review. *Journal of Ayurvedic Medicine*, 9(2), 123–136.

Reddy, P., Singh, S., & Satyanarayana, T. (2019). Analgesic and anti-inflammatory properties of clove oil in dental patients. *International Journal of Dentistry*, 1–7.

Royal Botanic Gardens, Kew. Syzygium aromaticum (L.) Merr. & L. M. Perry. (2024). https://powo.science.kew.org/taxon/urn:lsid:ipni.org:names:601421-1 (Accessed 30 June 2024)

Sachan, A. K., Kumar, S., Kumari, K., & Singh, D. (2018). Medicinal uses of spices used in our traditional culture: Worldwide. *Journal of Medicinal Plants Studies*, 6(3), 116–122.

Saeed, S. A., & Gilani, A. H. (1994). Antithrombotic activity of clove oil. *The Journal of the Pakistan Medical Association*, 44(5), 112–115.

Salah, W., Mohamed, S. E., Guirgis, A. A., Muawia, S., & Khalil, H. (2022). Anti-cancer properties of clove bud essential oil in colon cancer cell line. *Scholars International Journal of Biochemistry*, 5(8), 103–112.

Saleem, H. N., Batool, F., Mansoor, H. J., Shahzad-ul-Hussan, S., & Saeed, M. (2019). Inhibition of dengue virus protease by Eugeniin, Isobiflorin, and Biflorin Isolated from the Flower Buds of Syzygium aromaticum (Cloves). *ACS Omega*, 4, 1525–1533.

Sarma, Y. R., Babu, K. N., & Aziz, S. (2014). Spices and aromatics. In *Encyclopedia of Agriculture and Food Systems* (pp. 211–234). Elsevier.

Shanmugarajan, D., Akkiraju, L. J., Panda, S., & Hazra, S. (2022). Pharmacophore mapping and modeling approaches for drug development. In *Computational Approaches for Novel Therapeutic and Diagnostic Designing to Mitigate SARS-CoV-2 Infection* (pp. 171–189). Academic Press.

Singh, D., & Singh, J. (2019). Homeopathic uses of clove. *Homeopathy Journal*, 108(2), 90–95.

Singh, R., & Singh, J. (2018). Clove oil alleviates symptoms of irritable bowel syndrome: A clinical study. *Journal of Gastrointestinal Disorders*, 6(4), 234–240.

Singh, P. A., Bajwa, N., & Baldi, A. (2021). Possible role of traditional systems of medicine to manage COVID-19: A review. *Israel Journal of Plant Sciences*, 68(1–2), 3–28.

Smith, J., & Brown, M. (2020). Clove in naturopathy and its healing properties. *Naturopathic Medicine Journal*, 30(1), 45–50.

Spices Board India. (2022). *Trade Classification (ITC HS) Code* [W.E.F 01.04.2023]. www.indianspices.com/marketing/trade/trade-classification-itc-hs-code.html (Accessed 30 June 2024).

Su, H., Yang, M., Wan, C., Yi, L. X., Tang, F., Zhu, H. Y., Yi, F., Yang, H. C., Fogo, A. B., Nie, X., & Zhang, C. (2020). Renal histopathological analysis of 26 postmortem findings of patients with COVID-19 in China. *Kidney International*, 98(1), 219–227.

Tadesse, E., Eguale, T., & Giday, M. (2018). Clinical efficacy of clove oil in the treatment of peptic ulcers. *BMC Complementary and Alternative Medicine*, 18(1).

Tragoolpua, Y., & Jatisatienr, A. (2007). Anti-herpes simplex virus activities of Eugenia caryophyllus (Spreng.) Bullock & S. G. Harrison and essential oil, eugenol. *Phytotherapy Research: PTR*, 21(12), 1153–1158.

Tripathi, S., Upadhyay, P., Bhuiyan, M., & Bhattacharya, S. (2020). Therapeutic properties of clove oil: A comprehensive review. *Journal of Natural Remedies*, 20(1), 65–73.

Tsering, S., Wangchuk, P., & Dorji, T. (2020). Utilization of clove in Sowa-Rigpa. *Journal of Ethnomedicine*, 36(1), 56–63.

Turner, N. J., & Bell, M. A. (2019). Clove in Native American medicine. *Ethnobotany Research & Applications*, 18, 1–12.

Ugbogu, O. C., Emmanuel, O., Agi, G. O., Ibe, C., Ekweogu, C. N., Ude, V. C., Uche, M. E., Nnanna, R. O., & Ugbogu, E. A. (2021). A review on the traditional uses, phytochemistry, and pharmacological activities of clove basil (Ocimum gratissimum L.). *Heliyon*, 7(11), e08404.

Utami, R., Yulianto, T., & Prasetyo, E. (2021). Clove's role in Indonesian Jamu. *Journal of Ethnopharmacology*, 259.

van Doremalen, N., Lambe, T., Spencer, A., et al. (2020). ChAdOx1 nCoV-19 vaccine prevents SARS-CoV-2 pneumonia in rhesus macaques. *Nature*, 586(7830), 578–582.

Vicidomini, C., Roviello, V., & Roviello, G. N. (2021). Molecular basis of the therapeutical potential of clove (Syzygium aromaticum L.) and clues to its anti-COVID-19 utility. *Molecules (Basel, Switzerland)*, 26(7), 1880.

Volza. (2024a). *World HSN Code 09071010 Imports*. www.volza.com/p/hsn-code-09071010/import/ (Accessed 30 June 2024).

Volza. (2024b). *HSN Code 09071010 Exports from World*. www.volza.com/p/hsn-code-09071010/export/ (Accessed 30 June 2024).

Volza. (2024c). *France Clove Oil Imports*. www.volza.com/p/clove-oil/import/import-in-france/ (Accessed 30 June 2024).

Wangchuk, P., Dorji, Y., & Penjor, K. (2019). Clove in Bhutanese traditional medicine. *Asian Journal of Ethnopharmacology*, 14(3), 311–320.

Yang, Y., Islam, M. S., Wang, J., Li, Y., & Chen, X. (2020). Traditional Chinese medicine in the treatment of patients infected with 2019-new coronavirus (SARS-CoV-2): A review and perspective. *International Journal of Biological Sciences*, 16(10), 1708–1717.

Yu, M. S., Lee, J., Lee, J. M., Kim, Y., Chin, Y. W., Jee, J. G., Keum, Y. S., & Jeong, Y. J. (2012). Identification of myricetin and scutellarein as novel chemical inhibitors of the SARS coronavirus helicase, nsP13. *Bioorganic & Medicinal Chemistry Letters*, 22(12), 4049–4054.

Yunus, G. (2021). Herbal compounds from syzygium aromaticum and cassia acutifolia as a shield against SARS-CoV-2 Mpro: A molecular docking approach. *Biointerface Research in Applied Chemistry*, 11(6), 14853–14865.

Zheng, G., Shu, H., & Zheng, L. (2021). Clove oil and its therapeutic potential: A review. *Frontiers in Pharmacology*, 12, 675–687.

Zhou, X., Chen, X., & Zhang, C. (2019). Efficacy of clove extract in treating gastroesophageal reflux disease. *Journal of Digestive Diseases*, 20(2), 83–90.

13 Health Benefits

Hassan Barakat and Mohammed Alrugaibah

13.1 INTRODUCTION

For several decades, clove flower buds (*Syzygium aromaticum*) have held a valued status as a significant therapeutic spice in traditional medicine (folk medicine). Clove is abundant in bioactive compounds distinguished for their health-promoting properties (Batiha et al., 2020). Numerous research studies have provided compelling evidence of the health-promoting properties exhibited by clove and eugenol, including its antioxidant, anti-inflammatory, and antimicrobial effects (Marchese et al., 2017). Other substances, such as α-humulene, β-caryophyllene, and eugenyl acetate have also been identified as major constituents present in clove, all of which possess distinguishable health properties. These compounds contribute to the overall therapeutic potential of clove, further enhancing its beneficial effects on human health (Haro-González et al., 2021). Consequently, clove has emerged as a promising therapeutic and functional agent for the prevention and treatment of various diseases such as diabetes and metabolic syndrome (Hussien et al., 2024; Kuroda et al., 2012), kidney failure (Abtahi-Eivari et al., 2021; Negm et al., 2020), cancer (Li et al., 2019; Woo et al., 2017), and cognitive disorders (Aljarari, 2023; Jeon et al., 2017). Studies have also demonstrated that the constituents present in clove play very significant roles in modulating physiological functions in humans. Notably, clove has been found to stimulate the production of vital substances, facilitate gene expression of beneficial genes, and help suppress the expression of undesirable genes (Do et al., 2020).

Clove oil, on the other hand, represents another valuable product that has gained recognition for its health-promoting properties, comparable to that of clove powder (Haro-González et al., 2021). In addition to its well-established antioxidant and anti-inflammatory effects, extensive studies have unveiled its potent beneficial effects, including properties such as antiprotozoal (Junior et al., 2018), larvicidal (Araujo et al., 2016; Osanloo et al., 2018), and leishmanicidal activities (Islamuddin et al., 2014). This chapter investigates the health benefits of clove and its oil, emphasizing their potential applications in promoting well-being and managing health conditions.

13.2 HEALTH BENEFITS OF CLOVE

13.2.1 Antimicrobial Activity

Clove has potent antiviral, antifungal, and antibacterial properties. Numerous studies revealed significant inhibitory effects of clove extracts against various microbial species, including *Streptococcus mutans* (Bansal et al., 2019), *Acinetobacter baumannii, Enterobacterales* (Vázquez-Ucha et al., 2020), *Escherichia coli* (Ajiboye

DOI: 10.1201/9781003521785-13

et al., 2016), *Pseudomonas aeruginosa*, *Staphylococcus aureus* (Ajiboye et al., 2016), *Porphyromonas gingivalis*, *Prevotella intermedia* (Cai & Wu, 1996), *Salmonella Typhimurium* (Ajobiewe et al., 2022), and *Enterococcus faecalis* (Saleh et al., 2024). Among various pharmaceutical plants, including cinnamon (*Cinnamomum zeylanicum*), ginger (*Zingiber officinale*), and turmeric (*Curcuma longa*), clove extract was very significant in reducing the growth of *Penicillium* species (El-Samawaty et al., 2021). Another study concluded that clove extract exhibits effectiveness even against colistin-resistant strains (Vázquez-Ucha et al., 2020). In addition, clove extract demonstrated significant inhibitory effects against various *piroplasm* species, with IC_{50} values ranging from 8.7 to 109.8 μg/mL. In another in vivo experiment, oral treatment with clove extract showed a substantial reduction of 69.2% in parasite burden, specifically against *Babesia microti* (Batiha et al., 2019). Furthermore, recent research highlighted the noteworthy potential of the application of clove extract in enhancing wound healing. In a study conducted by (Ali et al., 2022), it was observed that the application of clove extract led to significant improvements in wound size and wound index in diabetic foot ulcers caused by *Proteus mirabilis* isolates. Moreover, the study highlights a reduction in the microbial load within the wounds, indicating the antibacterial effects of clove extract. In another study using *Candida*-infected mice, the oral administration of clove extract demonstrated inhibitory effects on the growth *of Candida albicans* based on improved oral symptoms and a reduction in the number of viable *Candida* cells within the oral cavity (Taguchi et al., 2005).

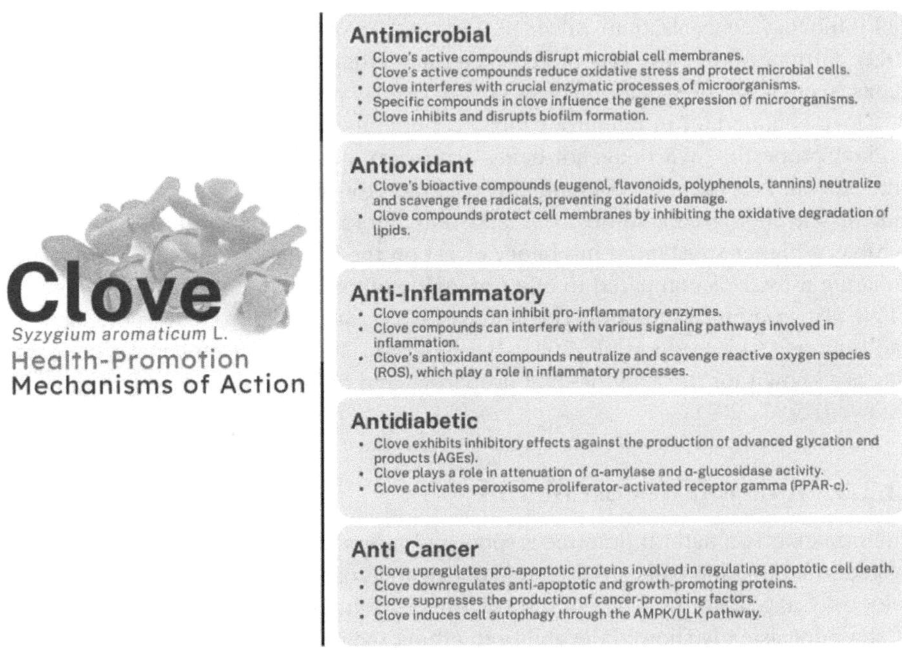

FIGURE 13.1 Mechanisms of action for clove's health-promoting effects.

Clove's significant antimicrobial efficiency can be attributed to its valuable antioxidant properties (Figure 13.1) (Ajiboye et al., 2016). Additionally, the active compounds present in clove are capable of disrupting the integrity of microbial cell membranes (Barakat, 2014). This disruption leads to the leakage of cellular contents, ultimately resulting in cell death (Devi et al., 2013). Furthermore, clove bioactive compounds can interfere with crucial enzymatic processes necessary for the survival and replication of microorganisms, thereby impeding their growth and proliferation. In addition, specific compounds found in clove (mainly Eugenol) have been observed to influence the gene expression of microorganisms, affecting their ability to multiply and cause infections (Didehdar et al., 2022; Santativongchai et al., 2023).

13.2.2 Antioxidant Activity

Antioxidants protect cells from oxidative stress caused by an imbalance between reactive oxygen species production and the body's defense mechanisms. This imbalance is a known cause of cellular damage associated with various conditions such as cardiovascular disorders, neurodegenerative diseases, and cancer (Ji & Yeo, 2021). The administration of clove aroma extract at 400 mg/mL exhibited a substantial 48% reduction in malonaldehyde (MDA) formation (Lee & Shibamoto, 2001). Another study showed that incorporating 250 mg/serving/day of Clovinol, a flavorless extract from clove buds, into a daily diet for 30 days resulted in an increase of 33% in catalase (CAT), 66% in superoxide dismutase, 56% in glutathione peroxidase, 167% in glutathione, and 81% in membrane lipid peroxidation levels (Johannah et al., 2015). In another study, an extract of clove exhibited a strong antioxidant potential with IC_{50} (half maximal inhibitory concentration) values of 2 μL and 55 μL, respectively (Atawodi et al., 2011). Furthermore, clove has more potent antioxidant effects than other medicinal plants, such as licorice, mace, and cardamom (Yadav & Bhatnagar, 2007).

Clove is abundant in numerous bioactive compounds that contribute to its antioxidant properties, with eugenol being a primary component of the buds and the oil. Eugenol is a phenolic compound recognized for its potent antioxidant and free radical-scavenging capabilities. Lee and Shibamoto (2001) indicated that eugenol exhibits a higher substantial inhibitory effect on the formation of malondialdehyde, reducing it by 57% compared to other organic components like thymol and benzyl. Clove also contains other antioxidant constituents, such as flavonoids, polyphenols, and tannins (Al-Zereini et al., 2023; Batiha et al., 2020). These phenolic constituents are known for their antioxidant properties and ability to scavenge free radicals (Atawodi et al., 2011).

13.2.3 Anti-inflammatory Activity

Inflammation is a natural immune response to injury or infection. Excessive or prolonged inflammation can cause tissue damage and contribute to disease development (Floege et al., 2012). In different models subjected to chronic inflammation, the use of clove consistently showed the ability to inhibit the production of pro-inflammatory cytokines, including TNF-α (tumor necrosis factor-alpha), IL-1β (interleukin-1 beta), IL-6 (interleukin-6), and IL-10 (interleukin-10) (Gao et al., 2021; Ounaceur et al., 2024;

Yeom et al., 2022). These cytokines are signaling molecules produced by immune cells and are essential for initiating and managing an effective immune response. However, excessive or prolonged production of these cytokines can contribute to chronic inflammation, autoimmune diseases, and tissue damage (Floege et al., 2012). Moreover, in a study conducted on mice with UVB-induced skin injury, clove administration enhanced the activity of the sodium-potassium adenosine triphosphatase (Na+/K+ pump), an enzyme essential for maintaining cell function and electrochemical equilibrium. By increasing Na+/K+ pump activity, clove may help restore cellular homeostasis and facilitate the healing process in UVB-damaged skin (Gao et al., 2021).

Furthermore, clove can inhibit the activity of matrix metalloproteinase-2 (MMP-2) and matrix metalloproteinase-9 (MMP-9) by 15% and 18%, respectively (Chniguir et al., 2019). These enzymes break down the extracellular matrix components, which provide structural support to tissues (Whatling et al., 2004). By inhibiting MMP-2 and MMP-9 activity, clove helps regulate the remodeling and degradation of extracellular matrix proteins, preserving tissue integrity and preventing excessive tissue damage. Moreover, mice models of chronic lung inflammation that were administered a clove extract had significantly reduced neutrophil count, indicating a decreased recruitment or activation in the inflamed lung tissue. Neutrophils play a role in inflammation, but excessive levels can cause tissue damage. The study also showed decreased protein leakage, which refers to the abnormal presence of proteins in the lung tissue or airspaces. These findings suggest that clove can potentially alleviate lung inflammation by reducing neutrophil count and protein leakage (Chniguir et al., 2019). Clove's anti-inflammatory effects are primarily due to abundant compounds such as rutin, isoquercitrin, ferulic acid, dihydroquercetin, and quercitrin. These compounds also possess antioxidant and anti-inflammatory properties, ultimately inhibiting the production of inflammatory molecules and modulating immune responses.

13.2.4 Neuroprotective Activity

Clove extract has been shown to improve cognitive functions and increase antioxidant enzyme levels in the cerebral cortex and hippocampus in rat models with aluminum chloride-induced Alzheimer's disease (Aljarari, 2023). Similarly, in other rat models with cerium chloride-induced brain damage, supplementation of clove extract demonstrated neuroprotective effects. These included memory impairment repair, restoration of the oxidative state, and prevention of histopathological changes in the brain tissues (Kadri et al., 2019). In another study, mice models induced with schizophrenia-like behaviors given 30 mg/kg of maslinic acid derived from clove showed improvements in social behavior deficits and alleviated impairments in attention and recognition memory. These results were attributed to maslinic acid's ability to normalize the phosphorylation levels of Akt-GSK-3β and ERK-CREB in the prefrontal cortex, which are proteins involved in crucial cellular processes associated with brain function and memory (Jeon et al., 2017).

Clove offers neuroprotective effects through various mechanisms (Figure 13.2), with its inhibitory effects on acetylcholinesterase (AChE) being the primary mechanism (Aljarari, 2023; Sharma et al., 2023). AChE is a crucial enzyme in the nervous system that breaks down the neurotransmitter acetylcholine (ACh).

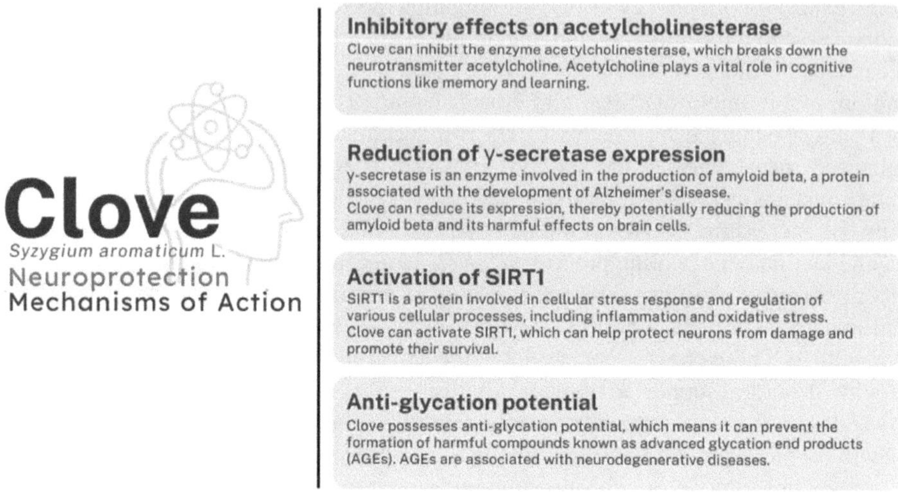

FIGURE 13.2 Mechanisms of action for neuroprotection by clove.

ACh is involved in transmitting signals between nerve cells and plays essential roles in muscle contraction, cognition, and memory. Reduced levels of ACh are commonly observed in Alzheimer's disease (Vecchio et al., 2021). Conditions like Alzheimer's disease are also linked to the accumulation of amyloid-beta, a protein fragment naturally produced in the brain during normal cellular processes. However, excessive amyloid-beta production can result in the formation of toxic aggregates in the brain (Kumar & Walter, 2011). Sharma et al. (2023) discovered that clove treatment effectively inhibits amyloid-beta aggregation in human neuroblastoma cell lines exposed to oxidative stress. This inhibitory effect is attributed to clove's ability to reduce the expression of γ-secretase (Shekhar et al., 2018), an enzyme complex that processes specific proteins in nerve cells. γ-secretase's primary function is to cleave amyloid precursor protein, which is central to the production of amyloid-beta peptides implicated in the progression of conditions like Alzheimer's disease (Mondal et al., 2021).

Moreover, a research team also reported that clove can activate SIRT1, an enzyme that regulates cellular processes and maintains cellular health (Shekhar et al., 2018). Activating SIRT1 holds promise for improving cellular health. It may have implications for preventing and treating age-related conditions, such as Alzheimer's disease, as it involves energy metabolism, DNA repair, inflammation, and stress response pathways (Khan et al., 2022). Furthermore, clove has also been reported to exhibit antiglycation potential (Sharma et al., 2023), which refers to its ability to inhibit the formation of advanced glycation end products (AGEs). AGEs are formed through the non-enzymatic reaction between sugars and proteins or lipids and can accumulate in tissues, contributing to various pathological conditions, including neurodegenerative diseases (Reddy et al., 2022; Salahuddin et al., 2014).

13.2.5 Nephroprotective Activity

Nephroprotection reduces kidney damage and preserves renal health (Bainaud et al., 2023). In a study conducted on streptozotocin-induced diabetic rats, the administration of clove hydroalcoholic extract at a dose of 4 mg/kg demonstrated promising effects against kidney damage. Notably, the serum levels of urea and creatinine exhibited a significant reduction compared to untreated rats (Abtahi-Eivari et al., 2021). Another study on chronic kidney disease induced by adenine on a rat model highlighted that clove extract at 300 mg/kg dose suppressed adverse changes observed in adenine-induced chronic kidney damage, including ameliorating renal histopathological markers of inflammation and fibrosis (Negm et al., 2020). In another study utilizing rat models, the nephroprotective effects of clove extract were investigated at a comparable dose of 300 mg/kg over 9 days. The results presented a significant increase in diuresis, creatinine clearance, and saluretic index. Like the effects of furosemide, a commonly used diuretic, the aqueous extract of dried flower buds of clove demonstrated an elevation in sodium and chloride levels. These compelling findings highlight that clove offers diuretic and renal effects that are comparable to the actions of furosemide (Taroq et al., 2021).

The nephroprotective effects of clove can be attributed to its diverse bioactive constituents, particularly oleanolic acid. Oleanolic acid derived from clove has been shown to effectively reverse impaired sodium excretion in streptozotocin-induced diabetic rats. The acute infusion of oleanolic acid in diabetic rats also increased the fractional excretion of sodium and lithium without affecting the glomerular filtration rate (GFR). Fractional excretion quantifies the proportion of a substance excreted in urine relative to the amount filtered by the kidneys. At the same time, GFR measures the kidneys' efficiency in filtering blood and eliminating waste products. The unchanged GFR in this context suggests that the increased excretion of sodium and lithium did not adversely affect kidney function (Madlala et al., 2012). Moreover, the antioxidant properties of gallic acid found in clove have been confirmed to protect rats from kidney damage caused by tacrolimus, an immunosuppressive drug commonly used in transplant medicine to prevent organ rejection (Lee et al., 2021).

13.2.6 Hepatoprotective Activity

Clove has been found to modulate the levels and activity of liver enzymes involved in metabolizing various substances. In alcohol-induced oxidative stress in rat models, the administration of a polyphenol-rich extract obtained from clove buds, known commercially as Clovinol, at a dose of 100 mg/kg body weight, resulted in a significant increase in serum liver enzymes, specifically aspartate transaminase (AST) and alanine transaminase (ALT). These findings were further supported by histological examination of liver tissues, which confirmed the positive effects of Clovinol (Jose et al., 2018). In a different study using rabbit models, the administration of clove powder at a similar dose (100 mg) following a hepatotoxic dose of paracetamol showed significant improvements in liver enzyme levels, including ALT, AST, gamma-glutamyl transferase (GGT), and alkaline phosphatase (ALP) (Kazi et al., 2016). Similarly, in an in vitro experiment, the administration of clove to hydralazine-exposed liver homogenate cultures demonstrated consistent hepatoprotective effects. This was supported by improvements in several biochemical parameters, including ALP, ALT, and AST (Patel et al., 2019).

13.2.7 Antidiabetic Activity

Clove exhibits nephroprotective and hepatoprotective effects, which may directly impact glycemic response, potentially improving or mitigating diabetic complications. A study on type 2 diabetic rats showed that administering dietary clove bud powder at 20–40 g/kg doses for 30 days gradually lowered blood glucose levels compared to non-supplemented diabetic rats (Adefegha et al., 2014). Consistent hypoglycemic results were observed in diabetic rodent models when supplemented with clove extract (Hussien et al., 2024; Kuroda et al., 2012). Moreover, in a pilot study involving healthy individuals, the supplementation of clove extract at a dosage of 250 mg once daily for 30 days showed a significant reduction in postprandial glucose levels (Mohan et al., 2019).

Clove's hypoglycaemic effects could be attributed to its inhibitory effects against the production of AGEs (Do et al., 2020), which are compounds formed due to the reaction of sugars with proteins or lipids, leading to an accumulation in tissues. This accumulation can promote oxidative stress, inflammation, and tissue damage, leading to aging-related diseases, including diabetes. Inhibiting the formation of AGEs represents a promising therapeutic approach for managing diabetes complications (Reddy et al., 2022; Salahuddin et al., 2014). Numerous studies have demonstrated that clove extract possesses inhibitory properties against AGE production, suggesting its potential to mitigate the detrimental impacts of AGE accumulation (Do et al., 2020; Sharma et al., 2023). Furthermore, clove has been found to play a role in attenuating the levels of α-amylase activity (Hussien et al., 2024; Mohan et al., 2019). Another study demonstrated that clove's free phenolic extract exhibited higher inhibitory effects against α-glucosidase activity compared to α-amylase inhibitory activity (Adefegha & Oboh, 2012). These two enzymes are involved in the digestion and absorption of complex carbohydrates into glucose molecules. Inhibiting their activity can decelerate the breakdown and absorption of carbohydrates, gradually releasing glucose into the bloodstream. This can be particularly advantageous for individuals with type 2 diabetes who encounter challenges in managing their postprandial glucose levels, as it aids in averting rapid spikes in blood sugar levels following meals (Li et al., 2022).

Furthermore, research has shown that clove may upregulate gene expression. In a study by Kuroda et al. (2012), a clove extract was found to activate PPAR-c (human peroxisome proliferator-activated receptor), which is a nuclear receptor that plays a crucial role in energy homeostasis, glucose metabolism, and lipid handling. PPAR-c is considered an important target for therapeutic interventions in metabolic disorders, including diabetes (Ruscica et al., 2019). In addition, oleanolic acid and maslinic acid derived from clove indicated inhibitory effects on the expression of ghrelin in the gastrointestinal tract. A study demonstrated that administering these derivatives decreased food intake, blood glucose levels, and ghrelin concentrations in diabetic rats (Luvuno et al., 2016). Ghrelin, an endogenous hormone predominantly synthesized in the gastrointestinal tract, is involved in metabolic disorders like obesity and type 2 diabetes. Ghrelin stimulates appetite, affects energy balance and weight regulation, and may impact insulin sensitivity and glucose metabolism (Pereira et al., 2017).

13.2.8 Anticancer Activity

Clove's pharmaceutical properties have been further investigated against the proliferation of tumors. In vitro studies have shown promising effects of clove in inhibiting the proliferation of different cancer cell lines, including pancreatic and colon cancer cells(Li et al., 2019), breast cancer cell lines (Begum et al., 2020; Kubatka et al., 2017), ovarian cancer cells (Woo et al., 2017), and cervical cancer cell lines (Das et al., 2018). Li et al. (2019) concluded that clove's anticancer activity relies on its ability to induce cell autophagy through the AMPK/ULK pathway, which is involved in cellular energy sensing and autophagy initiation. This pathway regulates autophagy, energy metabolism and cell cycle control and sensitizes cancer cells to treatments (Guo et al., 2021). Further studies have demonstrated that clove can upregulate the expression of specific pro-apoptotic proteins that regulate apoptotic cell death. Simultaneously, clove has been found to downregulate the expression of specific antiapoptotic and growth-promoting proteins. By modulating the expression of these proteins, clove shows potential in promoting apoptosis, limiting cell proliferation, and potentially hindering the progression of diseases like cancer (JECFA, 2005). Additional studies have reported that clove can suppress the production of cancer-promoting factors. These factors may include various molecules such as cytokines, growth factors, and enzymes that contribute to the development and progression of cancer (Woo et al., 2017). These findings highlight the promising chemopreventive potential of clove due to its multifaceted effects on cellular processes associated with cancer development.

13.3 HEALTH BENEFITS OF CLOVE OIL

Clove oil also has comparable health effects to clove powder, including notable antioxidant and anti-inflammatory activities. Moreover, extensive research focused on the efficacy of clove oil in various domains, such as antiprotozoal (Junior et al., 2018), larvicidal (Elzayyat et al., 2018; Osanloo et al., 2018), and leishmanicidal properties (Islamuddin et al., 2014). This section provides an overview of some of the advantageous aspects of clove oil, revealing its wide-ranging applications in fostering health and well-being.

13.3.1 Bio-Pesticide Activity

Clove can be utilized as a bio-pesticide due to its natural insecticidal properties. The essential oil derived from clove contains active compounds such as eugenol and β-caryophyllene, which have shown potent pesticidal effects against various pests. These compounds disrupt the physiological activities of pests, leading to mortality, inhibition of egg development, and repellency (Andrade-Ochoa et al., 2018; Czarnobai De Jorge et al., 2022). Research on the effectiveness of clove essential oil, mainly eugenol, has been proven against fleas as a bio-pesticide (*Ctenocephalides felis felis*). The essential oil demonstrated an LC_{50} of 5.70 μg cm^{-2} at 24 hours and 3.91 μg cm^{-2} at 48 hours for flea mortality, while eugenol exhibited an LC_{50} of 2.40 μg cm^{-2} at 24 hours and 1.40 μg cm^{-2} at 48 hours. Both treatments also displayed inhibitory effects on flea egg maturation, suggesting their potential as effective agents against fleas and hindering their egg development (Lambert et al., 2020). Another study demonstrated the insecticidal toxicity of clove essential oil against the maize weevil, *Sitophilus zeamais*. The exposed weevils exhibited notable avoidance behavior, altered mobility, and reduced respiratory rates (Haddi et al., 2015). In another study, clove oil formulations exhibited excellent efficacy, ranging from 92.47% to 100%, against the cattle tick (*Rhipicephalus microplus*), comparable to commercial chemical products (de Mello et al., 2014). Furthermore, clove essential oil showed contact toxicity against the poultry red mite (*Dermanyssus gallinae*), with a prolonged toxic effect lasting up to 4 days (Tabari et al., 2020).

13.3.2 Larvicidal Activity

The larvicidal potential of clove essential oil has been proven against various mosquito species, such as *Anopheles stephensi* (Osanloo et al., 2018), *Culex pipiens* (Elzayyat et al., 2018), *Aedes aegypti*, and *Culex quinquefasciatus* (Araujo et al., 2016; Fayemiwo et al., 2014). A gel formulation containing 20% clove essential oil demonstrated long-lasting complete protection of 4–5 hours against various other mosquito species, including *Mansonia uniformis*, *Anopheles dirus*, *Culex Gelidus*, and *Culex tritaeniorhynchus*. The gel formulation exhibited excellent repellency with an average of 97%, although a slightly lower repellency of 89% was explicitly observed against *Culex* species (Trongtokit et al., 2004). Fayemiwo et al. (2014) reported a lower repellency of the essential oil, with an 85% decrease observed within 24 hours. On the other hand, Araujo et al. (2016) found that the essential oil exhibited larvicidal activity against both resistant and susceptible populations. The larvicidal activity of clove essential oil has been attributed to the presence of eugenol. According to Osanloo et al. (2018), eugenol exhibited higher LC_{50} and LC_{90} values compared to clove essential oil. The LC_{50} value for clove essential oil was 57.49 ppm, while eugenol had an LC_{50} value of 86.96 ppm. Similarly, the LC_{90} value for clove essential oil was 93.14 ppm, whereas eugenol had an LC_{90} value of 128.18 ppm.

The effectiveness of clove as a mosquito repellent is believed to stem from various combined mechanisms (Figure 13.3). Among these mechanisms, clove's strong and distinct odor can mask or camouflage the attractant signals emitted by humans or other mosquito-attracting targets. This characteristic of clove can make it challenging for mosquitoes to locate and target their hosts by masking these attractive odors. Another possible mechanism relies on volatile compounds found in *S. aromaticum*, which can stimulate mosquitoes' sensory receptors, resulting in discomfort and disturbance to their regular feeding and host-seeking activities. This sensory irritation can effectively discourage mosquitoes from landing on or biting individuals. Furthermore, certain compounds found in clove have been shown to elicit toxic or physiological effects on mosquitoes upon direct contact with their bodies. These effects can potentially impair vital functions, disrupt the nervous system, or interfere with reproductive capabilities

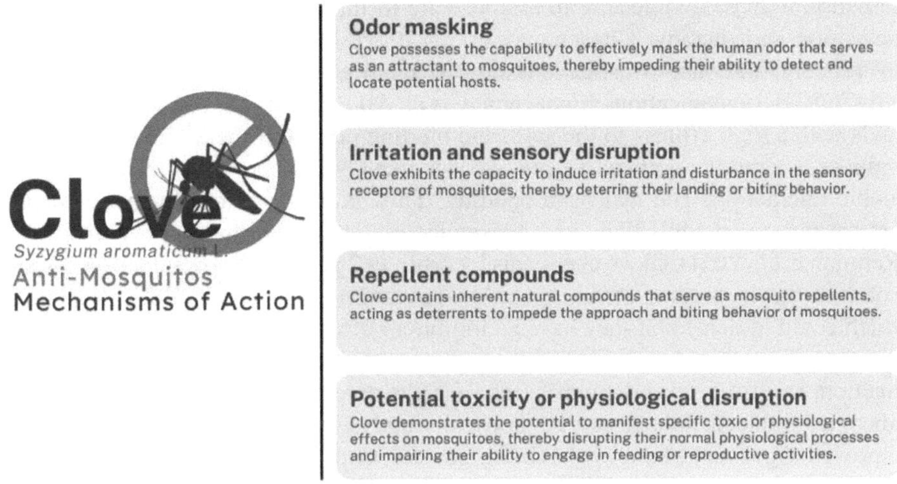

FIGURE 13.3 Mechanisms of action for clove as an antimosquito agent.

(Andrade-Ochoa et al., 2018; Czarnobai De Jorge et al., 2022). In particular, some of these mechanisms may be attributed to the anti-AChE activity of *S. aromaticum*, as suggested by Affonso et al. (2018). By inhibiting AchE, the breakdown of the neurotransmitter acetylcholine is prevented, leading to its accumulation. This accumulation can disrupt the normal functioning of the mosquitos' nervous system, ultimately causing paralysis and leading to their eventual death.

13.4 CLOVE IN COVID-19 OUTBREAK

Clove's anti-inflammatory, immunostimulatory, and antithrombotic qualities, including its traditional use in respiratory illnesses and its action against several viruses, make it a promising COVID-19 treatment. Clove, *Eucalyptus globulus*, *Cymbopogon citratus*, *Zingiber officinale*, and other economically and widely available medicinal plants prevent and treat SARS-CoV-2 (Mputu Kanyinda, 2020). As a medicinal plant, cloves can prevent and treat COVID-19, according to Kanyinda, who reported a demonstrated impact when used early (Mputu Kanyinda, 2020). The process involved boiling cloves and other plant materials in water for 15 minutes to make a decoction. Thus, patients breathe these volatile active compounds resulting from the decoction for 5 minutes. Notably, studies have been carried out in Morocco and India, two countries that have seen relatively minor pandemic effects (Samaddar et al., 2020), to determine the several COVID-19 home cures that the local population has applied, which have comprised a wide variety of spices and herbs. Among the most intriguing findings, almost 93% of the Indians surveyed held the belief that spices could aid in the treatment of COVID-19 and other viral diseases, as well as strengthen the immune system. Along with cinnamon, ginger, black pepper, garlic, neem, and basil, cloves are among the most often utilized spices and herbs in the area under examination during the present COVID-19 pandemic (Singh et al., 2021).

Herbalists from Salé Prefecture in Morocco are further utilizing cloves for COVID-19 prevention and therapy (Chaachouay et al., 2021). From a molecular perspective, several computational investigations suggested clove's phytocompounds as effective anti-COVID-19 medications (Chaachouay et al., 2021; Joshi et al., 2020); kaempferol binds with a high affinity to the substrate binding pocket of the SARS-CoV-2 main protease, interacting with active site residues like Cys145 and His41 through hydrophobic interactions and hydrogen bonding. This finding provides further evidence that natural compounds like clove flavonoids may be able to inhibit SARS-CoV-2. (Rehman et al., 2021). Clove chemicals bicornin (9.2 kcal/mol) and biflorin (8.5 kcal/mol) have strong binding affinities to Mpro, according to molecular docking studies, which could indicate that they have an inhibitory effect (Rehman et al., 2021).

After controlling baseline scores, employment status, biomass index, olfactory function, and time since diagnosis, inhaling the essential oil blend for 2 weeks significantly reduced fatigue scores. Subscale analysis showed that the intervention improved vigor and global, behavioral, general, and mental weariness. This study shows that a customized aromatherapy combination can boost energy in COVID-19-exhausted women (Hawkins et al., 2022).

Clove's traditional use to treat respiratory ailments, experimentally proven activity against various viruses and anti-inflammatory, immunostimulatory, and anti-thrombotic properties suggest that cloves and their phytochemical constituents may benefit the fight against COVID-19. Additionally, clove essential oil has shown significant antibacterial activity against infections in immunosuppressed hospitalized patients (Chaieb et al., 2007), proposing its use to prevent subsequent bacterial infections in COVID-19 patients (Bahramsoltani & Rahimi, 2020). Finally, cloves, a valued spice used in locations where the novel coronavirus is less prevalent than in Western countries, may help prevent and treat COVID-19. We need more molecular insights into clove phytochemicals' specific interactions with SARS-CoV-2 protein targets and future clinical data on cloves and their constituents' activity in COVID-19 patients to develop new drugs and effective treatment protocols.

13.5 SAFETY AND TOXICOLOGY ISSUES

The U.S. Food and Drug Administration approved clove and its derivatives (i.e., eugenol) as generally recognized as safe (GRAS) for use as a food additive (GRAS–184.1257) (FDA, 2021). Numerous studies have investigated the safety of clove. Consistent findings have indicated its promising safety profile. Consequently, clove oil was reported to be safe as high as 1,500 ppm. Acute and sub-chronic toxicity studies on rat models at 1,000 mg/kg body weight showed no toxicological alterations in biological parameters (Batiha et al., 2020). Additional non-genotoxic results were also observed, supporting the favorable safety profile of cloves (Vijayasteltar et al., 2016). Another study examined mouse models' 35-day spermatogenic cycles after prolonged oral clove hexane extract treatment. The therapy did not cause systemic toxicity at 15 mg, 30 mg, or 60 mg/kg body weight (Mishra & Singh, 2008). Consistent findings across multiple studies demonstrate the absence of any toxicological effects associated with clove administration (Hu et al., 2022; Issac et al., 2015). Nevertheless, numerous studies have demonstrated the overall safety of eugenol when used in appropriate concentrations and applications (Chaieb et al., 2007; Haro-González et al., 2021).

However, excessive consumption or intravenous administration of clove oil can cause lethal adverse effects such as Acute Respiratory Distress Syndrome, Fulminant Hepatic Failure, and Central Nervous System Decline, as well as harming human cells (Haddi et al., 2015). Eugenol has been found to exhibit cytotoxic effects on human dermal fibroblasts, endothelial cells, and dental pulp fibroblasts (Aburel et al., 2021). Usually, there is severe hypoglycemia, respiratory depression, and acidosis. The deadly oral dose is 3.75 g per kg of body weight. However, it is worth noting that contrasting findings have also indicated its toxicity at lower doses, while no such incidence was observed at higher doses (Gooderham et al., 2020). Eugenol excesses in herbal medicines have caused serious liver and kidney damage, mostly in children. Overdoses (10–30 mL clove oil) cause agitation, loss of consciousness, and coma (LiverTox, 2019).

Furthermore, eugenol may exert a more pronounced influence on the growth and viability of beneficial bacteria in the intestinal microbiota when compared to specific harmful pathogens. The beneficial commensal bacterium *Faecalibacterium prausnitzii* was found to exhibit greater sensitivity to eugenol, even at comparable or lower concentrations, compared to pathogens including *Escherichia coli*, *Salmonella enteritidis*, *Salmonella typhimurium*, *Clostridium difficile*, and *Clostridium perfringens* (Hu et al., 2018). These findings highlight the importance of considering the potential effects of eugenol on the delicate balance of the gut microbiota.

Eugenol is listed as a flavoring substance in the European Commission Implementing Regulation (EUCIR, 2012) besides the Joint FAO/WHO Expert Committee on Food Additives (JECFA, 2005). The European Food Safety Authority (EFSA) explains that the use of clove oils in animal feed is of low risk as follows: up to 50 mg/kg for clove bud oil, up to 100 mg/kg for clove leaf oil, and up to 20 mg/kg for a β-caryophyllene-rich fraction of clove leaf oil (Additives et al., 2023). Nevertheless, it is advised to be cautious regarding the consumption of clove buds and their derivatives.

13.6 CONCLUSIONS

Clove and its oil as GRAS offer various health benefits, including antimicrobial, anti-inflammatory, antioxidant, analgesic, antidiabetic, and anticancer properties. They are recognized for their complex chemical composition, containing eugenol, eugenol acetate, β-caryophyllene, α-humulene, and other bioactive compounds. In addition, the neuroprotection mechanism of action concluded that clove and its essential oil have an inhibitory effect on acetylcholinesterase, reduction of γ-secretase expression, activation of SIRT1, and antiglycation potential. Moreover, clove possesses bio-pesticide activity via odor-masking capability, which attracts mosquitoes, exhibits irritation and disturbance in mosquitoes' sensory receptors, contains repellent components, and has potential toxicity or physiological disruption. Furthermore, cloves were effectively used in COVID-19 treatments. But still, more molecular insights into clove phytochemicals' specific interactions with SARS-CoV-2 protein targets and future clinical data on cloves and their constituents' activity in COVID-19 patients to develop new drugs and effective treatment protocols are urgently needed.

REFERENCES

Abtahi-Eivari, S.-H., Shokoohi, M., Ghorbani, M., Halimi, M., Hajizadeh, H., Pourlak, T., Bahrami, J., & Ghoreishi, Z. (2021). Effects of hydroalcoholic extracts of cloves (*Syzygium aromaticum*) on the serum biomarkers, antioxidant status, and histopathological changes of kidneys in diabetic rats. *Crescent Journal of Medical & Biological Sciences*, *8*(4).

Aburel, O. M., Pavel, I. Z., Dănilă, M. D., Lelcu, T., Roi, A., Lighezan, R., Muntean, D. M., & Rusu, L. C. (2021). Pleiotropic effects of eugenol: The good, the bad, and the unknown. *Oxidative Medicine and Cellular Longevity*, *2021*, 3165159. https://doi.org/10.1155/2021/3165159

Additives, E. P. O., Products or Substances used in Animal, F., Bampidis, V., Azimonti, G., Bastos, M. d. L., Christensen, H., Durjava, M., Kouba, M., López-Alonso, M., López Puente, S., Marcon, F., Mayo, B., Pechová, A., Petkova, M., Ramos, F., Sanz, Y., Villa, R. E., Woutersen, R., Brantom, P.,. . . Dusemund, B. (2023). Safety and efficacy of feed additives consisting of essential oils derived from the flower buds or the leaves of *Syzygium aromaticum* (L.) Merr. & L.M. Perry (clove bud oil and clove leaf oils) for all animal species (FEFANA asbl). *EFSA Journal*, *21*(7), e08183. https://doi.org/https://doi.org/10.2903/j.efsa.2023.8183

Adefegha, S. A., & Oboh, G. (2012). In vitro inhibition activity of polyphenol-rich extracts from *Syzygium aromaticum* (L.) Merr. & Perry (Clove) buds against carbohydrate hydrolyzing enzymes linked to type 2 diabetes and Fe2+-induced lipid peroxidation in rat pancreas. *Asian Pacific Journal of Tropical Biomedicine*, *2*(10), 774–781.

Adefegha, S. A., Oboh, G., Adefegha, O. M., Boligon, A. A., & Athayde, M. L. (2014). Antihyperglycemic, hypolipidemic, hepatoprotective and antioxidative effects of dietary clove (*Szyzgium aromaticum*) bud powder in a high-fat diet/streptozotocin-induced diabetes rat model. *Journal of the Science of Food and Agriculture*, *94*(13), 2726–2737.

Affonso, R. S., Lima, J. A., Lessa, B. M., Caetano, J. V., Obara, M. T., Nóbrega, A. B., Nepovimova, E., Musilek, K., Kuca, K., & Slana, G. B. (2018). Quantification through TLC-densitometric analysis, repellency and anticholinesterase activity of the homemade extract of Indian cloves. *Biomedical Chromatography*, *32*(2), e4096.

Ajiboye, T., Mohammed, A., Bello, S., Yusuf, I., Ibitoye, O., Muritala, H., & Onajobi, I. (2016). Antibacterial activity of *Syzygium aromaticum* seed: Studies on oxidative stress biomarkers and membrane permeability. *Microbial Pathogenesis*, *95*, 208–215.

Ajobiewe, H., Elisha, E., Ibrahim, A., Ajobiewe, J., Salami, A., UmejiLC, U. P., Yashim, A., & Alau, K. (2022). Antimicrobial activity of clove plant flower bud extract (*Syzygium aromaticum*) on *Escherichia coli*. *Scholars Journal of Applied Medical Sciences*, *5*, 673–684.

Ali, R., Khamis, T., Enan, G., El-Didamony, G., Sitohy, B., & Abdel-Fattah, G. (2022). The healing capability of clove flower extract (CFE) in streptozotocin-induced (STZ-induced) diabetic rat wounds infected with multidrug resistant bacteria. *Molecules*, *27*(7), 2270.

Aljarari, R. (2023). Neuroprotective effects of a combination of Boswellia papyrifera and *Syzygium aromaticum* on AlCl 3 induced Alzheimer's disease in male albino rat. *Brazilian Journal of Biology*, *83*, e272466.

Al-Zereini, W. A., Al-Trawneh, I. N., Al-Qudah, M. A., TumAllah, H. M., Abudayeh, Z. H., & Hijazin, T. (2023). Antibacterial, antioxidant, and cytotoxic activities of *Syzygium aromaticum* (L.) Merr. & Perry essential oil with identification of its chemical constituents. *Zeitschrift für Naturforschung C*, *78*(3–4), 105–112.

Andrade-Ochoa, S., Sánchez-Aldana, D., Chacón-Vargas, K. F., Rivera-Chavira, B. E., Sánchez-Torres, L. E., Camacho, A. D., Nogueda-Torres, B., & Nevárez-Moorillón, G. V. (2018). Oviposition deterrent and larvicidal and pupaecidal activity of seven essential oils and their major components against *Culex quinquefasciatus* Say (Diptera: Culicidae): Synergism–antagonism effects. *Insects*, *9*(1), 25.

Araujo, A. F. D. O., Ribeiro-Paes, J. T., Deus, J. T. D., Cavalcanti, S. C. D. H., Nunes, R. D. S., Alves, P. B., & Macoris, M. D. L. D. G. (2016). Larvicidal activity of *Syzygium aromaticum* (L.) Merr and *Citrus sinensis* (L.) Osbeck essential oils and their antagonistic effects with temephos in resistant populations of *Aedes aegypti. Memórias do Instituto Oswaldo Cruz, 111*, 443–449.

Atawodi, S., Atawodi, J., Pfundstein, B., Spiegelhalder, B., Bartsch, H., & Owen, R. (2011). Assessment of the polyphenol components and in vitro antioxidant properties of *Syzygium aromaticum* (L.) Merr. & Perry. *Electronic Journal of Environmental, Agricultural & Food Chemistry, 10*(3).

Bahramsoltani, R., & Rahimi, R. (2020). An evaluation of traditional persian medicine for the management of SARS-CoV-2. *Frontiers in Pharmacology, 11*, 571434. https://doi.org/10.3389/fphar.2020.571434

Bainaud, M., Try, M., & Zaidan, M. (2023). Nephroprotection: General principles and application to the patients with cancers-when nephroprotection is essential for oncological care plan. *Bulletin du Cancer*, In Press.

Bansal, V., Gupta, M., Bhaduri, T., Shaikh, S. A., Sayed, F. R., Bansal, V., & Agrawal, A. (2019). Assessment of Antimicrobial Effectiveness of Neem and Clove Extract Against *Streptococcus mutans* and *Candida albicans*: An: In vitro: Study. *Nigerian Medical Journal, 60*(6), 285–289.

Barakat, H. (2014). Composition, antioxidant, antibacterial activities and mode of action of clove (Syzygium aromaticum L.) buds essential oil. *British Journal of Applied Science & Technology, 4*(13), 1934–1951.

Batiha, G. E.-S., Alkazmi, L. M., Wasef, L. G., Beshbishy, A. M., Nadwa, E. H., & Rashwan, E. K. (2020). *Syzygium aromaticum* L.(Myrtaceae): Traditional uses, bioactive chemical constituents, pharmacological and toxicological activities. *Biomolecules, 10*(2).

Batiha, G. E.-S., Beshbishy, A. M., Tayebwa, D. S., Shaheen, H. M., Yokoyama, N., & Igarashi, I. (2019). Inhibitory effects of *Syzygium aromaticum* and *Camellia sinensis* methanolic extracts on the growth of Babesia and Theileria parasites. *Ticks and Tick-Borne Diseases, 10*(5), 949–958.

Begum, S., Ali, S. N., Farooq, A. D., Siddiqui, F., Siddiqui, B. S., Ayub, A., & Fatima, M. (2020). Volatile constituents and in vitro activity of *Syzygium aromaticum* flower buds (clove) against human cancer cell lines. *Pakistan Journal of Pharmaceutical Sciences, 33*(6).

Cai, L., & Wu, C. D. (1996). Compounds from *Syzygium aromaticum* possessing growth inhibitory activity against oral pathogens. *Journal of Natural Products, 59*(10), 987–990.

Chaachouay, N., Douira, A., & Zidane, L. (2021). COVID-19, prevention and treatment with herbal medicine in the herbal markets of Salé Prefecture, North-Western Morocco. *European Journal of Integrative Medicine, 42*, 101285. https://doi.org/10.1016/j.eujim.2021.101285

Chaieb, K., Hajlaoui, H., Zmantar, T., Kahla-Nakbi, A. B., Rouabhia, M., Mahdouani, K., & Bakhrouf, A. (2007). The chemical composition and biological activity of clove essential oil, *Eugenia caryophyllata (Syzigium aromaticum* L. Myrtaceae): A short review. *Phytotherapy Research, 21*(6), 501–506. https://doi.org/10.1002/ptr.2124

Chniguir, A., Zioud, F., Marzaioli, V., El-Benna, J., & Bachoual, R. (2019). *Syzygium aromaticum* aqueous extract inhibits human neutrophils myeloperoxidase and protects mice from LPS-induced lung inflammation. *Pharmaceutical Biology, 57*(1), 55–63.

Czarnobai De Jorge, B., Hummel, H. E., & Gross, J. (2022). Repellent activity of clove essential oil volatiles and development of nanofiber-based dispensers against pear psyllids (Hemiptera: Psyllidae). *Insects, 13*(8), 743.

Das, A., Harshadha, K., SK, D. K., & Jayaprakash, B. (2018). Evaluation of therapeutic potential of eugenol-a natural derivative of Syzygium aromaticum on cervical cancer. *Asian Pacific Journal of Cancer Prevention: APJCP, 19*(7), 1977.

de Mello, V., Prata, M. C. D. A., da Silva, M. R., Daemon, E., da Silva, L. S., Guimarães, F. D. G., de Mendonça, A. É., Folly, E., Vilela, F. M. P., & do Amaral, L. H. (2014). Acaricidal properties of the formulations based on essential oils from *Cymbopogon winterianus* and *Syzygium aromaticum* plants. *Parasitology Research, 113*, 4431–4437.

Devi, K. P., Sakthivel, R., Nisha, S. A., Suganthy, N., & Pandian, S. K. (2013). Eugenol alters the integrity of cell membrane and acts against the nosocomial pathogen Proteus mirabilis. *Archives of Pharmacal Research, 36*, 282–292.

Didehdar, M., Chegini, Z., & Shariati, A. (2022). Eugenol: A novel therapeutic agent for the inhibition of Candida species infection. *Frontiers in Pharmacology, 13*, 872127.

Do, M. H., Choi, J., Kim, Y., Ha, S. K., Yoo, G., & Hur, J. (2020). Syzygium aromaticum reduces diabetes-induced glucotoxicity via the NRF2/Glo1 pathway. *Planta Medica, 86*(12), 876–883.

El-Samawaty, A. E.-R. M., El-Wakil, D. A., Alamery, S., & Mahmoud, M. M. (2021). Potency of plant extracts against Penicillium species isolated from different seeds and fruits in Saudi Arabia. *Saudi Journal of Biological Sciences, 28*(6), 3294–3302.

Elzayyat, E., Elleboudy, N., Moustafa, A., & Ammar, A. (2018). Insecticidal, oxidative, and genotoxic activities of *Syzygium aromaticum* and *Eucalyptus globulus* on Culex pipiens adults and larvae. *Türkiye Parazitolojii Dergisi, 42*(3), 213.

EUCIR. (2012). Commission Implementing Regulation (EU) No 872/2012 of 1 October 2012 adopting the list of flavouring substances provided for by Regulation (EC) No 2232/96 of the European Parliament and of the Council, introducing it in Annex I to Regulation (EC) No 1334/2008 of the European Parliament and of the Council and repealing Commission Regulation (EC) No 1565/2000 and Commission Decision 1999/217/EC Text with EEA relevance. In: *EUR-Lex*. https://eur-lex. europa. eu/legal-content/EN/TXT.

Fayemiwo, K. A., Adeleke, M. A., Okoro, O. P., Awojide, S. H., & Awoniyi, I. O. (2014). *Larvicidal efficacies* and chemical composition of essential oils of *Pinus sylvestris* and *Syzygium aromaticum* against mosquitoes. *Asian Pacific Journal of Tropical Biomedicine, 4*(1), 30–34.

FDA, U. (2021). *Guidance for industry: Preparation of food contact notifications for food contact substances (toxicology recommendations) 2002*. US Food & Drug Administration.

Floege, J., Lüscher, B., & Müller-Newen, G. (2012). Cytokines and inflammation. *European Journal of Cell Biology, 91*(6–7), 427–427.

Gao, X., Luo, F., & Zhao, H. (2021). Cloves regulate Na+-K+-ATPase to exert antioxidant effect and inhibit UVB light-induced skin damage in mice. *Oxidative Medicine and Cellular Longevity, 2021*.

Gooderham, N. J., Cohen, S. M., Eisenbrand, G., Fukushima, S., Guengerich, F. P., Hecht, S. S., Rietjens, I. M., Rosol, T. J., Davidsen, J. M., & Harman, C. L. (2020). FEMA GRAS assessment of natural flavor complexes: Clove, cinnamon leaf and West Indian bay leaf-derived flavoring ingredients. *Food and Chemical Toxicology, 145*, 111585.

Guo, L., Cui, J., Wang, H., Medina, R., Zhang, S., Zhang, X., Zhuang, Z., & Lin, Y. (2021). Metformin enhances anti-cancer effects of cisplatin in meningioma through AMPK-mTOR signaling pathways. *Molecular Therapy-Oncolytics, 20*, 119–131.

Haddi, K., Oliveira, E. E., Faroni, L. R., Guedes, D. C., & Miranda, N. N. (2015). Sublethal exposure to clove and cinnamon essential oils induces hormetic-like responses and disturbs behavioral and respiratory responses in *Sitophilus zeamais* (Coleoptera: Curculionidae). *Journal of Economic Entomology, 108*(6), 2815–2822.

Haro-González, J. N., Castillo-Herrera, G. A., Martínez-Velázquez, M., & Espinosa-Andrews, H. (2021). Clove essential oil (*Syzygium aromaticum* L. Myrtaceae): Extraction, chemical composition, food applications, and essential bioactivity for human health. *Molecules, 26*(21), 6387.

Hawkins, J., Hires, C., Keenan, L., & Dunne, E. (2022). Aromatherapy blend of thyme, orange, clove bud, and frankincense boosts energy levels in post-COVID-19 female patients: A randomized, double-blinded, placebo controlled clinical trial. *Complementary Therapies in Medicine, 67*, 102823. https://doi.org/https://doi.org/10.1016/j.ctim.2022.102823

Hu, Q., Zhou, M., & Wei, S. (2018). Progress on the antimicrobial activity research of clove oil and eugenol in the food antisepsis field. *Journal of Food Science, 83*(6), 1476–1483.

Hu, Z.-J., Yang, J.-W., Chen, Z.-H., Chang, C., Ma, Y.-P., Li, N., Deng, M., Mao, G.-L., Bao, Q., & Deng, S.-Z. (2022). Exploration of clove bud (*Syzygium aromaticum*) essential oil as a novel attractant against *Bactrocera dorsalis* (Hendel) and its safety evaluation. *Insects, 13*(10), 918.

Hussien, A. G., Fyiad, A. A. E.-A., Ali, M. M., & Mahmoud, A. E. E.-D. (2024). The influence of *Syzygium aromaticum* ethanolic extract against streptozotocin-induced diabetic rats. *Egyptian Journal of Chemistry, 67*(2), 357–369.

Islamuddin, M., Sahal, D., & Afrin, F. (2014). Apoptosis-like death in *Leishmania donovani* promastigotes induced by eugenol-rich oil of *Syzygium aromaticum. Journal of Medical Microbiology, 63*(1), 74–85.

Issac, A., Gopakumar, G., Kuttan, R., Maliakel, B., & Krishnakumar, I. (2015). Safety and anti-ulcerogenic activity of a novel polyphenol-rich extract of clove buds (*Syzygium aromaticum* L). *Food & Function, 6*(3), 842–852.

JECFA. (2005). *Joint FAO/WHO Expert Committee on Food Additives. Specifications for Flavourings, Eugenol.* www.fao.org/food/food-safety-quality/scientific-advice/jecfa/jecfa-flav/details/en/c/1520/(1460–2180).

Jeon, S. J., Kim, E., Lee, J. S., Oh, H. K., Zhang, J., Kwon, Y., Jang, D. S., & Ryu, J. H. (2017). Maslinic acid ameliorates NMDA receptor blockade-induced schizophrenia-like behaviors in mice. *Neuropharmacology, 126*, 168–178.

Ji, L. L., & Yeo, D. (2021). Oxidative stress: An evolving definition. *Faculty Reviews, 10.*

Johannah, N., Renny, R., Gopakumar, G., Maliakel, B., Sureshkumar, D., & Krishnakumar, I. (2015). Beyond the flavour: A de-flavoured polyphenol rich extract of clove buds (*Syzygium aromaticum* L) as a novel dietary antioxidant ingredient. *Food & Function, 6*(10), 3373–3382.

Jose, S. P., Ratheesh, M., Asha, S., Krishnakumar, I., Sandya, S., & Girish, K. B. (2018). Hepato-protective effect of clove bud polyphenols (*Syzygium aromaticum* L.) (Clovinol®) by modulating alcohol induced oxidative stress and inflammation. *Journal of Field Robotics, 7*, 10–20.

Joshi, T., Joshi, T., Sharma, P., Mathpal, S., Pundir, H., Bhatt, V., & Chandra, S. (2020). In silico screening of natural compounds against COVID-19 by targeting Mpro and ACE2 using molecular docking. *European Review for Medical and Pharmacological Sciences, 24*(8), 4529–4536. https://doi.org/10.26355/eurrev_202004_21036

Junior, G. Z., Massago, M., Kian, D., & de Ornelas Toledo, M. J. (2018). Efficacy of essential oil of *Syzygium aromaticum* alone and in combination with benznidazole on murine oral infection with Trypanosoma cruzi IV. *Experimental Parasitology, 185*, 92–97.

Kadri, Y., Nciri, R., Bardaa, S., Brahmi, N., Saber, S., Harrath, A. H., Aldahmash, W., Alwasel, S., Mohany, M., & El Feki, A. (2019). *Syzygium Aromaticum* alleviates cerium chloride-induced neurotoxic effect in the adult mice. *Toxicology Mechanisms and Methods, 29*(1), 26–34.

Kazi, S., Abbasi, P., & Arain, A. A. (2016). *Syzygium aromaticum*: A potential hepatoprotective agent. *JSZMC, 7*(2), 956–959.

Khan, I., Preeti, K., Kumar, R., Khatri, D. K., & Singh, S. B. (2022). Activation of SIRT1 by silibinin improved mitochondrial health and alleviated the oxidative damage in experimental diabetic neuropathy and high glucose-mediated neurotoxicity. *Archives of Physiology and Biochemistry*, 1–17. https://doi.org/10.1080/13813455.2022.2108454

Kubatka, P., Uramova, S., Kello, M., Kajo, K., Kruzliak, P., Mojzis, J., Vybohova, D., Adamkov, M., Jasek, K., & Lasabova, Z. (2017). Antineoplastic effects of clove buds (*Syzygium aromaticum* L.) in the model of breast carcinoma. *Journal of Cellular and Molecular Medicine*, *21*(11), 2837–2851.

Kumar, S., & Walter, J. (2011). Phosphorylation of amyloid beta (Aβ) peptides–A trigger for formation of toxic aggregates in Alzheimer's disease. *Aging (Albany NY)*, *3*(8), 803.

Kuroda, M., Mimaki, Y., Ohtomo, T., Yamada, J., Nishiyama, T., Mae, T., Kishida, H., & Kawada, T. (2012). Hypoglycemic effects of clove (Syzygium aromaticum flower buds) on genetically diabetic KK-A y mice and identification of the active ingredients. *Journal of Natural Medicines*, *66*, 394–399.

Lambert, M. M., Campos, D. R., Borges, D. A., de Avelar, B. R., Ferreira, T. P., Cid, Y. P., Boylan, F., Scott, F. B., de Almeida Chaves, D. S., & Coumendouros, K. (2020). Activity of *Syzygium aromaticum* essential oil and its main constituent eugenol in the inhibition of the development of Ctenocephalides felis felis and the control of adults. *Veterinary Parasitology*, *282*, 109126.

Lee, J. H., Park, M., Jung, K., Hong, G., Lee, H. L., Kim, D.-W., Kim, C.-E., & Kang, K. S. (2021). Identification of gallic acid as an active ingredient of *Syzygium aromaticum* against tacrolimus-induced damage in renal epithelial LLC-PK1 cells and rat kidney. *Bioorganic & Medicinal Chemistry Letters*, *41*, 128012.

Lee, K.-G., & Shibamoto, T. (2001). Inhibition of malonaldehyde formation from blood plasma oxidation by aroma extracts and aroma components isolated from clove and eucalyptus. *Food and Chemical Toxicology*, *39*(12), 1199–1204.

Li, C., Xu, H., Chen, X., Chen, J., Li, X., Qiao, G., Tian, Y., Yuan, R., Su, S., & Liu, X. (2019). Aqueous extract of clove inhibits tumor growth by inducing autophagy through AMPK/ULK pathway. *Phytotherapy Research*, *33*(7), 1794–1804.

Li, X., Bai, Y., Jin, Z., & Svensson, B. (2022). Food-derived non-phenolic α-amylase and α-glucosidase inhibitors for controlling starch digestion rate and guiding diabetes-friendly recipes. *Lwt-Food Science and Technology*, *153*, 112455.

LiverTox. (2019). *Clinical and Research Information on Drug-Induced Liver Injury* [Internet]. Bethesda (MD): National Institute of Diabetes and Digestive and Kidney Diseases; 2012–. Eugenol (Clove Oil) [Updated 2019 Oct 28]. Available from: www.ncbi.nlm.nih.gov/books/NBK551727. In.

Luvuno, M., Mbongwa, H. P., & Khathi, A. (2016). The effects of *Syzygium aromaticum*-derived triterpenes on gastrointestinal ghrelin expression in streptozotocin-induced diabetic rats. *African Journal of Traditional, Complementary and Alternative Medicines*, *13*(4), 8–14.

Madlala, H. P., Masola, B., Singh, M., & Musabayane, C. T. (2012). The effects of *syzygium aromaticum*-derived oleanolic acid on kidney function of male sprague–dawley rats and on kidney and liver cell lines. *Renal Failure*, *34*(6), 767–776.

Marchese, A., Barbieri, R., Coppo, E., Orhan, I. E., Daglia, M., Nabavi, S. F., Izadi, M., Abdollahi, M., Nabavi, S. M., & Ajami, M. (2017). Antimicrobial activity of eugenol and essential oils containing eugenol: A mechanistic viewpoint. *Critical Reviews in Microbiology*, *43*(6), 668–689.

Mishra, R. K., & Singh, S. K. (2008). Safety assessment of *Syzygium aromaticum* flower bud (clove) extract with respect to testicular function in mice. *Food and Chemical Toxicology*, *46*(10), 3333–3338.

Mohan, R., Jose, S., Mulakkal, J., Karpinsky-Semper, D., Swick, A. G., & Krishnakumar, I. (2019). Water-soluble polyphenol-rich clove extract lowers pre-and post-prandial blood glucose levels in healthy and prediabetic volunteers: An open label pilot study. *BMC Complementary and Alternative Medicine*, *19*, 1–9.

Mondal, A., Bose, S., Banerjee, S., & Pal, D. (2021). Role of γ-secretase inhibitors for the treatment of diverse disease conditions through inhibition of notch signaling pathway. *Current Drug Targets, 22*(15), 1799–1807.

Mputu Kanyinda, J.-N. (2020). Coronavirus (COVID-19): A protocol for prevention and treatment (Covalyse®). *European Journal of Medical and Health Sciences, 2*(3). https://doi.org/10.24018/ejmed.2020.2.4.340

Negm, A., Abbas, R. A., Awadalla, A., Abd El-Azim, A. O., & Elkhawaga, O.-A. (2020). *Syzygium Aromaticum* ameliorates oxidative stress and fibrosis in adenine-induced chronic kidney disease in rats. *Journal of Applied Biological Sciences, 14*(3), 322–338.

Osanloo, M., Sedaghat, M. M., Esmaeili, F., & Amani, A. (2018). Larvicidal activity of essential oil of *Syzygium aromaticum* (Clove) in comparison with its major constituent, eugenol, against *Anopheles stephensi. Journal of Arthropod-Borne Diseases, 12*(4), 361.

Ounaceur, L. S., Messarah, M., Lankar, A., Touaibia, D., Ghadab, H., Garri, Z., Boudjil, S., Djeffel, K. B., Atoui, L., & Ounaceur, A. (2024). In vivo determination of the anti-inflammatory and antioxidant effects of the aqueous extract of *Syzygium aromaticum* (clove) in an asthmatic rat model. *Cellular and Molecular Biology, 70*(3), 29–39.

Patel, U. D., Shah, F. C., & Jain, N. K. (2019). Hepatoprotective effect of methanolic extract of *Syzygium aromaticum* against hydralazine induced toxicity: An in vitro study. *The Pharma Innovation Journal, 8*(6), 537–541.

Pereira, J. A. D. S., Silva, F. C. D., & de Moraes-Vieira, P. M. M. (2017). The impact of ghrelin in metabolic diseases: An immune perspective. *Journal of Diabetes Research, 2017.*

Reddy, V. P., Aryal, P., & Darkwah, E. K. (2022). Advanced glycation end products in health and disease. *Microorganisms, 10*(9), 1848.

Rehman, M. T., AlAjmi, M. F., & Hussain, A. (2021). Natural compounds as inhibitors of SARS-CoV-2 main protease (3CLpro): A molecular docking and simulation approach to combat COVID-19. *Current Pharmaceutical Design, 27*(33), 3577–3589.

Ruscica, M., Busnelli, M., Runfola, E., Corsini, A., & Sirtori, C. R. (2019). Impact of PPAR-alpha polymorphisms—the case of metabolic disorders and Atherosclerosis. *International Journal of Molecular Sciences, 20*(18), 4378.

Salahuddin, P., Rabbani, G., & Khan, R. H. (2014). The role of advanced glycation end products in various types of neurodegenerative disease: A therapeutic approach. *Cellular & Molecular Biology Letters, 19*, 407–437.

Saleh, R. O., Dheyab, A. S., Hadi, B. H., Hasan, R. N., & Jasim, S. A. (2024). Effect of ethanolic extract of *Syzygium aromaticum* plant against *Enterococcus faecalis* isolated from women with urinary tract infections. *Archives of Clinical Infectious Diseases, 19*(1).

Samaddar, A., Gadepalli, R., Nag, V. L., & Misra, S. (2020). The enigma of low COVID-19 fatality rate in India [Perspective]. *Frontiers in Genetics, 11*, 854. https://doi.org/10.3389/fgene.2020.00854

Santativongchai, P., Tulayakul, P., & Jeon, B. (2023). Enhancement of the antibiofilm activity of nisin against *Listeria monocytogenes* using food plant extracts. *Pathogens, 12*(3), 444.

Sharma, H., Kim, D. Y., Shim, K. H., Sharma, N., & An, S. S. A. (2023). Multi-targeting neuroprotective effects of *Syzygium aromaticum* bud extracts and their key phytocompounds against neurodegenerative diseases. *International Journal of Molecular Sciences, 24*(9), 8148.

Shekhar, S., Yadav, Y., Singh, A. P., Pradhan, R., Desai, G. R., Dey, A., & Dey, S. (2018). Neuroprotection by ethanolic extract of *Syzygium aromaticum* in Alzheimer's disease like pathology via maintaining oxidative balance through SIRT1 pathway. *Experimental Gerontology, 110*, 277–283.

Singh, N. A., Kumar, P., Jyoti, & Kumar, N. (2021). Spices and herbs: Potential antiviral preventives and immunity boosters during COVID-19. *Phytotherapy Research, 35*(5), 2745–2757.

Tabari, M. A., Rostami, A., Khodashenas, A., Maggi, F., Petrelli, R., Giordani, C., Tapondjou, L. A., Papa, F., Zuo, Y., & Cianfaglione, K. (2020). Acaricidal activity, mode of action, and persistent efficacy of selected essential oils on the poultry red mite (*Dermanyssus gallinae*). *Food and Chemical Toxicology, 138*, 111207.

Taguchi, Y., Ishibashi, H., Takizawa, T., Inoue, S., Yamaguchi, H., & Abe, S. (2005). Protection of oral or intestinal candidiasis in mice by oral or intragastric administration of herbal food, clove (*Syzygium aromaticum*). *Nippon Ishinkin Gakkai Zasshi, 46*(1), 27–33.

Taroq, A., Bakour, M., El Atki, Y., El Kamari, F., Aouam, I., & Zahra, F. (2021). Comparative study of the effects of *Laurus nobilis* and *Syzygium Aromaticum* aqueous extracts on urine volume and renal function in rats. *International Journal of Pharmaceutical Sciences and Research, 12*(2), 776–782.

Trongtokit, Y., Rongsriyam, Y., Komalamisra, N., Krisadaphong, P., & Apiwathnasorn, C. (2004). Laboratory and field trial of developing medicinal local Thai plant products against four species of mosquito vectors. *Southeast Asian Journal of Tropical Medicine and Public Health, 35*, 325–333.

Vázquez-Ucha, J. C., Martínez-Guitián, M., Lasarte-Monterrubio, C., Conde-Pérez, K., Arca-Suárez, J., Álvarez-Fraga, L., Pérez, A., Crecente-Campo, J., Alonso, M. J., & Bou, G. (2020). *Syzygium aromaticum* (clove) and Thymus zygis (thyme) essential oils increase susceptibility to colistin in the nosocomial pathogens *Acinetobacter baumannii* and *Klebsiella pneumoniae. Biomedicine & Pharmacotherapy, 130*, 110606.

Vecchio, I., Sorrentino, L., Paoletti, A., Marra, R., & Arbitrio, M. (2021). The state of the art on acetylcholinesterase inhibitors in the treatment of Alzheimer's disease. *Journal of Central Nervous System Disease, 13*, 11795735211029113.

Vijayasteltar, L., Nair, G. G., Maliakel, B., Kuttan, R., & Krishnakumar, I. (2016). Safety assessment of a standardized polyphenolic extract of clove buds: Subchronic toxicity and mutagenicity studies. *Toxicology Reports, 3*, 439–449.

Whatling, C., McPheat, W., & Hurt-Camejo, E. (2004). Matrix management: Assigning different roles for MMP-2 and MMP-9 in vascular remodeling. *Arteriosclerosis, Thrombosis, and Vascular Biology, 24*, 10–11.

Woo, J.-H., Ahn, J.-H., Jang, D. S., Lee, K.-T., & Choi, J.-H. (2017). Effect of Kumatakenin Isolated from cloves on the apoptosis of cancer cells and the alternative activation of tumor-associated macrophages. *Journal of Agricultural and Food Chemistry, 65*(36), 7893–7899. https://doi.org/10.1021/acs.jafc.7b01543

Yadav, A. S., & Bhatnagar, D. (2007). Free radical scavenging activity, metal chelation and antioxidant power of some of the Indian spices. *Biofactors, 31*(3–4), 219–227.

Yeom, J. E., Kim, S.-K., & Park, S.-Y. (2022). Regulation of the gut microbiota and inflammation by β-caryophyllene extracted from cloves in a dextran sulfate sodium-induced colitis mouse model. *Molecules, 27*(22), 7782.

14 Regulatory Guidelines

Nisha Khatri, Preet Amol Singh, Neha Bajwa, and Subhajit Hazra

14.1 INTRODUCTION

For thousands of years, people have been interested in using plants that promote health and serve as natural remedies (Singh et al., 2021a, 2021b, 2022; Singh & Baldi, 2023) An essential kind of medicinal plants utilized in traditional Chinese and Indian medicine are spices, which are aromatic plants used to flavour and preserve food. The therapeutic qualities of spices have been uncovered by contemporary science, together with the natural substances (phytonutrients) that underlie these advantages (Vijayasteltar et al., 2016; Singh & Bajwa, 2024).

Syzygium aromaticum, commonly referred to as clove (synonym: *Eugenia caryophyllata*), is a medium-sized (8–12 m) Myrtaceae tree (Cortés-Rojas et al., 2014). With roots in the 1st century BC, cloves are among the most valuable and ancient of all the spices. Clove was initially used during the Chinese Han dynasty (207 BC to 220 AD). By the 4th century AD, Arab traders had brought clove from Asia to Europe. Clove flowers and oil have been widely utilized in Indian and Chinese traditional medicine for more than 2,000 years. The East India Company brought clove to India approximately 1,800 years ago when they planted it in their spice garden in Court Allam, Tamil Nadu. The Indonesian Molucca Islands are the original home of cloves. Madagascar and Zanzibar are the primary exporters, with Madagascar's cloves having been seen to be superior in the past, even though Indonesia is the biggest producer (Yadav et al., 2021). Coastal regions are common places to cultivate clove plants. Four years after planting, they begin to produce flower buds, a component that is valuable to businesses. Before they bloom, during the maturity stage, these buds are collected. Either a natural phytohormone that naturally releases ethylene in plant tissue to promote early maturation can be used for the collection, or it can be done manually (Cortés-Rojas et al., 2014).

Regarding uses of clove, it blossoms are often used in British Columbia to treat pigs and dogs with roundworms and tapeworms. In China, Japan, and Korea, the flower buds of *E. caryophyllata* are traditionally used as a vermifuge, an antibacterial agent, and a treatment for toothaches, asthma, and other allergens. Clove flower buds are used in several Asian and Australian traditional remedies to flavour food and treat digestive system disorders, headaches, sore throats, and dental and respiratory problems. Clove also has antipyretic, aphrodisiac, appetite-stimulant, expectorant, antiemetic, anxiolytic, myorelaxant, analgesic, decongestant, anti-inflammatory, and hypnotic properties. It is used to treat dyspepsia, gastritis, and diarrhoea (Singh, 2012). Western herbal medicine, Chinese medicine, Unani medicine, and Indian

DOI: 10.1201/9781003521785-14

Ayurvedic medicine all make substantial use of clove. Warm and scented, it is used as a carminative, for toothaches, as an aphrodisiac, to strengthen kidney yang (hypochlorhydria), and to treat male sexual abnormalities. Clove oil is regarded as a natural anthelmintic and has been shown to improve peristalsis, function as an anodyne, and induce anaesthesia. Clove is a prevalent ingredient in toothpastes and mouth fresheners in India since it is frequently used for dental disorders. Clove has several applications in the culinary arts, food processing, pharmaceuticals, perfumes, and cosmetics in addition to medicine (Singh, 2012).

As far as the trade dynamics of clove is concerned, it is designated under the international trade code "0907: clove: whole fruit, cloves, and stems", and its global production is valued at more than $500 million annually, accounting for around 4% of the world's spice trade and placing it between fifth and seventh in the world in terms of value. With large hubs in the United Arab Emirates and Singapore, as well as important routes connecting Madagascar and Sri Lanka to Asia and Europe and Brazil to the Americas, cloves are mostly traded in Asia and consumed worldwide, with the Netherlands playing a significant role in this latter region. The intricate clove business in Indonesia is sourced from Tanzania and Madagascar through Singapore, with exports to other Asian nations as well as the Indian subcontinent. The production of kretek cigarettes, which started in Java in the 1870s and employs over 300,000 people despite increasing mechanization, uses the majority of Indonesia's clove harvest. Ninety-one percent of Indonesian smokers prefer kretek cigarettes, and by 2025, that number is expected to reach 95 million. The second-largest market for cloves is the agri-food sector, which uses them in whole and crushed form in a variety of cuisines—Japanese cuisine excluded. India, which topped the world spice market in 2016 with 21% of exports, buys a significant amount of cloves from Tanzania, Madagascar, and Sri Lanka and has emerged as Madagascar's main customer in recent years (Danthu et al., 2020).

Due to a lack of statistics, evaluating the production of clove essential oil is more difficult than evaluating cloves themselves. With an annual production of almost 4,000 tonnes, Indonesia is the world's top producer, contributing 60% of the total output. About 2,000 tonnes are produced annually in Madagascar, but none are produced in the Comoros since they would rather protect their timber supplies. Eugenol and clove essential oil are prized for their ease of manufacture, low cost of manufacturing, and natural abundance; this is especially true in tiny, inexpensive distilleries in Madagascar and Indonesia. These oils are used for a variety of purposes, such as antibacterial and antifungal treatments and fragrance. Due to its wide-spectrum antibacterial qualities, clove oil is frequently used in aromatherapy, dental treatments for its analgesic effects, and even in agriculture to stop potato sprouting (Danthu et al., 2020).

Established under the Food Safety and Standards Act, the Food Safety and Standards Authority of India (FSSAI) is the primary regulatory organization operating under the purview of the Indian government. The FSSAI is in charge of creating food-related standards and regulations and seeing to it that they are followed correctly across the nation. In addition, two more regulatory authorities in India that assist in establishing norms and recommendations pertaining to quality standards, the use of clove in ayurvedic remedies, storage, transportation, packaging, import, and export are the Indian Spice Board and AYUSH. These rules also ensure adherence to the labelling, quality, and safety regulations set forth by the

FSSAI. International standards like those established by the Codex Alimentarius Commission also have an impact on these regulations. Similar to this, the Food and Drug Administration (FDA) in the U.S. has approved oleoresins, clove buds, clove essential oil, and eugenol as permissible food additives and established guidelines for their labelling and packaging. For the clove industry to maintain consistency in quality, all of these rules are necessary. Regulations pertaining to product labelling and traceability guarantee supply chain transparency. Following global rules makes clove goods more competitive by bringing them into compliance with international standards. Regulatory oversight and enforcement monitor adherence to these standards in order to ensure consistency and safety in clove products.

14.2 GLOBAL REGULATORY FRAMEWORK

Cloves, like other spices, are governed by a number of international regulations that are related to their production, commerce, and consumption. Codex Alimentarius Commission is one such international organization responsible for establishing guidelines that guarantee the safety, quality, and authenticity of spices like clove. Additionally, the World Trade Organization (WTO) agreements are also very important since they make sure that the worldwide practices related to trade adhere to the principles of fair competition and non-discrimination.

14.2.1 International Standards for Spices

The Food and Agriculture Organization (FAO) and the World Health Organization (WHO) work together to form the Codex Alimentarius Commission, with the intention of safeguarding the health of consumers and making sure that the food trade is conducted fairly (Varadi, 2007). In order to establish guidelines for spices and herbs, it established the Codex Committee on Spices and Culinary Herbs (CCSCH) (Press Information Bureau, 2021a, 2021b).

"Codex standard for dried floral parts: Cloves (CXS 344–2021)", provides science-based standard that includes definitions, essential compositions, and quality characteristics for cloves. According to these standards, cloves are made from the dried flower parts of the clove tree and can be ground or powdered without the need for additional ingredients by specifying requirements for their chemical and physical features, such as moisture content, acid insoluble ash, and appropriate labelling. It also emphasizes the importance of the requirement of efficient control methods to guard against contamination and guarantee clove authenticity (Codex Alimentarius Commission, 2021).

14.2.2 World Trade Organization (WTO) Agreements

Trade in products, services, and intellectual property is covered by the WTO accords. They define permissible deviations and lay out the fundamentals of trade liberalisation. The accords specify each nation's pledges to uphold open services markets and lower customs tariffs and other trade obstacles. They also set up conflict resolution processes. These are flexible agreements that can be expanded with new ones and are renegotiable on a regular basis. These agreements give member nations a framework to adhere to particular guidelines and requirements, guaranteeing impartial competition and antidiscrimination (World Trade Organization, 2024). For example, there was a case in which WTO was involved in disputes about flavoured clove cigarettes. This issue was concerned with Indonesia's challenge against the U.S. government's Section 907 of the Food, Drug, and Cosmetic Act (as amended by the Tobacco Control Act) prohibiting clove cigarettes. Indonesia said that, while domestically made menthol cigarettes were excluded, the ban discriminated against imported flavoured tobacco products, such as clove cigarettes. The United States' restriction on clove cigarettes was found to be a violation of the WTO Agreement on Technical Barriers to Trade (TBT), according to the WTO's Dispute Settlement Body (DSB), which ruled in favour of Indonesia (American Society of International Law, 2012; Centre for WTO studies, 2012).

14.3 REGULATIONS RELATED TO CLOVE

14.3.1 United States

14.3.1.1 Trade Regulations Related to Clove

Regarding cloves and clove-related items, the United States has a number of important restrictions. As long as they adhere to certain requirements and specifications, clove and its derivatives are confirmed as generally recognized as safe (GRAS) for use as adjuvants and flavouring agents in food (Code of Federal Regulations, 2024). Clove bud oil and clove leaf oil are duty-free items under subheading 3301.29.5113 for essential oils of clove in the Harmonised Tariff Schedule of the United States (HTSUS). The Food, Drug, and Cosmetic Act, as amended by the Tobacco Control Act, notably Section 907, forbids the manufacture and distribution of cigarettes with "characterising flavours", such as clove, in the United States (Harmonized Tariff Schedule, 2010; Code of Federal Regulations, 2024).

The World Trade Organization's Agreement on Technical Barriers to Trade was deemed to be in conflict with this restriction because it gave preference to indigenous menthol cigarettes over imported clove cigarettes. The United States may outlaw clove cigarettes, but it cannot treat imports differently from comparable domestic goods. The United States has maintained that Section 907 will comply with trade agreements if further scientific research is done and regulations on menthol cigarettes are implemented.

Additionally, clove is one of the spices for which the American Spice Trade Association (ASTA) is responsible for developing industry standards and procedures. It has established comprehensive clove quality standards that take into account a number of variables. According to these guidelines, detached clove stems may include up to 5% less weight in extraneous matter than other permitted contaminants. Clove essential oil, especially clove leaf essential oil, must also comply with FDA and ASTA criteria in terms of its physico-chemical characteristics. In order to preserve the essential standards of the oil and prevent adulteration, particularly with synthetic eugenol, ASTA requirements are required. Additionally, limits for observable extraneous matter have been established, along with cleaning criteria. In order to maintain the quality and authenticity of clove products, ASTA makes sure that ISO standards are followed. These criteria cover factors including non-volatile ether extract, total ash, acid-insoluble ash, volatile oil concentration, crude fibre, and moisture (Guidance from the American Spice Trade Association, 2011).

A crucial role is played by the Flavour and Extract Manufacturers Association (FEMA) in determining the safety and approving the usage of flavouring components derived from cloves in the food and beverage industry. The following crucial components are essential to FEMA's operation: using its established approach, the FEMA Expert Panel evaluates the safety of natural flavour compounds made from West Indian Bay leaf, clove, and cinnamon. Notably, several clove-derived flavouring components, such as clove bud oil (FEMA 2323), clove leaf oil (FEMA 2325), clove stem oil (FEMA 2328), clove bud extract (FEMA 2322), and clove bud oleoresin (FEMA 2324), have been given "generally recognised as safe" (GRAS) status by the FEMA Expert Panel, confirming their suitability for intended uses. The FEMA Expert Panel's thorough safety assessment, which included the chemical makeup, anticipated dietary exposure, and known toxicological data about these compounds produced from cloves, led to the GRAS designation. Notably, estimated per capita intakes for clove bud oil (FEMA 2323), clove leaf oil (FEMA 2325), and clove stem oil (FEMA 2328) range from 190 to 2,350 µg/person/day; the FEMA Expert Panel determined these values to be safe after careful examination (Gooderham et al., 2020; Baker et al., 2018).

14.3.1.2 Safety and Efficacy Regulations Related to Clove

To protect consumer safety and advance ethical business practices in the food industry, the Food and Drug Administration (FDA) provides rules for the labelling and quality standards for spices, including clove. The dried flower buds of *Syzygium aromaticum* (L.) Merr. & L.M. Perry are referred to as cloves. These entire cloves have a powerful, aromatic scent and a strong, slightly numbing flavour. They are dark brown and fashioned like nails. The volatile oil content, total ash, acid-insoluble ash, and moisture content of cloves are used to determine their quality (Food and Drug Administration, 2018; Code of Federal Regulations, 2024).

The U.S. Food and Drug Administration (FDA) classifies cloves and their derivatives as direct food ingredients recognized as generally safe (GRAS) under 21 CFR Part 184, Subpart B, Section 184.1257. According to 170.3(0)(12), clove and its derivatives (CAS Reg. No. 8000–34–8) are used as adjuvants and flavouring agents in food and are extracted from the dried flower buds of the evergreen clove tree, Eugenia caryophyllata Thunberg (Myrtaceae family). According to § 184.1(b)(1), these ingredients are used in food at amounts that do not exceed good manufacturing practices (Code of Federal Regulations, 2024). Before the flowers open, cloves, which are the dried, unopened flower buds and calyx tubes of the clove tree, are plucked. Oleoresins, natural extractives, and essential oils extracted from buds, leaves, and stems are some of their derivatives. Clove and its derivatives are subject to a number of requirements laid forth in FDA regulations. These include requirements for volatile oil concentration, optical rotation, refractive index, specific gravity, and phenols (as eugenol), as well as the absence of residual solvents unless GRAS or within predetermined tolerance limits. According to the "Food Chemicals Codex", fourth edition (1996), clove bud oil, clove leaf oil, clove stem oil, and eugenol (CAS Reg. No. 97–53–0) must all adhere to the rules. For these substances, there are no current punishments or suspended penalties that deviate from the usage listed in this section (Code of Federal Regulations, 2024; Electronic Code of Federal Regulations (eCFR), 2024a, 2024b).

14.3.1.3 Agronomy Regulations Related to Clove

The USDA Organic certification verifies that a product meets the USDA's organic requirements. These standards are designed to ensure that products that are certified organic are grown without the use of artificial pesticides, herbicides, or fertilizers and that the way they are handled and processed preserves their organic integrity. Claiming USDA Organic status for cloves has several advantages. In the first place, it expands market access by drawing in customers who respect organic goods and are prepared to pay more for them. Second, it boosts sales by satisfying customers' growing desire for organic options. Additionally, it improves the operation's reputation by showcasing a dedication to manufacturing premium organic products. By enabling the selling of clove products at premium rates in comparison to non-organic competitors, it also promotes higher profitability (Agriculture Marketing Service, U.S. Department of Agriculture, 2024). Clove must follow a number of steps (Figure 14.1) as specified by the USDA National Organic Programme (NOP) and certifying agents in order to receive USDA organic certification (Organic, 2018).

14.3.1.4 Regulations Related to Products of Clove

Clove and related goods must comply with FDA rules, quality standards, and labelling requirements in order to be approved in the U.S. Clove oil and clove buds are typically regarded by the FDA as safe (GRAS) food additives. According to 21 CFR 101.22(a)(2), clove must be labelled as "clove" when used in food. Important details such the product name, manufacturer, batch number, date of manufacturing, grade, producing nation, and net weight must be included on labels and packaging. Packed in fresh, clean, dry bags composed of materials such as high-density polyethylene, polypropylene, or jute, whole cloves should be kept dry.

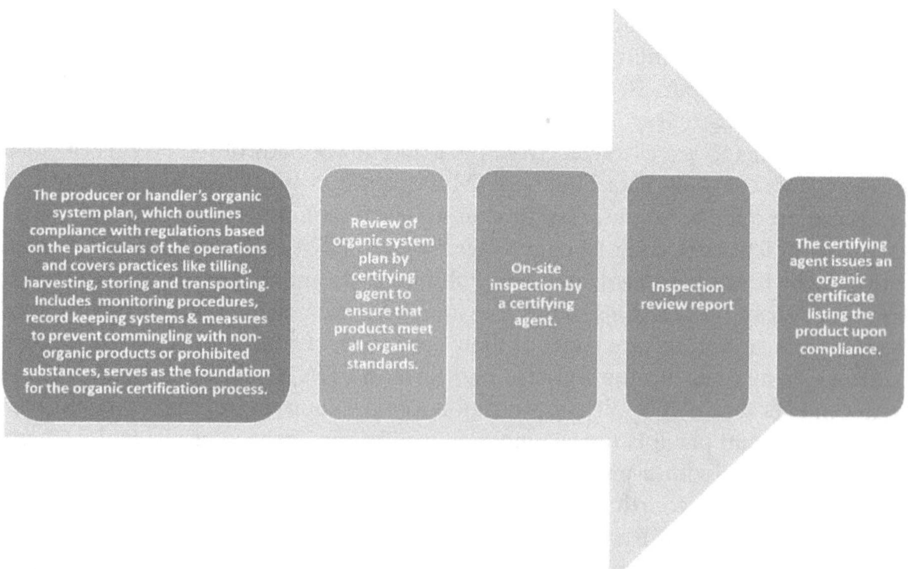

FIGURE 14.1 A comprehensive flow chart representing organic USDA certification process.

Glass, tin, aluminium, or food-grade plastic pouches are good materials to use as fresh, dry, and clean containers to keep clove powder. A high-quality clove should meet certain colour and appearance standards and be whole, unbroken, oil-rich, and stalkless. They are ranked in accordance with ISO standard 2254–2004 and the national standards of the countries that produced them. Cloves must adhere to sanitation standards set by the American Spice Trade Association (ASTA), which exclude macroscopic foreign objects like rodent and insect debris. Clove products must abide by European laws, especially those pertaining to food safety and traceability, in order to be exported to Europe. Exporters are required to make sure they adhere to all applicable laws and acquire the required licences or certificates.

To ensure the quality and safety of their products, manufacturers are required to follow Good Manufacturing Practices (GMP). The FDA's post-market surveillance programme guarantees continued adherence to quality and safety requirements. Additionally, pharmacopeial requirements for purity and quality must be met by clove products. To prove a product's safety and efficacy, clinical trials can be required, depending on the product and its intended usage. Based on the evaluation of the application and any accompanying materials, the FDA will decide whether to grant or refuse market authorization. To retain the validity of their market licence, manufacturers must maintain compliance with FDA laws and guidelines after gaining market permission and have mechanisms in place for recalls related to safety concerns or regulatory non-compliance (Food and Drug Administration, 2021, FDA Advisory No. 2021–0729; Tania SL Voon et al., 2012).

14.3.2 Indian Regulations

Clove is widely used and recognized throughout India and is known by several Indian names, including "Laung" in Hindi, "Lawang" in Bengali, "Lavang" in Gujarati, "Lavanga" in Kannada, "Grambu" in Malayalam, "Luvang" in Marathi, "Labang" in Oriya, "Laung" in Punjabi, "Lavanga" in Sanskrit, and "Kirambu" or "Lavangam" in Tamil, with "Lavangalu" in Telugu and "Laung" in Urdu (Spices Board India, 2024, Clove).

Additionally, in India, the Indian Trade Clarification based on the Harmonised System of Coding (ITC HS Codes) provides a classification system for cloves and clove products that is given as follows:

HS Code 0907: **STEMS, WHOLE FRUIT, AND CLOVES**

This category includes the following goods made using cloves:

- **09071010:** cloves extracted.
- **09071020:** whole cloves (stems excluded).
- **09071030:** stems of cloves.
- **09071090:** any further clove goods not listed elsewhere.
- Crushed or crushed cloves **(09072000).**

Specific rules described in Policy Condition No. 1 of Chapter-09 of the ITC HS Policy apply to the import of cloves. Regarding food items, importers must adhere to the ports of entry defined in Appendix-V List-A to Schedule-I of the Import Policy, ITC (HS) (How to Import Cloves into India in 2023: An Essential Guide for Traders—Exim SEO—EIG, 2023).

14.3.2.1 Product-related Specifications of Clove

Furthermore, in charge of creating guidelines and guaranteeing clove and clove product quality control is the Bureau of Indian Standards (BIS). To guarantee the safety and purity of cloves, clove oil, and clove stem oil, BIS has established strict guidelines. The following essential criteria have been established by BIS for cloves and related products:

IS 6/1972: clove specification. The requirements for whole cloves are described in this standard, together with information on acceptable levels of pollutants and adulterants, packaging and labelling guidelines, and physical and chemical characteristics.

IS 7809/1975: clove oil specification. Clove oil's colour, odour, relative density, optical rotation, refractive index, and total eugenol concentration are all specified in this standard.

IS 7810/1975: clove stem oil specification. the specifications described in this standard for clove oil also apply to clove stem oil (Bureau of Indian Standards, 2013a, 2013b).

These guidelines guarantee that cloves and clove-derived products are safe to eat and satisfy certain quality standards. BIS also offers multiple certification programmes. A few of them are:

Certification with ISI mark: regulations for items containing cloves marketed in India. Products that satisfy BIS criteria are awarded the ISI mark, which attests to their adherence to the required specifications.

Certification from the export inspection council (EIC): applicable to products made with cloves that are destined for export. In order to make sure these goods satisfy importer needs as well as global standards, BIS works with the EIC.

Acclaim certification: according to quality standards established by the Agricultural and Processed Food Products Export Development Authority (APEDA), for spices, including cloves.

Throughout the supply chain, these certification programmes aid in preserving the quality and safety of cloves and clove-related goods. This alignment guarantees that Indian cloves and clove-related products satisfy the demands of the global market (Bureau of Indian Standards, 2013).

14.3.2.2 Trade-related Specifications of Clove

The Spices Board of India functions as the export promotion and regulatory organisation for Indian spices. Comprehensive rules pertaining to the export of cloves and products connected to cloves from India have been put in place by the Indian Spice Board. These rules are intended to support standards of quality, safety, and compliance while promoting the development of the spice sector (Spice Board India, 2024; The Economic Times, 2024).

Manufacturer exporters and merchant exporters alike must have a Registration-cum-Membership Certificate (RCMC) from the Spice Board in order to export cloves from India. This certificate attests to the exporter's legal status and compliance with board rules. Obtaining the Spice Board RCMC entails the following steps:

(1) Application submission: exporters must submit an online application via the Spice Board portal, together with all required certifications and documentation.
(2) Document Upload and verification: the Spice Board's Regional Office verifies the uploaded documentation. In case of any noncompliance, the application can be sent back for revisions.
(3) Issuance of certificate: the exporter obtains the Spice Board RCMC following a successful verification. Whether the exporter is a manufacturer or a merchant, as well as whether the application is for a new registration, renewal, or amendment, will determine the registration fees.

To protect product integrity and avoid contamination during storage and transportation, exported cloves must be packaged in hygienic, durable, and airtight containers made of appropriate materials. Labels on the container need to state product name and grade, net weight, lot/batch number, packing date, best before date, exporter's name and address and Spice Board RCMC number. Phytosanitary Certificates are also necessary to guarantee that the exported cloves are free of illnesses and pests. These certifications attest to the cloves' compliance with the phytosanitary laws of the destination nation. They are provided by the Plant Quarantine and Storage Department of the Ministry of Agriculture and Farmers Welfare. At the port of entry, phytosanitary certificates are essential for the efficient processing of clove exports (How to Get Spice Board Registration to Export Clove from India, 2024).

In order to guarantee compliance with legal requirements and preserve the quality and safety of the product, exporting (Figure 14.2) and importing (Figure 14.3) cloves and clove-related items from India requires a number of crucial actions.

14.3.2.3 Traditional Medicine and Food-related Specifications of Clove

In traditional Ayurvedic and Siddha medicine, cloves, also known as "*Lavanga*" in Sanskrit, are an important component. They are essential parts of several AYUSH concoctions that are prized for their medicinal qualities. Cloves are classified as having a sharp flavour (Katu Rasa), having a heating power (Ushna Virya), and being light and dry (Laghu and Ruksha) in Ayurveda (Svastha Ayurveda, 2023; Plant Ayurveda, 2019).

Various Guidelines issued by the Ministry of AYUSH under the National AYUSH Mission (NAM) emphasized the importance of incorporating various spices like cloves in one's daily regimen (Ministry of AYUSH, 2018a, 2018b). The Ministry

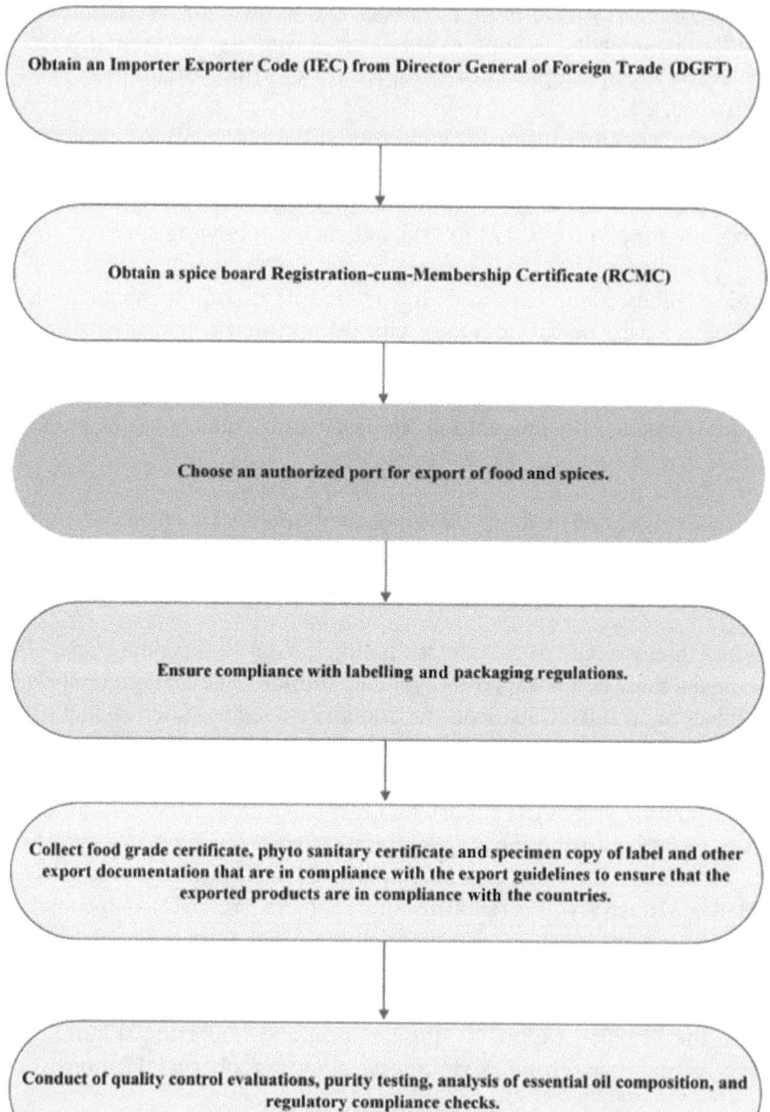

FIGURE 14.2 Clove export process of India.

of AYUSH in consultation with FSSAI formulated the Ayurveda aahar products under the Food Safety and Standards (Ayurveda Aahara) Regulations, 2022. They have also defined various methods for clove to be used in traditional food recipes (Ayushman Bharat Traditional Food Recipies, 2020) (Press Information Bureau, 2022, Ayurvedic Aahar Products). The Ministry of AYUSH also emphasizes adopting good agricultural practices in general so as to maintain the quality of products throughout the supply chain. NAM provides funding to support the establishment of

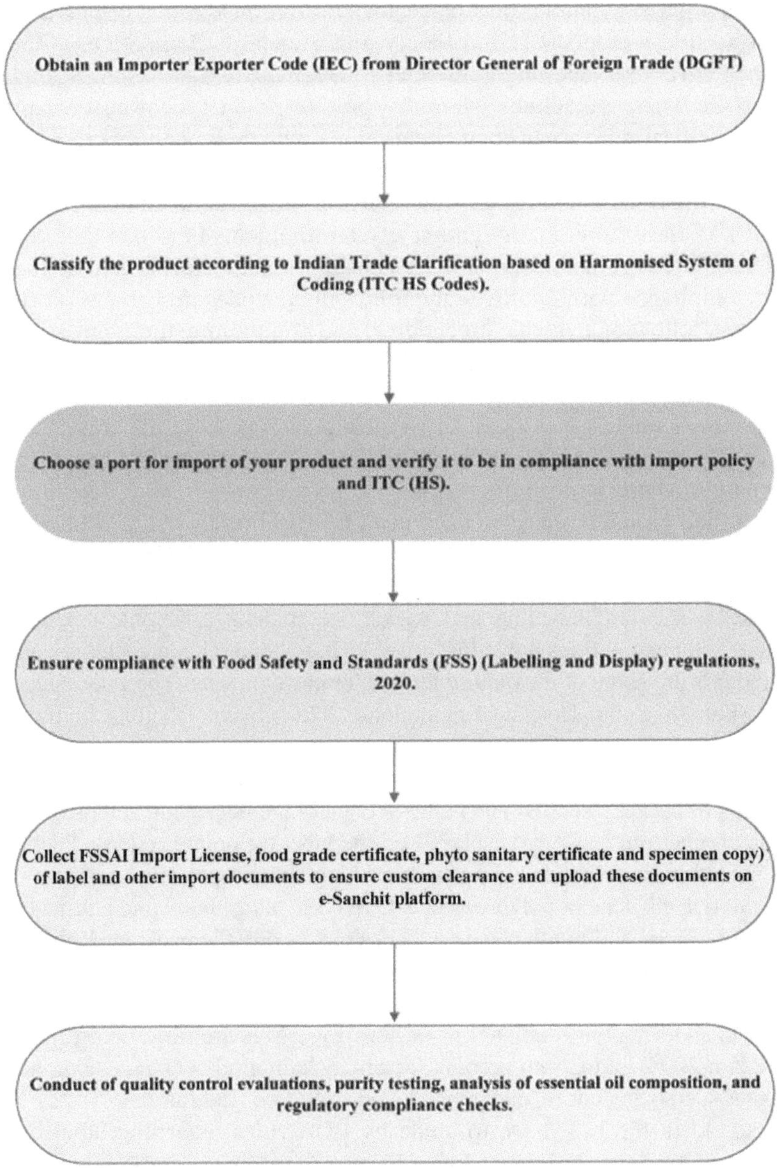

FIGURE 14.3 Clove import process of India.

nurseries, the acquisition of planting supplies, and the promotion of organic farming practices. To promote sustainable agriculture, the mission also involves programmes like teaching farmers and setting up demonstration plots (Press Information Bureau, 2021; Ministry of AYUSH, 2023).

Likewise, in Food Safety and Standards (Health Supplements, Nutraceuticals, Food for Special Dietary Use, Food for Special Medical Purpose, Functional Food and Novel Food) Regulations, 2016, the Food Safety and Standards Authority of

India (FSSAI) has put regulations into place for controlling the use of cloves in herbal and nutraceutical products (Food Safety and Standards Regulation, 2009). It has established particular labelling guidelines for herbal and nutraceutical goods that contain cloves. These guidelines guarantee precise product identification and thorough consumer disclosure about components along with their quantity present in a particular product (Kumar Pandey et al., 2022).

Furthermore, FSSAI has also established rules covering good manufacturing practices (GMP) to ensure the quality and safety requirements of goods that contain cloves. These GMP guidelines guarantee constant manufacturing and control of products in compliance with quality standards. This includes things like proper storage conditions, following hygienic procedures, and maintaining thorough paperwork (Food Standards and Safety Authority of India, 2022). Moreover, the FSSAI regulations limit several claims made about the medicinal qualities of herbal and nutraceutical products made with cloves; these include claims about the prevention, treatment, or cure of diseases as well as claims suggesting the product can be used as a dietary alternative. Manufacturers are responsible for making sure that any claims they make about their products are accurate, free of false information, and supported by empirical data (Food Standards and Safety Authority of India, 2021).

To ensure proper identification, quality control, and regulatory compliance for cloves and clove products, standard labelling and packaging guidelines become essential. Moisture content, total ash, ash insoluble in diluted HCl, volatile oil concentration, and crude fibre standards are some of the quality factors for clove powder. The Food Safety and Standards (Labelling and Display) Regulations, 2020 govern labelling in India. Along with the country of origin, labels should have the following information: product name, batch number, net weight in metric units, best before date, storage instructions, and the manufacturer's or packer's details. The name or code of the inspection authority and the certification number must also be displayed on the label for organic cloves. Packed whole, cloves should be in fresh, dry bags made of high-density polyethylene, jute, or fabric coated with polyethylene or polypropylene. Glass, tin, aluminium, and laminated, metallized, multi-layered food-grade plastic pouches are among the packaging options for clove powder. Ensuring the quality and integrity of the cloves during storage and transportation requires packaging free from dangerous substances, fungal infection, unwanted aromas, and insect infestation. For uniformity, cloves are rated according to ISO standard 2254–2004, taking into account factors including oil content, colour, and physical attributes such being entire, undamaged, and stalk-free (Indiramma, 2012).

Clove items sold in the U.S. have to abide by FDA rules regarding labelling, quality standards, and regulations. Clove products and non-retail containers must follow the General Standard for the Labelling of Non-Retail Containers of Foods (CXS 346–2021) as provided by the Codex alimentarius commission. This standard has also been mandated in India by the FSSAI. By guaranteeing that all pertinent information is given in an understandable manner on the container or in related documentation, this standard promotes transparency and complies with FDA rules (Codex Alimentarius, 2023). According to 21 CFR 101.22(a)(2), clove must be labelled as "clove" when used in food. Important information such the product name, manufacturer, batch number, date of manufacturing, product grade, producing nation, and net weight must be included on labels and packaging.

Packed whole, cloves should be in fresh, dry bags made of materials such as high-density polyethylene, jute, or fabric bonded with polyethylene or polypropylene. It is recommended to package clove powder in fresh, dry, and clean glass, tin, or aluminium containers or food-grade plastic pouches. The product name, batch code, net weight, shelf life, suggested storage conditions, lot identification number, country of origin, and manufacturer information must all be listed on export package labels. FDA standards pertaining to food information, such as minimum font size requirements for required information, allergen labelling, nutrition, and provenance, must be followed by consumer packaging. The name, code, and certification number of the inspection organisation must appear on labels for organic cloves. Furthermore, according to U.S. Department of Transportation (DOT) regulations, packages containing cloves must meet the standards for combustible materials. These standards include labelling single packages 119 U.S. gallons/450 litres with the combustible symbol and proper shipping name and allowing a maximum net quantity per outer packaging of 1,000 ml Bel-Art Products, 2015) (Code of Federal Regulations [C.F.R.], 2023, Part 101). Respecting these comprehensive guidelines aids in preserving the integrity, safety, and quality of clove goods all the way through the supply chain.

14.4 COMPARISON OF REGULATIONS BETWEEN USA AND INDIA

Despite the goal being nearly the same, which is, to ensure the quality, safety, and purity of clove and clove products, there are certain differences in the national regulations of the U.S. and India. Table 14.1 summarizes the key differences that lie between the U.S. and Indian regulations of clove and clove products.

TABLE 14.1
Key differences between the U.S. and Indian regulations with respect to clove and clove products.

Aspect	United States regulations	Indian regulations	References
Regulatory bodies	FDA, USDA, ASTA, FEMA	FSSAI, BIS, Spice Board of India, Ministry of AYUSH	(Commissioner, 2020)
GRAS status	Clove and its derivatives are generally recognized as safe (GRAS) for use as adjuvants and flavoring agents (21 CFR 184.1257).	Not specifically mentioned as GRAS but subject to FSSAI safety and quality standards	(National Archives and Reports Administration, The Federal Register, 2016; Code of Federal Regulations, 2024, Part 184.1257)
Tariff classification	Clove bud oil and clove leaf oil are duty-free under HTSUS subheading 3301.29.5113.	Clove and clove products classified under ITC HS Codes (e.g., 09071010 for extracted cloves)	(Harmonized Tariff Schedule, 2010; Code of Federal Regulations, 2024, Part 184.1257).

(Continued)

TABLE 14.1 *(Continued)*
Key differences between the U.S. and Indian regulations with respect to clove and clove products.

Aspect	United States regulations	Indian regulations	References
Quality standards	ASTA sets quality standards for clove products, including limits on extraneous matter and prevention of adulteration.	BIS sets standards for cloves and clove products, such as IS 6/1972 (Clove Specification), IS 7809/1975 (Clove Oil Specification), and IS 7810/1975 (Clove Stem Oil Specification).	(Guidance from the American Spice Trade Association, 2011; Bureau of Indian Standards, 2013)
Labelling requirements	Product name, manufacturer, batch number, date of manufacturing, product grade, producing nation, net weight, lot identification number, shelf life, storage conditions, inspection organization name/code, certification number for organic cloves	Product name, batch number, net weight in metric units, best before date, storage instructions, manufacturer details, inspection authority name/code, certification number for organic cloves	(Food Safety and Standards Regulation, 2009; Codex Alimentarius Commission, 2021)

14.5 EXAMPLES OF DIFFERENT APPROVED PRODUCTS OF CLOVE

Clove is used in dental cement and as an addition in food with approval of FDA. It is available in a variety of formulations, including dental goods, over-the-counter medicines like clove oil liquid, and combo products like Arhtri Plus Clove oil spray and Amol Clove oil. Moreover, clove buds are used as key materials in the medicinal and perfume industries and to flavour food whether they are dried intact or powdered for culinary purposes (Baker et al., 2018). Due to its high eugenol content, clove oil is highly valued for its antibacterial and antimicrobial qualities, which make it very helpful in tooth care, and so clove is integrated in toothpaste; an example of such a toothpaste is the "Colgate Swarna Vedshakti" that is certified by the Indian Dental Association. Other products present in the current market that comprise clove as a main constituent include chewing gum such as the Adams Sons Company's clove gum (Kamandloo et al., 2023), shaving soap, organic weed killers, perfumes, and insect repellents, in addition to its traditional uses. Potential future goods could include cloves in dairy products such as ice cream, yoghurt, butter, milk, and cheese. Furthermore, new developments such as the benzethonium chloride-containing clove bud essential oil emulsion show antibacterial activity against microorganisms, underscoring its importance in food safety. These illustrations highlight the wide range of uses for clove in the culinary, medicinal, and aromatic domains while guaranteeing adherence to strict quality and safety guidelines required by regulatory bodies such as the FDA and FEMA (Jackson-Davis et al., 2022).

14.6 CONCLUSION

Clove has a complex and multifaceted regulatory system that reflects its important position in global trade and its many uses in traditional medicine, food, and pharmaceuticals. There are regulatory bodies both at national and international levels ensuring that the quality and standards of clove and clove products are maintained at all times during the import, export, and production process. To guarantee the safety, purity, and authenticity of cloves, the Codex Alimentarius Commission establishes standards at a global level that address several issues such as chemical characteristics and processing. These guidelines along with those provided by the WTO agreements promote fair commerce by guaranteeing non-discriminatory practices worldwide. Also, different nations like the U.S. and India have different organizations to guarantee safety and quality compliance. Organizations such as the USDA and FDA implement stringent regulations and certification procedures for cloves and clove products in the U.S, whereas in India the regulatory framework for cloves and other spices is supervised by the Indian Spice Board, AYUSH, and FSSAI, which attempts to protect cloves' market integrity and encourage their use in traditional and nutraceutical products. By comparing these regulations, one can see how different and similar their regulatory strategies are, underscoring the necessity for unified standards to promote global trade and guarantee high-quality products.

REFERENCES

Agriculture Marketing Service, U.S. *Department of Agriculture. Organic Certification and Accreditation.* Retrieved May 20, 2024, from: www.ams.usda.gov/grades-standards/organic-standards (Accessed 03 July 2024).

American Society of International Law. (2012). *The WTO Appellate Body Outlaws Discrimination in U.S. Flavored Cigarette Ban.* Retrieved from: www.asil.org/insights/volume/16/issue/15/wto-appellate-body-outlaws-discrimination-us-flavored-cigarette-ban (Accessed 03 July 2024).

Baker, B. P., Grant, J. A., & Malakar-Kuenen, R. (2018). *Cloves & Clove Oil Profile Cloves & Clove Oil Profile Active Ingredient Eligible for Minimum Risk Pesticide Use.* Retrieved from: http://hdl.handle.net/1813/56120 (Accessed 03 July 2024).

Bel-Art Product. (2015). *Odo-Clave (Clove) Safety Data Sheet.* Retrieved from: www.belart.com/corporate/SDS/Odo-Clave%20(Clove).pdf (Accessed 03 July 2024).

Bureau of Indian Standards. (2013a). *Indian Standard Specification for Clove Oil.* Retrieved from: https://law.resource.org/pub/in/bis/S11/is.6698.1972.pdf (Accessed 03 July 2024).

Bureau of Indian Standards. (2013b). *Indian Standard Spices and Condiments—Cloves, Whole And Ground—Specification (Third Revision).* Retrieved from: https://law.resource.org/pub/in/bis/S06/is.4404.2010.pdf (Accessed 03 July 2024).

Centre for WTO Studies. (2012). *United States- Measures Affecting The production and Sale of Clove Cigarettes.* Retrieved from: https://wtocentre.iift.ac.in/DisputeAnalysis/Dispute%20Analysis_US-Clove%20Cigarettes%28AB%29.pdf (Accessed 03 July 2024).

Code of Federal Regulations. (2024). *Food and Drug Administration, Department of Health and Human Service. Subchapter B-Food for Human Consumption.* 21(3) US Food and Drug Administration. 21CFR184.1257.

Codex Alimentarius Commission. (2021). *Appendix VI: Proposed Draft for Dried Floral Parts- Dried Cloves).* Retrieve from: www.fao.org (Accessed 03 July 2024).

Codex Alimentarius Commission. (2023). *Standard for Dried Floral Parts: Cloves (CXS 344–2021).* Retrieved from: www.codexalimentarius.org (Accessed 03 July 2024).

Commissioner, O. of the (2020). *USDA, FDA Strengthen U.S. Food Supply Chain Protections, U.S. Food and Drug Administration.* Retrieved from: www.fda.gov/news-events/press-announcements/usda-fda-strengthen-us-food-supply-chain-protections (Accessed 04 May 2024).

Cortés-Rojas, D. F., de Souza, C. R. F., & Oliveira, W. P. (2014). Clove (Syzygium aromaticum): A precious spice. *Asian Pacific Journal of Tropical Biomedicine*, 4(2), 90.

Danthu, P., Simanjuntak, R., Fawbush, F., Leong Pock Tsy, J. M., Razafimamonjison, G., Abdillahi, M. M., Jahiel, M., & Penot, E. (2020). The clove tree and its products (clove bud, clove oil, eugenol): Prosperous today but what of tomorrow's restrictions? *Fruits*, 75(5), 224–242.

The Economic Times. (2024). *Spice Board Formulates Guidelines for Exporters to Prevent ETO Contamination.* Retrieved from: https://economictimes.indiatimes.com/industry/cons-products/food/spices-board-formulates-comprehensive-guidelines-for-exporters-to-prevent-eto-contamination/articleshow/110017387.cms (Accessed 04 July 2024).

Electronic Code of Federal Regulations (eCFR). (2024a). *Title 21, Chapter I, Subchapter B, Part 101.* Retrieved from: www.ecfr.gov/current/title-21/chapter-I/subchapter-B/part-101 (Accessed 03 July 2024).

Electronic Code of Federal Regulations (eCFR). (2024b). *Title 21, Chapter I, Subchapter B, Part 184, Subpart B.* Retrieved from: www.ecfr.gov/current/title-21/chapter-I/subchapter-B/part-184/subpart-B (Accessed 03 July 2024).

The Federal Register (2016) *Federal Register: Request Access.* Retrieved from: www.ecfr.gov/current/title-21/chapter-I/subchapter-B/part-184/subpart-B/section-184.1257 (Accessed 04 July 2024).

Food and Drug Administration. (2018). *CPG Sec 525.750 Spices—Definitions.* Retrieved from: www.fda.gov/regulatory-information/search-fda-guidance-documents/cpg-sec-525750-spices-definitions (Accessed 03 July 2024).

Food and Drug Administration. (2021). *FDA Advisory No.2021–0729, Against the Purchase and Use of the Unregistered Drug Product "Arefielo Botanical Essense with Turmeric and Clove Oil 60mL"—Food and Drug Administration.* Retrieved May 20, 2024, from: www.fda.gov.ph/fda-advisory-no-2021–0729-against-the-purchase-and-use-of-the-unregistered-drug-product-arefielo-botanical-essense-with-turmeric-and-clove-oil-60ml/ (Accessed 03 July 2024).

Food Safety and Standards Authority of India. (2009). *Food Safety and Standards Regulation.* Retrieve from: https://fssai.gov.in/upload/uploadfiles/files/FSSAI-regulations.pdf (Accessed 03 July 2024).

Food Safety and Standards Authority of India. (2022). *Food Safety and Standards (Food Products Standards and Food Additives) Regulations, 2011.* Retrieve from: www.fssai.gov.in/upload/uploadfiles/files/Compendium_Food_Additives_Regulations_20_12_2022.pdf (Accessed 03 July 2024).

Food Standards and Safety Authority of India. (2021). *Food Safety And Standards (Health Supplements, Nutraceuticals, Food For Special Dietary Use, Food For Special Medical Purpose, Functional Food And Novel Food) Regulations, 2016.* Retrieved from: www.fssai.gov.in/upload/uploadfiles/files/Compendium_Nutra_29_09_2021.pdf (Accessed 03 July 2024).

Gooderham, N. J., Cohen, S. M., Eisenbrand, G., Fukushima, S., Guengerich, F. P., Hecht, S. S., Rietjens, I. M. C. M., Rosol, T. J., Davidsen, J. M., Harman, C. L., Murray, I. J., & Taylor, S. V. (2020). FEMA GRAS assessment of natural flavor complexes: Clove, Cinnamon leaf and West Indian bay leaf-derived flavoring ingredients. *Food and Chemical Toxicology*. Retrieved from: https://doi.org/10.1016/j.fct.2020.111585 (Accessed 03 July 2024).

Guidance from the American Spice Trade Association. (2011). *American Spice Trade Association*, 1–34. Retrieved from: https://astaspice.org/safe-view/57815

Harmonized Tariff Schedule. (2010). *N087879: The Tariff Classification of Clove Bud Oil, CAS No. 8000–34–8 and Clove Leaf Oil, CAS No. 8015–97–2 from India*. Retrieved May 18, 2024, from: https://hts.usitc.gov/search?query=3301.29.5113 (Accessed 03 July 2024).

How to get spice board registration to export clove from India. (n.d.). Retrieved May 20, 2024, from: www.professionalutilities.com/blogs/how-to-export-clove-from-india.php (Accessed 03 July 2024).

How to Import Cloves into India in 2023: An Essential Guide for Traders—Exim SEO—EIG. (2023). Retrieved May 20, 2024, from: https://exportimport.guru/Exim_SEO/Single/Exporting/how-to-import-cloves-into-india-2023 (Accessed 04 July 2024).

Indiramma, A. R. (2012). *Packaging Aspects Of Spices & Spice Products. Indian Center for Plastics in The Environment*. Retrieved from: www.icpe.in/Plastics%20in%20Food%20Packaging/pdf/12-Final.pmd.pdf (Accessed 04 July 2024).

Jackson-Davis, A., White, S., Kassama, L. S., Coleman, S., Shaw, A., Mendonca, A., Cooper, B., Thomas-Popo, E., Gordon, K., & London, L. (2022). A review of regulatory standards and advances in essential oils as antimicrobials in foods. *Journal of Food Protection*, 86(2), 100025.

Kamandloo, F., Takbirgou, H., Gashtasbi, S., Zaghari, N., Ghamari, F., Fotouhi, L., & Salami, M. (2023). Exploring the potential of chewing gums containing plant extracts (clove, saffron, and clove-mint) as a natural solution for controlling obesity and associated health issues. *Functional Food Science*, 3(12), 288.

Kumar Pandey, V., Shams, R., Singh, R., Dar, A. H., Pandiselvam, R., Rusu, A. V., & Trif, M. (2022). A comprehensive review on clove (Caryophyllus aromaticus L.) essential oil and its significance in the formulation of edible coatings for potential food applications. *Frontiers in Nutrition*, 9, 987674.

Ministry of AYUSH. (2018a). *Central Council for Research in Ayurvedic Sciences, Ayurveda Based Diet & Life Style Guidelines for Prevention and Management of Skin Diseases*. Retrieved from: www.ccras.nic.in/sites/default/files/ebooks/24052018_CCRAS_HQ_Ayurvedabaseddiet%26lifeStyleGuidelinesSkinDiseases.pdf (Accessed 04 July 2024).

Ministry of AYUSH. (2018b). *General Standards for Drug Development of Ayurvedic Formulations*. Retrieved from: www.ccras.nic.in/sites/default/files/viewpdf/Publication/CCRAS_Guideline%20of%20Drug%20Development.pdf (Accessed 04 July 2024).

Ministry of AYUSH. (2020). *Ayushman Bharat Traditional Food Recipes from AYUSH System of Medicine*. Retrieved from: https://ayushportal.nic.in/pdf/Food_Recipes_From_AYUSH.pdf (Accessed 04 July 2024).

Ministry of AYUSH. (2023). *National AYUSH Mission Operational Guidelines*. Retrieved from: https://cghealth.nic.in/ehealth/dishm/National_AYUSH_Mission2021-22to2025-26.pdf.

Organic (2018). *Organic | Agricultural Marketing Service*. Retrieved from: www.ams.usda.gov/grades-standards/organic-standards (Accessed 20 May 2024).

Plant Ayurveda. (2019). *Lavanga, Clove (Syzgium Aromaticum)*. Retrieved from: www.planetayurveda.com/library/lavanga-syzygium-aromaticum/ (Accessed 04 July 2024).

Press Information Bureau. (2021a). *Codex Committee on Spice and Culinary Herbs (CCSCH) Finalize Quality Standards for 4 More Spice*. Retrieved from: https://pib.gov.in/PressReleaseIframePage.aspx?PRID=1715125 (Accessed 04 July 2024).

Press Information Bureau. (2021b). *Promotion of Ayush*. Retrieved May 25, 2024, from: https://pib.gov.in/PressReleaseIframePage.aspx?PRID=1695617 (Accessed 04 July 2024).

Press Information Bureau. (2022). *Ayurvedic Aahar Products*. Retrieved May 25, 2024, from: https://pib.gov.in/PressReleasePage.aspx?PRID=1846185 (Accessed 04 July 2024).

Singh, J., Baghotia, A., & Goel, S. P. (2012). Eugenia caryophyllata thunberg (family myrtaceae): A review. *International Journal of Research in Pharmaceutical and Biosciences*, 3, 1469–1475.

Singh, P. A., & Bajwa, N. (2024). Is Tinospora cordifolia responsible for drug-induced liver injury? *Current Drug Safety*, 19(1), 8–10.

Singh, P. A., Bajwa, N., & Baldi, A. (2021a). A comparative review on the standard quality parameters of turmeric. *Indian Journal of Natural Products*, 35(1).

Singh, P. A., Bajwa, N., & Baldi, A. (2021b). Possible role of traditional systems of medicine to manage COVID-19: A review. *Israel Journal of Plant Sciences*, 68(1–2), 3–28.

Singh, P. A., Bajwa, N., Chinnam, S., Chandan, A., & Baldi, A. (2022). An overview of some important deliberations to promote medicinal plants cultivation. *Journal of Applied Research on Medicinal and Aromatic Plants*, 31, 100400.

Singh, P. A., & Baldi, A. (2023). Quality risk assessment on traditional immunity booster plants cultivators in Punjab, India. *Indian Journal of Traditional Knowledge*, 22(2), 273–282.

Spices Board India, Clove. Retrieved May 20, 2024, from: www.indianspices.com/spice-catalog/clove.html (Accessed 04 July 2024).

Svastha Ayurveda. (2023). *The Powerful Healing Properties of Clove*. Retrieved from: https://svasthaayurveda.com/the-powerful-healing-properties-of-clove/ (Accessed 04 July 2024).

Tania SL Voon. (2012). United States-Measures Affecting the Production and Sale of Clove Cigarettes (pp 824–830). *American Journal of International Law*. Retrieved from: www.researchgate.net/publication/297451573_United_States-Measures_Affecting_the_Production_and_Sale_of_Clove_Cigarettes (Accessed 04 July 2024).

Varadi, M. (2007). The international regulation of chemical toxicants in food. Codex Alimentarius. In *Food Toxicants Analysis* (pp. 1–10). Elsevier.

Vijayasteltar, L., Nair, G., Maliakel, B., & Kuttan, R. (2016). Safety assessment of a standardized polyphenolic extract of clove buds: Subchronic toxicity and mutagenicity studies. *Toxicology Reports*, 3, 439–449.

World Trade Organization. (2024). *What We Do*. Retrieve May 25, 2024 from: www.wto.org/english/thewto_e/whatis_e/what_we_do_e.htm (Accessed 04 July 2024).

Yadav, M., Yadav, P., Sahu, S., Yadav, V., & Narayan Gupta, S. (2021). Review literature on clove. *International Journal of Creative Research Thoughts*, 9(1), 2320–2882.

Index

Note: Page numbers in *italics* indicate a figure and page numbers in **bold** indicate a table on the corresponding page.

D

E

F

G

H

Q

R